# Test Bank

Raymond F. X. Williams
*Howard University*

# CHEMISTRY

## The Study of Matter and Its Changes

**Second Edition**

James E. Brady
*St. John's University, New York*

John R. Holum
*Augsburg College (Emeritus), Minnesota*

**John Wiley & Sons, Inc.**
**New York / Chichester / Brisbane / Toronto / Singapore**

ISBN 0-471-12088-X

Printed in the United States of America

10 9 8 7 6 5 4 3 2 1

Printed and bound by Malloy Lithographing, Inc.

# CONTENTS


| | | |
|---|---|---|
| Chapter 1 | Building a Foundation | 1 |
| Chapter 2 | The Structure of Matter: Atoms, Molecules and Ions | 15 |
| Chapter 3 | Stoichiometry: Quantitative Chemical Relationships | 32 |
| Chapter 4 | Chemical Reactions in Solution | 46 |
| Chapter 5 | Energy and Thermochemistry | 63 |
| Chapter 6 | Atomic and Electronic Structure | 74 |
| Chapter 7 | Chemical Bonding I | 90 |
| Chapter 8 | Chemical Bonding II | 104 |
| Chapter 9 | Chemical Reactions: Periodic Correlations | 121 |
| Chapter 10 | Properties of Gases | 129 |
| Chapter 11 | Intermolecular Attractions and the Properties of Liquids and Solids | 147 |
| Chapter 12 | Solutions | 158 |
| Chapter 13 | Thermodynamics | 179 |
| Chapter 14 | Kinetics: The Study of Rates of Reaction | 192 |
| Chapter 15 | Chemical Equilibrium - General Concepts | 205 |
| Chapter 16 | Acid Base Equilibrium | 218 |
| Chapter 17 | Solubility and Simultaneous Equilibria | 231 |
| Chapter 18 | Electrochemistry | 242 |
| Chapter 19 | Metal Complexes and Their Equilibria | 262 |
| Chapter 20 | Nuclear Reactions and Their Role in Chemistry | 271 |
| Chapter 21 | Organic Compounds, Polymers, and Biochemicals | 282 |
| Supplementary Chapter 1 | Simple Molecules and Ions of Nonmetals: Part I | 295 |
| Supplementary Chapter 2 | Simple Molecules and Ions of Nonmetals: Part II | 307 |
| Supplementary Chapter 3 | The Properties and Chemistry of Metals | 318 |

# Chapter 1 Building a Foundation

## Multiple Choice

1. A characteristic which distinguishes between a true science and a pseudoscience is:

   a. one deals with quantitative information only while the other deals with qualitative information only
   b. one deals with natural phenomena, while the other deals with artificial phenomena
   ! c. one deals with reproducible observations, while the other does not
   d. one deals with scientific facts, while the other deals with empirical facts
   e. one is developed in academic research laboratories, while the other is developed in non-academic laboratories

2. Complete the following sentence. A scientific law is:

   ! a. a statement describing a reproducible relationship between phenomena
   b. a tentative explanation for a set of observations that can be tested by further experimentation
   c. a tested explanation that explains a body of facts and relations
   d. a model used to visualize the invisible
   e. an absolutely correct explanation of a scientific principle to which no exception can be found

3. Which one of the following does not involve a chemical change?

   a. a fish that is left unrefrigerated decomposes
   b. grape juice which is left in an open bottle ferments
   c. bread rises when baking
   ! d. when heated, liquid water is transformed into steam
   e. when sugar is fermented, alcohol is produced

4. Which one of the following is a physical change?

   a. when ignited with a match in open air, paper burns
   ! b. in cold weather, water condenses on the inside surface of single pane windows
   c. when treated with bleach, some dyed fabrics change color
   d. when heated strongly, sugar turns dark brown
   e. in the absence of refrigeration, apple juice turns sour

5. Which one of the following is a chemical change?

   ! a. when blood is washed with 3% hydrogen peroxide solution, it changes color
   b. when water is boiled, it forms steam
   c. when a solid stick of butter is heated, it becomes a liquid
   d. when blue paint is mixed with yellow paint, a green color is obtained
   f. when a bar of gold metal is pounded with a hammer, it flattens out

6. Which one of the following properties is a chemical property?

! a. combustibility
b. volatility
c. viscosity
d. malleability
e. ductility

7. Which one of the following is an extensive property of matter?

a. density
b. specific gravity
c. electrical conductivity
d. refractive index
! e. mass

8. Which one of the following is an SI base unit?

a. dyne
b. newton
c. milliliter
! d. ampere
e. joule

9. The kilo is

a. a unit of mass
b. a unit employed in medical terminology
! c. a decimal multiplier in the metric system
d. a unit of speed
e. a volume unit employed by the DEA (drug enforcement agency)

10. The SI base units of temperature and mass respectively are

a. degree and gram
! b. kelvin and kilogram
c. celsius and milligram
d. °K and kilogram
e. kelvin and gram

11. The device used for measuring mass accurately is

! a. analytical balance
b. ballistic galvanometer
c. scale
d. buret
e. dynamometer

12. Which one of the following is not a unit of surface?

! a. gigabyte
b. acre
c. square mile
d. $dm^2$
e. square yard

13. Which one of the following is a unit of energy, but not an SI unit of energy?

a. joule
b. g/cm$^3$
! c. calorie
d. pascal
e. newton

14. The SI unit for force is the:

a. pascal
b. g/cm$^3$
c. calorie
d. joule
! e. newton

15. The boiling point of sulfur is 444 °C. Determine the equivalent value in °F

a. 279 °F
! b. 831 °F
c. 264 °F
d. 857 °F
e. 797 °F

16. The melting point of platinum is listed in one handbook as 1770 °C. Determine the equivalent value in °F

a. 3154 °F
b. 3128 °F
c. 1015 °F
! d. 3218 °F
e. 3244 °F

17. On a day in the summer of 1992, the temperature fell from 95 °F to 72° F in just three hours. The temperature drop expressed in celsius degrees (C°) was

! a. 13 C°
b. 41 C°
c. 45 C°
d. 9 C°
e. 75 C°

18. The melting point of aluminum was listed in one handbook as 1220 °F. Expressed in kelvins this temperature would be

a. 919 K
b. 2501 K
! c. 933 K
d. 798 K
e. 812 K

19. The melting point of nickel chloride, a yellow solid, is 1001 °C. Determine the melting point of this compound in units of °F.

a. 588 °F
b. 728 °F
c. 1274 °F
! d. 1834 °F
e. 2002 °F

20. The boiling point of chlorine is -34.6 C. This temperature expressed in Kelvins is

a. -30.3 K
b. 177.4 K
! c. 238.5 K
d. 243 K
e. 307.6 K

21. A special alloy melts at 601 °C. What is this temperature in °F?

a. 302 °F
b. 365 °F
c. 1050 °F
d. 1082 °F
! e. 1114 °F

22. Ammonia boils at -33.4 °C. What is this temperature in °F?

a. +13.5 °F
b. -18.5 °F
! c. -28.1 °F
d. -60.1 °F
e. -92.1 °F

23. The highest temperature recorded in the field house was 122.0 °F. Express this in °C

! a. 50.0 °C
b. 64.4 °C
c. 67.8 °C
d. 162 °C
e. 219.6 °C

24. An electronic scale used in the mailroom is calibrated (and displays) in tenths of a kilogram from 1 to 140 kg. Determine how many significant figures should be used to express the mass of anyone using the scale whose mass is between 80.2 and 83.5 kg.

! a. 3
b. 5
c. 4
d. 2
e. 1

25. Which response gives the correct number of significant figures for all three of the following measurements? #1) 7.103 cm #2) 0.00005 inch #3) $1.3400 \times 10^{-4}$ $dm^3$

a. 3, 5, and 4
b. 3, 1, and 3
c. 4, 1, and 3
! d. 4, 1, and 5
e. 4, 5, and 5

26. After carrying out the operations, (13.726 + 0.027)/8.221, how many significant figures are appropriate to show in the result?

a. 1
b. 2
c. 3
! d. 4
e. 5

27. How many significant figures should there be in result of the operation, 8.5201 + 1.93?

a. 1
b. 2
c. 3
! d. 4
e. 5

28. When the expression, 412.272 + 0.00031 - 1.00797 + 0.000024 + 12.8, is evaluated the result should be expressed as

a. 424
b. 424.0
! c. 424.1
d. 424.06
e. 424.064364

29. When the expression, 16.0200 + 0.00048 - 11.184 - 221.1, is evaluated the result should be expressed as

! a. - 216.3
b. - 216.26
c. - 216.2635
d. - 216.26352
e. - $2.2 \times 10^2$

30. When a student evaluates the expression below, the result should be expressed as

$$M = [4.268 \times 0.082057 \times 373.15]/[(744.6/760.0)(2.688)]$$

a. 49.623
b. 49.631
! c. 49.62
d. 49.63
e. 49.623202

31. When a student evaluates the expression below, the result should be expressed as

$$V = [0.04616 \times 0.082057 \times 293.30]/0.654$$

a. 1.69
! b. 1.70
c. 1.699
d. 1.6987
e. 1.69870

32. One radio station operates on an assigned frequency of 96.3 megahertz while another one operates on an assigned frequency of 1280 kilohertz. What is the ratio of the larger to the smaller value?

a. 13.3
b. 13.29
! c. 75.2
d. 75.23
e. 13.2918

33. The SI prefixes giga and micro represent, respectively:

! a. $10^{9}$ and $10^{-6}$
b. $10^{-9}$ and $10^{-6}$
c. $10^{6}$ and $10^{-3}$
d. $10^{3}$ and $10^{-3}$
e. $10^{-9}$ and $10^{-3}$

34. One radio station operates on an assigned frequency of 88.1 megahertz while another one operates on an assigned frequency of 1460 kilohertz. What is the ratio of the larger to the smaller value?

a. 16.5
b. 16.57
c. 16.57208
! d. 60.3
e. 60.342

35. A distance of $14 \times 10^{-3}$ meters in numerically equivalent to

a. 1.4 micrometers
b. 1.4 millimeters
c. 0.14 meters
! d. 1.4 centimeters
e. 14 kilometers

36. The diameter of an atom is approximately $1 \times 10^{-8}$ cm. How many nanometers is this?

a. $1 \times 10^{1}$ nm
b. $1 \times 10^{-19}$ nm
c. $1 \times 10^{-15}$ nm
! d. $1 \times 10^{-1}$ nm
e. $1 \times 10^{-10}$ nm

37. How many micrometers are there in 3.726 km?

a. $3.726 \times 10^{6}$
b. $2.684 \times 10^{-7}$
c. $2.684 \times 10^{-4}$
d. 3.726 x 103
! e. $3.726 \times 10^{9}$

38. How many mm (millimeters) are there in 3.6 km?

a. $3.6 \times 10^{-5}$ mm
b. 3600 mm
c. $3.6 \times 10^{4}$ mm
d. $3.6 \times 10^{5}$ mm
! e. $3.6 \times 10^{6}$ mm

39. How many $cm^3$ is 0.00400 $dm^3$?

a. 0.0400 $cm^3$
b. 0.400 $cm^3$
! c. 4.00 $cm^3$
d. 0.00000400 $cm^3$
e. 250 $cm^3$

40. How many cubic inches are in 1.00 $dm^3$?

! a. 61.0 $in^3$
b. 155 $in^3$
c. 394 $in^3$
d. $1.64 \times 10^{4}$ $in^3$
e. $383 \times 10^{2}$ $in^3$

41. The average distance from the earth to the sun is $9.30 \times 10^{7}$ miles. Light travels at a speed of $3.00 \times 10^{8}$ meters per second. If the sun were to be suddenly extinguished, how long would it take for the effect to be visible here on the earth?

a. 8.31 seconds
b. 5.38 minutes
c. 0.499 seconds
! d. 8.31 minutes
e. 0.310 seconds

42. Which one of the following represents the smallest distance?

! a. 4.5 mm
b. 0.20 inch
c. 0.83 cm
d. 0.73 m
e. 0.30 yard

43. How many square meters are there in a rectangular piece of carpet which measures 12.0 feet by 22.0 feet?

! a. 24.5 $m^2$
b. 28.4 $m^2$
c. 866 $m^2$
d. 80.5 $m^2$
e. 966 $m^2$

44. How many square meters of floor space are there in a room which has 225.0 square yards of floor space?

! a. 188.1 square meters
b. 269.0 square meters
c. 246.0 square meters
d. 205.7 square meters
e. 172.0 square meters

45. If a car has an EPA mileage rating of 30 miles per gallon, what is this rating in km/L (kilometers per liter)? 1 liter = 1 $dm^3$

! a. 13 km/L
b. 200 km/L
c. 180 km/L
d. 70 km/L
e. 11 km/L

46. The number, 0.0030600, is properly expressed in scientific notation as

a. $3.0600 \times 10^{-2}$
b. $0.30600 \times 10^{-2}$
c. $0.306 \times 10^{-2}$
d. $3.06 \times 10^{-3}$
! e. $3.0600 \times 10^{-3}$

47. The number, 0.02100, is properly expressed in scientific notation as

a. $0.21 \times 10^{-1}$
b. $2.1 \times 10^{-2}$
! c. $2.100 \times 10^{-2}$
d. $21.0 \times 10^{-3}$
e. $2.10 \times 10^{-2}$

48. An empty volumetric flask, weighing 60.42 grams, weighed 309.60 grams when filled with water. After emptying the water and drying the flask, it was filled with ethylene glycol. It now weighed 338.72 grams. What is the specific gravity of the ethylene glycol?

a. 0.8048
b. 0.9140
c. 1.094
! d. 1.1169
e. 1.2424

49. An empty volumetric flask, weighing 27.16 grams, has a volume of 100.4 $cm^3$. How much would it weigh when it filled with bromine (a liquid element) which has a density of 3.1028 $g/cm^3$?

a. 59.51 g
b. 284.4 g
c. 311.5 g
! d. 338.7 g
e. 395.8 g

50. How many $cm^3$ of water will overflow from a full container of water if a 52.5 gram sample of nickel (density = 8.90 $g/cm^3$) is carefully placed in the container so there is no splashing, just overflowing.

a. 467 $cm^3$
b. 0.170 $cm^3$
! c. 5.90 $cm^3$
d. 43.6 $cm^3$
e. 61.4 $cm^3$

51. How many $cm^3$ of water will overflow from a full container of water if a 66.7 gram sample of vanadium (density = 6.11 $g/cm^3$) is carefully placed in the container so there is no splashing, just overflowing.

a. 11.9 $cm^3$
b. 9.92 $cm^3$
c. 60.6 $cm^3$
! d. 10.9 $cm^3$
e. 0.0916 $cm^3$

52. A sample of zinc metal (density = 7.14 $g/cm^3$) was submerged in a graduated cylinder containing water. The water level rose from 162.5 $cm^3$ to 186.0 $cm^3$ when the sample was submerged. How many grams did the sample weigh?

! a. 168 g
b. 22.7 g
c. 26.1 g
d. 48.8 g
e. 3.29 g

53. A sample of an alloy (density = 9.62 g/cm$^3$) was submerged in a graduated cylinder containing water. The water level rose from 166.5 cm$^3$ to 182.0 cm$^3$ when the sample was submerged. How many grams did the sample weigh?

a. 0.621 g
b. 175 g
c. 18.9 g
d. 17.8 g
! e. 149 g

54. A compound, once named "Isooctane" and used as a fuel in gasoline engines because it had an antiknock octane rating of 100, has a density of 0.6919 g/cm$^3$. How many pounds would 20.4 gallons of this fuel (a typical full tank) weigh? 1 gallon = 3785 cm$^3$ = 4 quart 1 pound = 453.6 g

a. 77.2 pounds
b. 246 pounds
! c. 118 pounds
d. 25.2 pounds
e. 50.6 pounds

55. Given this data from lab, for mass and then the volume of water displaced in a graduated cylinder by introduction of a metal sample to the cylinder. Mass of weighing cup = 0.452 g. Mass of weighing cup + metal sample = 72.943 g. Volume of water in cylinder = 15.2 cm$^3$. Introduction of the metal sample causes the water above the submerged metal sample to rise to the 19.0 cm$^3$ mark. What is the density of the metal?

a. 2.15 g/cm$^3$
b. 3.84 g/cm$^3$
c. 4.80 g/cm$^3$
! d. 19 g/cm$^3$
e. 19.2 g/cm$^3$

56. Mercury, which has a density of 13.595 g/cm$^3$, is usually stored in iron vessels for shipment. Typically, these vessels have a capacity of 2.60 liters (1 liter = 1 dm$^3$). How many pounds of mercury would a filled container hold? 1 pound = 0.4536 kg.

! a. 77.9 pounds
b. 86.7 pounds
c. 11.5 pounds
d. 16.0 pounds
e. 42.6 pounds

57. Iron has a density of 7.86 g/cm$^3$. How many pounds does a block of iron with a volume of 1.65 cubic feet weigh? 1 pound = 0.4536 kg.

! a. 810 pounds
b. 0.871 pounds
c. 871 pounds
d. 491 pounds
e. 3.01 pounds

58. Gold has a density of 19.3 g/cm$^3$. How many pounds does a gold sphere (volume = 1.333333 x $\pi$ x radius$^3$) weigh if it has a diameter of 5.20 inches? 1 pound = 0.4536 kg.

a. 7.96 pounds
b. 16.3 pounds
c. 19.7 pounds
! d. 51.3 pounds
e. 63.7 pounds

59. A spherical cannon ball made of an iron alloy has a diameter of 9.55 inches and a density of 7.89 g/cm$^3$. (volume = 1.333333 x $\pi$ x radius$^3$). How many pounds does this cannon ball weigh? 1 pound = 0.4536 kg.

a. 59.0 pounds
! b. 130 pounds
c. 41.4 pounds
d. 124 pounds
e. 21.0 pounds

60. Iridium has a density of 22.65 g/cm$^3$. The vice president for research and development has an iridium paperweight on his desk which weighs 11.50 pounds. What is its volume, in cubic inches?

a. 5.533 cubic inches
b. 9.410 cubic inches
! c. 14.05 cubic inches
d. 35.70 cubic inches
e. 90.67 cubic inches

61. The metric equivalent of the popular 55 gallon drum has a volume of 0.200 cubic meters. One such drum was filled with a colorless liquid, Sukanol, which has a density of 1.168 g/cm$^3$. How many kg should this quantity of Sukanol weigh?

a. 24.3 kg
! b. 23.4 kg
c. 243 kg
d. 2.34 x $10^5$ kg
e. 500 kg

62. A coastal patrol boat which weighed 35.6 metric tons (1 metric ton = 1000 kg) was traveling at a speed of 15.0 knots (1 knot = 1.852 km/hr) when, due to navigational error, it collided with a very bulky submerged object. For the record (and the court of inquiry) how much kinetic energy did the boat possess prior to the impact?

a. 1.37 x $10^{13}$ joules
! b. 1.06 x $10^3$ kilojoules
c. 1.06 newtons
d. 13.7 megajoules
e. 17.8 kilojoules

63. A flatbed truck loaded with cinder blocks had a gross weight of 65.2 tons at the last weigh station. It lost its brakes and went off a curve on the interstate highway at a speed of 76.5 miles/hr, crashing into a bridge abutment. Calculate the kinetic energy possessed by the truck, in megajoules, prior to the impact?

a. 0.191 megajoules
b. 13.4 megajoules
! c. 34.6 megajoules
d. 64.9 megajoules
e. 168 megajoules

64. The density of iron is 7.86 g/cm$^3$. What is the mass (in kg) of 20.00 cubic inches of iron?

a. 6.46 x $10^{-3}$ kg
b. 4.17 x $10^{-2}$ kg
c. 0.393 kg
d. 2.54 kg
! e. 2.58 kg

## Fill in the Blanks

65. How many significant figures does the number 30.340 contain? ______ ( ! 5)

66. What will be the cost, in dollars, of gasoline for a 3710 mile trip in a car pulling a trailer that delivers 13.3 miles per gallon, if the average price of gas is 89.9 cents per gallon? ______ ( ! $250.77)

67. What will be the cost of gasoline for a 4710 mile automobile trip if the car gets 27.3 miles per gallon, and the average price of gas is $1.249 per gallon? ______ (! $215.49)

68. A kilometer is ______ times as long as a centimeter. ( ! $10^{+5}$)

69. Carbon tetrachloride, a colorless liquid has a density of 1.5867 g/cm$^3$ and a freezing point -22.9 °C, while water has a density of 0.99704 g/cm$^3$ and a freezing point of 0.00 °C. What is the specific gravity of carbon tetrachloride? ______ ( ! 1.5914)

70. A 201 g sample weighs ______ mg. ( ! 2.01 x $10^5$ mg)

71. A spot on a microchip which is 7500 nm (nanometers) in diameter is ______ pm (picometers) in diameter. ( ! 7.5 x $10^6$ pm)

72. An object weighing 450 kg, expressed in megagrams (Mg), is ______ Mg. ( ! 0.450 Mg)

73. Many home freezers maintain a temperature of 0.0 °F. Express this temperature in °C. ______ ( ! -17.8 °C)

74. How many significant digits are there in the number 1.050 x $10^9$? ______ ( ! 4)

75. Express the result of the operation, 8.520 + 2.7, to the proper number of significant digits. ______ ( ! 11.2)

## True and False

76. Scientific theories which do not require testing by observation are called hypotheses. ( ! F)

77. A sample of each of the seven base units in the SI system of measurement is kept by the French Academy of Sciences in a special temperature controlled room. ( ! F)

78. The international standard meter is preserved on a platinum alloy bar kept at the National Institute of Standards and Technology (NIST) in Gaithersburg, Maryland near Washington DC. ( ! F)

79. A large crowd attending a bicentennial celebration in a metropolitan area was described as 450,000 in the morning newspaper the next day. A chemistry student stated that, mindful of the purpose of using scientific notation, this should be correctly expressed in scientific notation as $4.50000 \times 10^5$. Is this true or false? ( ! F)

80. Any number known accurately to six or more significant digits is defined as an exact number. ( ! F)

81. In determining the number of significant digits in the result of a calculation, exact numbers are considered as having a value to six significant digits. ( ! F)

82. The ambiguity in the precision of a number being expressed can be eliminated by proper use of scientific notation. ( ! T)

83. The result of the following operation, 8.52010 x 7.9, should be expressed as 67.3088. ( ! F)

84. A 15 km distance run is a shorter run than a 10 mile distance run. ( ! T)

85. A 5.00 pound bag of sugar weighs more than a 2.50 kg bag of sugar. ( ! F)

86. A piece of carpet which measures 44.0 square yards is smaller that a piece of carpet which measures 44.0 square meters. ( ! T)

87. A liter of carbon tetrachloride is a smaller quantity than a quart of the same substance. ( ! F)

88. An increase of one kelvin in temperature is a smaller change than an increase of one degree Fahrenheit. ( ! F)

89. Since the large ocean going container vessel (200,000 tons) was drifting at just 1.5 miles per hour on the tide when it struck the bridge pylon, it possessed less kinetic energy than the 4500 pound speedboat which struck the same pylon last year while going 48.5 miles per hour. ( ! F)

90. Intensive properties of matter depend on only one property, color. ( ! F)

91. Specific gravity equals (mass/volume) divided by time. ( ! F)

## Critical Thinking

92. A spherical cannon ball made of an iron alloy has a specific gravity of 7.88, and weighs 22.12 pounds. It has a diameter of 13.46 cm. On the other hand, spent uranium (from processing which removes the commercially important rare isotope) has a specific gravity of 19.107. How many pounds would a uranium ball of the same dimensions as the iron cannon ball weigh? ______ ( ! 53.6 pounds)

93. A spherical cannon ball made of an iron alloy has a density of 7.86 g/cm$^3$, and weighs 22.12 pounds. On the other hand, spent uranium (from processing which removes the commercially important rare isotope) has a density of 19.05 g/cm$^3$. What would be the diameter, in cm, of a uranium round shot which has exactly twice the weight as the iron ball described above? ______ (! 12.62 cm)

94. Calculate a conversion factor to four significant digits, which will convert cubic feet into cubic meters directly. (1 m$^3$ = ______ ft$^3$) (! 35.31)

95. A young high school student has invented a new temperature scale, the Zuban scale (his last name, of course). In common with the Kelvin, Celsius, and Fahrenheit scales, it is a linear scale. According to Zuban, 1) the boiling point of water which is 100.00 °C = 373.15 K = 0.00 °Z; 2) the boiling point of sulfur which is 444.60 °C = 717.75 K = 250.00 °Z; 3) the melting point of silver which is 960.15 °C = 1233.30 K = 624.02 °Z. Calculate the value of absolute zero on the Zuban Scale. ______ (! -270.71 °Z)

96. Iron has a density of 7.86 g/cm$^3$. As part of their exam, students in the junior class at a boarding school were to weigh a metal sphere, measure its diameter and calculate its density. Some of the seniors sneaked into the lab the weekend before, took the iron sphere, took it downtown, and had the interior partially hollowed and the surface repaired so it wouldn't be noticed and replaced it in the cabinet late on Sunday. The junior students who had this iron sphere for their "unknown" obtained 9.30 cm for the diameter and 2.44 kg for the mass. What value should they have reported for the mass of the sphere if it had not been tampered with, and what was the volume of the hollowed out space in the interior of the sphere? ________________ (! 3.31 kg and 111 cm$^3$)

97. Kevin, a ninth grade whiz kid, has invented a new temperature scale, the Vitellan scale (his last name, of course). In common with the Kelvin, Celsius, and Fahrenheit scales, it is a linear scale. According to Kevin, 1) the melting point of benzene which is 5.53 °C = 0.00 °V; 2) the boiling point of benzene which is 80.10 °C = 100.00 °V; 3) the melting point of lead which is 327.50 °C = 431.77 °V. Calculate the value for the melting point of mercury ( -38.86 °C) on the Vitellan scale. ______ (! -59.53 °V)

# Chapter 2 The Structure of Matter: Atoms, Molecules, and Ions

## Multiple Choice

1. The two major types of pure substances are

! a. compounds and elements
b. compounds and solutions
c. elements and mixtures
d. mixtures and solutions
e. solutions and elements

2. All of the following properties of a sample of a pure substance can be used for identification except its

a. density
b. freezing point temperature
! c. mass
d. melting point temperature
e. solubility in 100 g of water (g solute/100 g water at 25°)

3. A chemical reaction must be used to separate

a. air into oxygen, nitrogen, and other components
! b. a compound into its elements
c. gases from liquids
d. a mixture into its components
e. solids from liquids

4. Distillation can readily be used to separate

a. the elements in a compound
b. a heterogeneous mixture of two solids
c. a homogeneous solution of two solids
d. iron filings from sugar and salt crystals
! e. a liquid solvent from a dissolved solid

5. The symbol "Si" is used to represent the element:

a. silver
! b. silicon
c. sodium
d. sulfur
e. tin

6. An example of a chemical compound is

a. air
b. brass
c. bronze
d. granite
! e. table salt

7. Which element below occurs in nature as a gas composed of diatomic molecules at ordinary temperatures and pressures

a. boron
b. gold
c. helium
! d. oxygen
e. sulfur

8. Which element below is a liquid which conducts electricity very well at ordinary temperatures and pressures

a. bromine
b. copper
c. krypton
! d. mercury
e. zinc

9. Which of the statements below is true?

a. When two atoms combine, they do so in definite proportions by weight
b. When two different compounds combine to form an element, they do so in definite proportions by weight
! c. When two different elements combine to form a compound, they do so in definite proportions by weight
d. When two molecules combine, they do so in definite proportions by weight
e. When two different elements combine to form a mixture, they do so in definite proportions by weight

10. The relative number of atoms of each element in a particular compound,

a. is always 1:1.
b. is the same as the density ratio
c. is the same as the weight ratio.
! d. is definite and constant.
e. cannot be determined experimentally.

11. The smallest unit of a compound that still maintains the chemical properties of that compound is the

a. atom
b. electron
! c. molecule
d. nucleon
e. nucleus

12. Naturally occurring boganium, Bo, has the following composition. $^{47}$Bo, 46.972 a.m.u., 69.472% $^{49}$Bo, 48.961 a.m.u., 21.667% $^{50}$Bo, 49.954 a.m.u., 8.8610%. What is the average atomic weight of naturally occurring boganium?

! a. 47.667
b. 47.699
c. 48.629
d. 48.667
e. 48.961

13. Naturally occurring engrium, En, has the following composition. $^{127}$En, 146.9672 a.m.u., 64.792% $^{149}$En, 148.9638 a.m.u., 26.117% $^{150}$En, 149.9592 a.m.u., 9.0910%. What is the average atomic weight of naturally occurring boganium?

a. 49.254
! b. 147.76
c. 148.63
d. 148.67
e. 147.80

14. Naturally occurring vitellium, Vi, has the following composition. $^{188}$Vi, 187.9122 a.m.u., 10.861% $^{191}$Vi, 190.9047 a.m.u., 12.428% $^{193}$Vi, 192.8938 a.m.u., 76.711%. What is the average atomic weight of naturally occurring vitellium?

a. 64.035
b. 190.57
c. 190.67
d. 192.08
! e. 192.11

15. Naturally occurring zubanium, Zb, has the following composition. $^{148}$Zb, 147.9554 a.m.u., 10.563% $^{151}$Zb, 150.9496 a.m.u., 70.811% $^{153}$Zb, 152.9461 a.m.u., 18.626%. What is the average atomic weight of naturally occurring zubanium?

a. 50.335
b. 150.62
c. 150.67
! d. 151.01
e. 151.08

16. Which of the statements below is reasonably accurate?

! a. a cation is formed by removal of an electron(s) from an atom
b. a cation is a positively charged electron
c. a cation is a positively charged neutron
d. a cation is a negatively charged neutron
e. a cation is a negatively charged proton

17. Which of the following contributes to the charge but does NOT contribute significantly to the mass of an atom?

! a. electrons
b. nuclei
c. nucleons
d. neutrons
e. protons

18. Uranium exists in nature in the form of several isotopes; the different isotopes have different

a. atomic numbers
b. charges
c. numbers of electrons
! d. numbers of neutrons
e. numbers of protons

19. Which answer below best describes all atoms of a given isotope of a particular element?

a. they possess the same mass, only
b. they possess the same chemical properties and the same mass, but nothing else in common
c. they possess the same atomic number and the same mass, but have nothing else in common
d. they possess the same number of electrons, the same atomic number, the same mass, but nothing else in common
! e. they possess the same number of electrons, the same atomic number, the same mass, and the same chemical properties

20. Which answer below best describes all atoms of a particular element?

a. they possess the same number of electrons, the same atomic number, the same mass, but nothing else in common
b. they possess the same mass and the same chemical properties, but nothing else in common
! c. they possess the same number of electrons, the same atomic number, the same chemical properties, but not the same mass
d. they possess the same chemical properties and the same mass, but nothing else in common
e. they possess the same atomic number and the same mass, but have nothing else in common

21. The nuclide below which has 24 neutrons is

a. $^{52}Cr$
b. $^{56}Fe^{2+}$
3. $^{24}Mg$
! d. $^{45}Sc$
e. $^{51}V$

22. Both $^{85}Rb^{+}$ and $^{80}Br^{-}$ have the same

! a. number of electrons
b. mass
c. number of neutrons
d. number of nucleons
e. number of protons

23. Consider the nuclides sodium-23 and magnesium-24. Both of these nuclides have the same

a. number of electrons
b. mass
! c. number of neutrons
d. number of nucleons
e. number of protons

24. Consider the nuclides iron-56 and iron-57. Both of these nuclides have the same

a. number of electrons
b. number of neutrons
c. number of nucleons
d. number of photons
! e. number of protons

25. Consider the nuclides copper-65 and zinc-65. Both of these nuclides have the same

a. number of electrons
b. mass
c. number of neutrons
! d. number of nucleons
e. number of protons

26. The oxide ion possesses,

a. 6 electrons
b. 8 electrons
! c. 10 electrons
d. 16 electrons
e. 18 electrons

27. Compare the sodium-23 atom and the sodium-23 ion. In what respect do these species differ?

! a. number of electrons
b. number of neutrons
c. number of nucleons
d. number of photons
e. number of protons

28. A neutral iodine atom has 78 neutrons. Which description below fits this atom?

a. 39 protons, 131 nucleons, 39 electrons
b. 53 protons, 78 nucleons, 53 electrons
c. 53 protons, 78 nucleons, 54 electrons
! d. 53 protons, 131 nucleons, 53 electrons
e. 53 protons, 131 nucleons, 54 electrons

29. What is the correct formula for an ionic compound formed between magnesium and chlorine?

a. $MgCl$
! b. $MgCl_2$
c. $Mg_2Cl$
d. $Mg_2Cl_7$
e. $Mg_7Cl_2$

30. A compound containing a cation and a complex oxyanion has the formula $CaFnO_4$. The complex oxyanion in this compound has a parent acid. What is the formula of the parent acid?

a. HFn
b. $H_2Fn$
c. $HFnO_4$
d. $H_2FnO_2$
! e. $H_2FnO_4$

31. A compound containing a cation and a complex oxyanion has the formula $Ca(ZbO_3)_2$. The complex oxyanion in this compound has a parent acid. What is the formula of the parent acid?

a. HZb
b. $H_2Zb$
! c. $HZbO_3$
d. $H_2ZbO_3$
e. $H_2ZbO_4$

32. Calcium karnellate, $Ca(KnO_3)_2$, contains an anion which has a parent acid. What is the name of that acid?

a. hydrokarnellic acid
b. hypokarnellous acid
c. karnellous acid
! d. karnellic acid
e. perkarnellic acid

33. Which one of the following named acids is the one which probably does not have ANY oxygen in it?

! a. hydrokarmic acid
b. hypokarmic acid
c. karmic acid
d. karmous acid
e. perkarmic acid

34. Which set of elements below are all in the same period?

a. Ba, Pb, As, Sn
b. Fr, U, Am, Ca
c. K, Na, Li, Cs
! d. Na, Al, P, Ar
e. Nd, Dy, Pu, Os

35. Which set of elements below includes only known elements which are metals?

a. lead, bismuth, bromine, magnesium
b. mobanium, nevadium, zeon, bromium, silanium
c. nitrogen, silicon, sulfur, arsenic
! d. uranium, americium, praseodymium, zinc
e. zinc, radon, barium, tin

36. Which set of elements below includes mostly non-metals?

a. barium, calcium, strontium
b. lanthanum, lutecium, rhodium
! c. oxygen, selenium, tellurium
d. silicon, zinc, strontium
e. sodium, lithium, nitrogen

37. What is the charge on all the simple ions of metals of group IA?

! a. +1
b. +2
c. -1
d. -2
e. +4

38. The set below featuring one alkali, one alkaline earth, one halogen, and one lanthanide is . . .

a. cesium, barium, bromine, erbium, samarium
! b. francium, beryllium, iodine, terbium, berkelium
c. lithium, manganese, fluorine, lanthanum, vanadium
d. potassium, radium, iodine, lutetium, platinum
e. rubidium, strontium, chlorine, thorium, plutonium

39. Which set below includes TWO and only two actinide elements?

a. gallium, germanium, iron, barium, tellurium
b. lithium, sodium, potassium, rubidium, francium
c. magnesium, gallium, fluronium, missourium, neptunium
d. radium, polonium, actinium, platinum, selenium
! e. uranium, francium, gallium, plutonium, titanium

40. Which set below includes only alkali metal elements?

a. gallium, germanium, iron, barium, tellurium
! b. lithium, sodium, potassium, rubidium, francium
c. magnesium, gallium, fluronium, missourium, neptunium
d. radium, polonium, actinium, platinum, selenium
e. uranium, francium, gallium, plutonium, titanium

41. Which set below does not include at least one transition metal element?

a. gallium, germanium, iron, barium, tellurium
b. lithium, radium, magnesium, palladium, boron
! c. magnesium, gallium, aluminum, missourium, neptunium
d. radium, polonium, actinium, platinum, selenium
e. uranium, francium, gallium, plutonium, titanium

42. Which of the sets below includes the largest number of elements?

a. alkali metals
b. alkaline earth elements
c. halogens
! d. lanthanides
e. noble gases

43. Which set below represents elements which are often found in nature as uncombined elements?

a. magnesium vanadium hydrogen tin chlorine
! b. neon gold platinum silver carbon
c. uranium iron argon silver manganese
d. sodium sulfur lithium barium helium
e. calcium chromium copper phosphorus zinc

44. The present form of the periodic table evolved from the pioneering work, in this area, of . . . .

a. John Dalton
! b. Dmitri Mendelyev
c. Isaac Newton
d. J. J. Thomson
e. Ernest Rutherford

45. What is the correct name for the compound $IBr_3$?

a. bromic iodide
b. iodine bromate
! c. iodine tribromide
d. iodine tribromine
e. monoiodine tribromite

46. What is the correct name for the compound $HF(aq)$?

a. fluorohydric acid
! b. hydrofluoric acid
c. hydrogen fluoride
d. hydrogen fluorate
e. monohydrogen monofluoride

47. What is the correct name for the compound $H_2Se(aq)$

a. dihydrogen monoselenium
b. dihydrogen selenide
! c. hydroselenic acid
d. selenic acid
e. hydrogen selenide

48. What is the correct name for the compound $S_2Cl_2$?

a. disulfur chlorate
! b. disulfur dichloride
c. disulfur dichlorine
d. sulfur(I) chloride
e. sulfur(II) chlorine(II)

49. What is the correct name for the compound $HI(g)$?

a. hydriodic acid
b. hydrogen monoiodide
! c. hydrogen iodide
d. iodic acid
e. monohydrogen monoiodide

50. An alkaline earth element, which we will call X, unites with a halogen, which we will call Q. What would be the correct formula of the resulting compound?

a. XQ
! b. $XQ_2$
c. $XQ_4$
d. $X_2Q$
e. $X_4Q$

51. Aluminum unites with a second element, which we will call E, to form a definite compound whose formula is $AlE_3$. Element E is most probably

a. an actinide element
b. an alkali metal
c. a chalcogen
! d. a halogen
e. a transition element

52. What is the correct name for the compound HI(*aq*)?

! a. hydriodic acid
b. hydrogen monoiodide
c. hydrogen iodide
d. iodic acid
e. monohydrogen monoiodide

53. What is the correct name for the compound $V_2O_5$? (Remember, for transition metals ......)

a. divanadium pentoxide
b. vanadic oxide
! c. vanadium(V) oxide
d. vanadium(V) pentoxide
e. vanadous oxide

54. What is the correct name for the compound $NaCl_3$?

a. sodium chlorate
b. sodium chlorite
c. sodium perchloride
d. sodium trichloride
! e. there is no such compound

55. What is the correct name for the compound $CuBr_2$? (Remember, for transition metals ......)

a. copper(I) bromide(II)
! b. copper(II) bromide
c. copper(II) bromite
d. copper dibromide
e. cuprous bromide

56. What is the correct name for the compound HCN(*g*)?

a. hydrocarbonitride
b. hydrocyanic acid
c. hydrogen carbonitride
d. hydrogen cyanate
! e. hydrogen cyanide

57. What is the correct name for the compound $Na_2O$?

a. disodium oxide
! b. sodium oxide
c. sodium(I) oxide
d. sodium peroxide
e. sodium superoxide

58. A typographical error on an exam produced the formula, $P_4Se_7$, in one of the questions. How would you name this compound?

a. tetraphosphorus hexaselenide
! b. tetraphosphorus heptaselenide
c. phosphorus heptaselenite
d. phosphorus(IV) selenide
e. phosphorus(VII) selenide

59. Which one of the following is a correct name for the compound $FeBr_3$?

a. ferrous bromide
! b. iron(III) bromide
c. iron bromite
d. iron tribromide
e. iron tribromine

60. Which one of the following is the correct formula for the compound ferrous sulfate?

! a. $FeSO_4$
b. $Fe(SO_4)_2$
c. $Fe_2SO_4$
d. $Fe_2(SO_4)_3$
e. $Fe_3(SO_4)_2$

61. Which one of the following is a correct name for the compound $Hg_2Cl_2$?

a. dimercury dichloride
b. mercuric chloride
! c. mercury(I) chloride
d. mercury(II) dichloride
e. there is no correct name, the formula should be HgCl

62. Which one of the following is a correct formula for mercury(I) phosphate?

a. $HgPO_3$
b. $HgPO_4$
c. $Hg_2PO_4$
d. $Hg_2PO_3$
! e. $(Hg_2)_3(PO_4)_2$

63. Which one of the following is a correct name for the compound $CoF_3$?

a. cobalt fluoride
b. cobalt trifluoride
! c. cobaltic fluoride
d. cobaltic trifluoride
e. cobaltous fluoride

64. A correct name for the compound, $SnF_4$, would be

a. stannic tetrafluoride
b. stannous fluoride
c. stannous(IV) fluoride
! d. tin(IV) fluoride
e. tin tetrafluoride

65. A correct formula for stannous nitrate would be

a. $Sn(NO_2)_2$
! b. $Sn(NO_3)_2$
c. $Sn(NO_3)_3$
d. $Sn(NO_3)_4$
e. $Sn_2NO_3$

66. What is the correct formula for the compound, magnesic chlorate?

a. $MgClO_3$
b. $Mg(ClO_3)_2$
c. $Mg_2ClO_3$
d. $MgO(ClO_3)_2$
! e. there is no such compound

67. What is the correct name for the compound $MgSeO_3$?

a. magnesium selenate
b. magnesium selenide
! c. magnesium selenite
d. magnesium selenium trioxide
e. magnesium selenoxate

68. What is the correct name for the compound $K_2Cr_2O_7$?

a. potassium chromium(VII)-ate
! b. potassium dichromate
c. potassium dichromium heptaoxide
d. potassium heptaoxochromate
e. potassium perchromate

69. The compound $Na_2S_2O_3$ is used extensively in photographic film processing. What is its chemical name?

a. sodium bisulfite
b. sodium disulfur trioxide
c. sodium oxosulfate(IV)
! d. sodium thiosulfate
e. sodium trioxosulfite

70. What is the correct name for the compound $H_2Cr_2O_7$?

a. bichromic acid
b. chromic acid
! c. dichromic acid
d. hydrogen dichromate
e. hydrogen dichromium heptaoxide

71. If the $NtO_4^{2-}$ ion is called nortonate, what is the correct name for the compound $H_2NtO_4$?

a. dihydrogen nortonium tetraoxide
b. dihydrogen nortonate
c. hydrogen nortonate
! d. nortonic acid
e. nortonous acid

72. What is the correct name for the compound $Cu_2SO_3$?

! a. copper(I) sulfite
b. copper(II) sulfite
c. copper thiosulfate
d. cuprous sulfate
e. dicopper sulfur trioxide

73. What is the correct name for the $C_2H_3O_2^-$ ion?

! a. acetate ion
b. hydrocarbonate ion
c. monocarbonate ion
d. oxalate ion
e. sucrose ion

74. What is the correct name for $H_3PO_2$?

a. hydrogen hypophosphite
b. hydrogen phosphite
! c. hypophosphorus acid
d. phosphoric acid
e. trihydrogen phosphorus dioxide

75. Which one of the following is a correct name for the $HCrO_4^-$ ion?

! a. bichromate ion
b. dichromate ion
c. hydrogen chromium tetraoxide ion
d. monochromate ion
e. monochromic acid

76. Which one of the following is a correct name for the compound $KHCr_2O_7$?

a. potassium bichromite
b. potassium bichromate
c. potassium dichromic acid
d. potassium monohydrogen chromite
! e. potassium monohydrogen dichromate

77. What is the correct name for the compound $BaHSeO_4$?

a. barium selenic acid
b. barium hydrogen selenium tetraoxide
c. barium monohydrogen selenite
d. barium monohydrogen selenoxate
! e. there is no compound with that formula

78. What is the correct name for the compound $LiHPO_4$?

a. lithium monohydrogen phosphate
b. lithium hydrogen phosphoric acid
c. lithium hydrogen phosphorus tetraoxide
d. lithium monohydrogen phosphite
! e. there is no compound with that formula

79. What is the correct name for the compound $Li_2HPO_4$?

! a. lithium monohydrogen phosphate
b. there is no compound with that formula
c. dilithium monohydrogen phosphate
d. lithium hydrogen phosphorus tetraoxide
e. lithium phosphoric acid

80. What is the correct name for the compound $CuHSO_4$?

! a. copper(I) bisulfate
b. copper(II) bisulfate acid
c. copper hydrogen sulfur tetraoxide
d. copper hydrogen sulfate
e. copper sulfuric acid

81. What is the correct formula for the oxalate ion?

a. $CO_3^{2-}$
b. $C_4O_2^{2-}$
c. $C_4O_4^{2-}$
! d. $C_2O_4^{2-}$
e. $C_2H_3O_2^-$

82. If the $NtO_4^{2-}$ ion is called nortonate, what is the correct name for the compound $CsHNtO_4$?

a. cesium monohydrogen nortonite
b. cesium nortonic acid
! c. cesium binortonate
d. cesium hydrogen nortonium tetraoxide
e. based on the information given, this compound couldn't exist with the formula as given

83. One of the compounds below has the formula written incorrectly. Which one?

a. $Al(H_2PO_4)_3$
b. $Al(HCO_3)_3$
c. $Ba(H_2AsO_4)_2$
d. $Ca(HCO_3)_2$
! e. $KHPO_4$

84. One of the compounds below has the formula written incorrectly. Which one?

! a. $Al(H_2CO_3)_3$
b. $Al(H_2PO_4)_3$
c. $Ba(HCO_3)_2$
d. $Ca(H_2AsO_4)_2$
e. $KH_2PO_4$

85. The correct name for the compound $Al(SO_4)_3$ is:

! a. there in no compound with that formula--it must be incorrectly written
b. aluminum sulfate
c. aluminum trisulfate
d. aluminum(III) sulfate
e. aluminum sulfite

86. Which one of the following chemical reactions will yield an acid salt?

a. $1\ H_2SO_4 + 2\ NaOH \longrightarrow$
b. $1\ HCl + 1\ KOH \longrightarrow$
c. $2\ H_3PO_4 + 3\ Ba(OH)_2 \longrightarrow$
d. $1\ H_2SO_4 + 1\ Ba(OH)_2 \longrightarrow$
! e. $1\ H_3PO_4 + 2\ NaOH \longrightarrow$

87. Which one of the following is not a polyprotic acid?

a. arsenic acid
b. sulfuric acid
! c. nitric acid
d. carbonic acid
e. phosphorous acid

## Fill in the Blanks

88. What is the correct name for the compound $CH_4$? ______ ( ! methane)

89. What is the correct name for the compound $H_3N(g)$? ______ ( ! ammonia)

90. How many atoms are there in one formula unit of $[NH_4]_4[Fe(CN)_6]$? ______ ( ! 33)

91. How many atoms are there in one formula unit of $NH_4Al(SO_4)_2 \cdot 24H_2O$? ______ ( ! 88)

92. How many atoms are there in one formula unit of $Al_2(SO_4)_3 \cdot 18H_2O$? ______ ( ! 71)

93. What is the charge on all the simple ions of metals of Group IIA? ______ ( ! +2)

94. What is the correct formula for the sulfide ion? ______ ( ! $S^{2-}$)

95. The correct formula for lithium phosphate is ______. ( ! $Li_3PO_4$)

96. The correct formula for barium selenite is ______. ( ! $BaSeO_3$)

97. The correct formula for barium iodite is ______. ( ! $Ba(IO_2)_2$)

98. What is the correct formula for the compound, chromium (III) dihydrogen phosphate? ______ ( ! $Cr(H_2PO_4)_3$)

99. What is the formula for calcium bicarbonate? ______ ( ! $Ca(HCO_3)_2$)

## True and False

100. A correct name for the compound $P_4Se_{10}$ is phosphorus(IV) selenide. ( ! F)

101. The correct name for the compound $CrBr_2$, is chromic bromide. ( ! F)

102. The correct name for the compound $RbClO_4$, is rubidium(I) perchlorate. ( ! F)

103. The compound, $HBrO_3$, is named hydrogen bromate acid. ( ! F)

104. The correct name for the $CrO_4^{2-}$ ion is perchromate. ( !F)

## Critical Thinking

105. Vitellium arsenate has the formula, $Vi_3(AsO_4)_2$, while sodium nortonate has the formula, $Na_2NtO_4$. Which of the following would be the expected formula for vitellium nortonate?

! a. $ViNtO_4$
b. $Vi_2NtO_4$
c. $Vi(NtO_4)_2$
d. $Vi_2(NtO_4)_3$
e. $Vi_3(NtO_4)_2$

106. Engrium sulfate has the formula, $En_2(SO_4)_3$, sodium nortonite has the formula $Na_2NtO_3$, and engrium arsenite has the formula, $EnAsO_3$. Based on these names and formulas, what would you expect for the formula of engrium nortonate?

a. $EnNtO_4$
b. $En_2NtO_4$
c. $En(NtO_4)_2$
! d. $En_2(NtO_4)_3$
e. $En_3(NtO_4)_2$

107. Two elements, Qr and E, combine to form an ionic compound whose formula is $QrE_2$. Qr also combines with element Z for form an ionic compound, $Qr_3Z_2$. Based on this information, what is a reasonable value for the charge on E?

a. +1
! b. -1
c. +2
d. -2
e. -3

## Chemicals in Use 2:

108. Most of the world's petroleum fuel reserves exist in the form of underground pools under both the land surface and the water surface ("offshore") of our planet. ( ! F)

109. Burning coal for production of heat energy releases considerable quantities of radioactive actinide elements into the atmosphere. ( ! T)

110. Petroleum was formed in an early geologic age from decaying vegetation. ( ! F)

111. Fossil "fuels" are more valuable as a major source of chemical feedstock for manufacturing many important chemical products for various industries than as energy sources !T

112. Which form of solid combustible fossil fuel contains the smallest amount of water and other volatile materials?

a. oil sand
b. lignite
c. peat
! d. anthracite
e. bituminous coal

113. The primary objective of refining crude oil is

a. to remove the methane and ethylene for which there is a large commercial market
! b. to separate it into various useful products, like gasoline, diesel oil, fuel oil, asphalt, and such
c. to drastically reduce the high sulfur content which is environmentally objectionable in liquid fuels
d. to remove the mercury compounds which can poison the atmosphere and waters
e. to remove solid particulate material which clogs injectors of diesel engines and large injector furnaces in energy generating stations

114. Oil shale

a. is a rich source of natural gas
b. contains up to 50% lignite
c. contains up to 50% peat
! d. contains around 4% to 5% petroleum
e. decomposes to form oil sand if allowed to stand in the open in bright sunlight

## Chapter 3 Stoichiometry: Quantitative Chemical Relationships

### Multiple Choice

1. A sample of phosphorus trifluoride, $PF_3$, contains 1.400 moles of the substance. How many atoms are there in the sample?

a) 4
b) 5.6
c) $8.431 \times 10^{23}$
d) $2.409 \times 10^{24}$
! e) $3.372 \times 10^{24}$

2. A sample of arabinose, $C_5H_{10}O_5$, contains 0.6000 moles of the substance. How many carbon atoms are there in the sample?

a) 3
b) 5
c) $3.613 \times 10^{23}$
! d) $1.807 \times 10^{24}$
e) $3.011 \times 10^{24}$

3. A sample of sulfolane, $C_4H_8O_2S$, contains $5.00 \times 10^{24}$ atoms. How many moles of sulfolane are there in the sample?

a) 0.120 moles
! b) 0.554 moles
c) 1.81 moles
d) 8.30 moles
e) $3.33 \times 10^{23}$ moles

4. A sample of trifluoromethanesulfonic acid, $CHF_3O_3S$, contains $4.62 \times 10^{23}$ oxygen atoms. How many moles of $CHF_3O_3S$ are there in the sample?

! a) 0.256 moles
b) 3.00 moles
c) 0.259 moles
d) 0.767 moles
e) $1.53 \times 10^{23}$ moles

5. How many moles of carbon atoms are combined with 11.2 moles of hydrogen atoms in a sample of the compound, $C_3H_8$?

a) 3.00
b) 5.60
! c) 4.20
d) $6.02 \times 10^{23}$
e) 29.9

6. A sample of $C_{12}H_{22}O_{11}$, contains 0.4662 moles of carbon atoms. How many moles of hydrogen atoms are there in the sample?

a) 0.2543 moles
b) 0.4662 moles
c) 10.26 moles
! d) 0.8547 moles
e) 0.9324 moles

7. A sample of $K_3Fe(CN)_6$ contains $1.084 \times 10^{24}$ carbon atoms. How many potassium atoms are there in the same sample?

a) $1.084 \times 10^{24}$ atoms
b) $3.252 \times 10^{24}$ atoms
c) $2.168 \times 10^{24}$ atoms
d) $3.613 \times 10^{23}$ atoms
! e) $5.420 \times 10^{23}$ atoms

8. A sample of $C_7H_5N_3O_4$, contains 0.4662 moles of carbon atoms. How many nitrogen atoms are there in the sample?

a) $2.807 \times 10^{23}$ atoms
! b) $1.203 \times 10^{23}$ atoms
c) $4.101 \times 10^{23}$ atoms
d) $1.998 \times 10^{22}$ atoms
e) $6.551 \times 10^{23}$ atoms

9. A sample of $(N_2H_5)_2C_3H_4O_4$ contains $1.084 \times 10^{24}$ carbon atoms. How many moles of hydrogen atoms are there in the same sample?

a) 4.200 moles
b) 4.725 moles
c) 7.000 moles
! d) 8.400 moles
e) 2.400 moles

10. The mass of 5.20 moles of glucose, $C_6H_{12}O_6$, is correctly expressed as

a) $1.56 \times 10^{-21}$ g
b) 31.2 g
c) 34.7 g
! d) 937 g
e) $6.43 \times 10^{20}$ g

11. How many molecules of carbon dioxide are there in 154.0 grams of carbon dioxide?

a) 3.499
! b) $2.107 \times 10^{24}$
c) $4.214 \times 10^{24}$
d) $9.274 \times 10^{25}$
e) $4.081 \times 10^{27}$

12. Which one of the following contains the greatest number of carbon atoms?

! a) 0.250 moles of glucose, $C_6H_{12}O_6$
b) 1.20 moles of carbon dioxide, $CO_2$
c) 0.500 moles of $CaC_2$
d) 0.450 moles of $Al_2(CO_3)_3$
e) 0.350 moles of $C_4H_8O_2S$

13. How many atoms are there in one formula unit of $(NH_4)_4Fe(CN)_6$?

a) 15
b) 25
c) 28
! d) 33
e) 35

14. How many atoms are there in one formula unit of $NiSO_4{\cdot}7H_2O$?

a) 9
b) 14
! c) 27
d) 28
e) 33

15. A gas sample contains 16.0 g of $CH_4$, 16.0 g of $O_2$, 16.0 g of $SO_2$, and 33.0 g of $CO_2$. The total number of moles, of ***everything***, is:

a) 2.25
! b) 2.50
c) 2.75
d) 3.00
e) 4.00

16. The formula mass of $(NH_4)_2SO_4$ is

a) 84.12
b) 116.12
c) 118.13
! d) 132.14
e) 221.53

17. The formula mass of $Co(NH_3)_6(ClO_4)_3$ is

a) 318.53
b) 389.43
c) 402.57
! d) 459.47
e) 754.13

18. The formula mass of $Ni(H_2O)_6Cl_2$ is

a) 157.69
b) 193.00
c) 227.61
! d) 237.69
e) 296.83

19. What is the percent, by weight, of calcium in $Ca(OCl)_2$, to the proper number of significant digits? Use the atomic weights from your text!

! a) 28.030
b) 28.571
c) 31.562
d) 43.787
e) 44.493

20. What is the percent, by weight, of boron in $Al(BF_4)_3$, to the proper number of significant digits? Use the atomic weights from your text!

a) 9.501 %
b) 9.385 %
c) 10.152 %
! d) 11.285 %
e) 23.713 %

21. What is the percent, by weight, of chromium in $K_2CrO_4$, to the proper number of significant digits? Use the atomic weights from your text!

! a) 26.776 %
b) 31.763 %
c) 40.268 %
d) 42.241 %
e) 51.996 %

22. What is the percent, by weight, of oxygen in $NiSO_4 \cdot 7H_2O$?

a) 14.846
b) 39.875
c) 43.273
d) 49.531
! e) 62.661

23. Magnetite is a binary compound containing only iron and oxygen. The percent, by weight, of iron is 72.360 %. What is the empirical formula of magnetite?

a) $FeO$
b) $FeO_2$
! c) $Fe_3O_4$
d) $Fe_2O_3$
e) $Fe_2O_5$

24. Which one of the following is definitely not an empirical formula?

a. $C_{12}H_{16}O_3$
b. $C_{12}H_{22}O_{11}$
c. $C_3H_8O_2$
d. $C_4H_{12}N_2O$
! e. $C_6H_{12}O_4$

25. A compound contains sodium, boron, and oxygen. The experimental analysis gave values of 53.976 % sodium and 8.461 % boron, by weight, the remainder is oxygen. What is the empirical formula of the compound?

a. $NaBO_2$
! b. $Na_3BO_3$
c. $Na_3BO_2$
d. $NaB_3O$
e. $Na_3B_3O_8$

26. A compound contains potassium, nitrogen and oxygen. The experimental analysis gave values of 45.942 % potassium and 16.458 % nitrogen, by weight, the remainder is oxygen. What is the empirical formula of the compound?

! a. $KNO_2$
b. $KNO_3$
c. $K_2N_2O_5$
d. $KN_3O_8$
e. $K_2N_2O$

27. A compound used in a polymer research project contains carbon, hydrogen, and nitrogen. The assay values are: carbon, 58.774%; hydrogen, 13.810%; nitrogen, 27.416% Determine the empirical formula of this compound.

a. $C_3H_7N$
b. $C_2H_7N$
c. $C_2H_8N$
! d. $C_5H_{14}N_2$
e. $C_7H_{16}N_4$

28. A well characterized compound contains potassium, sulfur, and oxygen. The assay values are: potassium, 49.410%; sulfur, 20.261%; Determine the empirical formula of this compound.

! a. $K_2SO_3$
b. $K_2SO_4$
c. $K_2S_2O_4$
d. $K_2S_2O_3$
e. $K_3S_2O_8$

29. A new compound contains nitrogen, hydrogen, boron, and fluorine. The assay values are: nitrogen, 13.360%; hydrogen, 3.8455%; boron, 10.312%; Determine its empirical formula.

a) $NH_3BF_3$
b) $NH_4B_3F$
c) $N_4HB_4F$
! d) $NH_4BF_4$
e) $NH_3BF_4$

30. A freshly prepared compound contains potassium, hydrogen, phosphorus, and oxygen. The assay values are: potassium, 44.895%; hydrogen, 0.5787%; phosphorus, 17.783%; Determine the empirical formula of this compound.

a) $K_2H_4P_4O$
! b) $K_2HPO_4$
c) $K_2H_2PO_4$
d) $K_2HP_2O_5$
e) $KH_2PO_4$

31. A 7.300 gram sample of aluminum combined quantitatively with some selenium to form a definite compound. The compound weighed 39.35 grams. What is the empirical formula for this compound?

a) AlSe
b) $Al_2Se$
c) $AlSe_2$
! d) $Al_2Se_3$
e) $Al_3Se_2$

32. A 4.626 gram sample of a hydrocarbon, upon combustion in a combustion analysis apparatus, yielded 6.484 grams of carbon dioxide. The percent, by weight, of carbon in the hydrocarbon is therefore:

! a) 38.25 %
b) 19.47 %
c) 71.35 %
d) 40.16 %
e) 42.16 %

33. A 6.789 gram sample of a hydrocarbon, upon combustion in a combustion analysis apparatus, yielded 9.883 grams of carbon dioxide. The percent, by weight, of carbon in the hydrocarbon is therefore

a) 18.75 %
! b) 39.73 %
c) 68.69 %
d) 45.57 %
e) 34.44 %

34. A 4.626 gram sample of a hydrocarbon, upon combustion in a combustion analysis apparatus, yielded 6.527 grams of water. The percent, by weight, of hydrogen in the hydrocarbon is therefore:

a) 14.11 %
! b) 15.79 %
c) 41.09 %
d) 66.22 %
e) 85.89 %

35. A 6.789 gram sample of a hydrocarbon, upon combustion in a combustion analysis apparatus, yielded 6.527 grams of water. The percent, by weight, of hydrogen in the hydrocarbon is therefore

a) 9.615 %
! b) 10.76 %
c) 28.00 %
d) 38.72 %
e) 58.53 %

36. A 6.789 gram sample of a compound was analyzed for nitrogen in a nitrogen analysis apparatus. In the procedure, all the nitrogen present was completely converted to ammonia ($NH_3$). 1.637 grams of ammonia were obtained. The percent, by weight, of nitrogen in the compound is therefore

a) 35.57 %
b) 24.11 %
c) 75.89 %
d) 31.77 %
! e) 19.83 %

37. A 4.927 gram sample of a compound was analyzed for nitrogen in a nitrogen analysis apparatus. In the procedure, all the nitrogen present was completely converted to ammonia ($NH_3$). 1.369 grams of ammonia were obtained. The percent, by weight, of nitrogen in the compound is therefore

a) 27.79 %
b) 38.48 %
! c) 22.85 %
d) 61.52 %
e) 36.71 %

38. In a quantitative analysis study, 4.624 grams of a hydrocarbon (which contains carbon and hydrogen only) sample yielded 13.84 g of $CO_2$ and 7.556 g of $H_2O$ in a combustion analysis apparatus. Determine the empirical formula of the hydrocarbon.

a) $CH_3$
b) $C_2H_5$
! c) $C_3H_8$
d) $C_{10}H_{26}$
e) $C_{10}H_{27}$

39. In a quantitative analysis study, 4.624 grams of a compound containing carbon, hydrogen and oxygen only yielded 6.557 g of $CO_2$ and 4.026 g of $H_2O$ in a combustion analysis apparatus. Determine the empirical formula of the compound.

a) $CH_2O$
! b) $CH_3O$
c) $CH_4O$
d) $C_2H_4O$
e) $C_4H_2O$

40. A compound has an empirical formula $CH_2O$. An independent analysis gave a value of 150.13 for its molecular mass. What is the correct molecular formula?

! a) $C_5H_{10}O_5$
b) $C_6H_{12}O_6$
c) $C_{11}H_2O$
d) $C_6H_6O_8$
e) $C_9H_{10}O_2$

41. A compound has an empirical formula $C_2H_4O$. An independent analysis gave a value of 132.16 for its molecular mass. What is the correct molecular formula?

a) $C_4H_4O_5$
b) $C_{10}H_{12}$
c) $C_7O_3$
! d) $C_6H_{12}O_3$
e) $C_4H_8O_5$

42. When the chemical equation, $AsF_3 + CCl_4 \longrightarrow AsCl_3 + CCl_2F_2$, is correctly balanced, the sum of the smallest set of integer coefficients will be

a) 4
b) 5
! c) 10
d) 14
e) 20

43. When the chemical equation, $BaCl_2 + K_3PO_4 \longrightarrow KCl + Ba_3(PO_4)_2(s)$ is correctly balanced, the sum of the smallest set of integer coefficients will be

a) 7
b) 8
c) 10
! d) 12
e) 16

44. When the remaining coefficients in the chemical equation,

$$1\ C_{10}H_8 + 6\ CrO_3 + \_\_H_2SO_4 \longrightarrow 1\ C_8H_6O_4 + 2\ CO_2 + 3\ Cr_2(SO_4)_3 + \_\_H_2O$$

are calculated, the coefficient for the $H_2O$ will be

a) 13
b) 5
c) 3
d) 20
! e) 10

45. When the remaining coefficients in the chemical equation,

$$1\ C_{10}H_8 + 9\ MnO_2 + \_\_H_2SO_4 \longrightarrow 1\ C_8H_6O_4 + 2\ CO_2 + \_\_MnSO_4 + \_\_H_2O$$

are calculated, the coefficient for the $H_2SO_4$ will be

! a) 9
b) 13
c) 8
d) 12
e) 10

46. When the coefficients in the chemical equation,

$$K_2Cr_2O_7 + BaCl_2 + H_2O \longrightarrow BaCrO_4 + KCl + HCl$$

are calculated, the sum of the coefficients will be

a) 8
! b) 10
c) 13
d) 16
e) 18

47. When $BaCl_2$ reacts with $Na_3PO_4$, $Ba_3(PO_4)_2$ and NaCl are formed. How many moles of $Ba_3PO_4$ are formed for each mole of $BaCl_2$ that is consumed?

a) 3
b) 1
! c) 0.3333
d) 2.3333
e) 1.5

48. The reactant and product in a chemical reaction are shown below,

$$P_4O_{10} + 6\ Ca(OH)_2 \longrightarrow 2\ Ca_3(PO_4)_2 + 6\ H_2O$$

If you start with 2.40 moles of $P_4O_{10}$ and 3.30 moles of $Ca(OH)_2$ and they react completely,

! a) 1.10 moles of $Ca_3(PO_4)_2$ are produced
b) 0.90 moles of $Ca(OH)_2$ are left over
c) 2.40 moles of $Ca_3(PO_4)_2$ are produced
d) 1.65 moles of $Ca_3(PO_4)_2$ are produced
e) 0.60 moles of $P_4O_{10}$ are consumed

49. The left side of a balanced chemical equation is shown below,

$$K_2Cr_2O_7 + 4\,H_2SO_4 + 3\,SeO_2 \longrightarrow \ldots..$$

Given 0.600 moles of $K_2Cr_2O_7$, 2.800 moles of $H_2SO_4$ and 1.500 moles of $SeO_2$?

a) $H_2SO_4$ is the limiting reagent
b) $K_2Cr_2O_7$ is the limiting reagent
c) there are 1.300 moles of $H_2SO_4$ in excess
! d) there are 0.100 moles of $K_2Cr_2O_7$ in excess
e) there are 0.300 moles of $SeO_2$ in excess

50. Given the balanced chemical equation, $C_3H_8 + 5\,O_2 \longrightarrow 3\,CO_2 + 4\,H_2O$. If one reacts 0.3818 moles of $C_3H_8$ with 1.718 moles of $O_2$, how many moles of water are produced?

a) 1.336 moles
! b) 1.374 moles
c) 1.527 moles
d) 1.718 moles
e) 3.426 moles

51. Given the balanced chemical equation, $C_4H_4 + 5\,O_2 \longrightarrow 4\,CO_2 + 2\,H_2O$. If one reacts 0.3618 moles of $C_4H_4$ with 1.818 moles of $O_2$, how many moles of carbon dioxide are produced?

! a) 1.447 moles
b) 1.454 moles
c) 1.456 moles
d) 2.180 moles
e) 0.3978 moles

52. Given the balanced chemical equation, $C_4H_4 + 5\,O_2 \longrightarrow 4\,CO_2 + 2\,H_2O$. If one reacts 0.3618 moles of $C_4H_4$ with 1.818 moles of $O_2$, how many grams of water should be produced?

a) 11.02 g
! b) 13.04 g
c) 13.20 g
d) 19.64 g
e) 65.50 g

53. Given the chemical reaction, $C_4H_8 + O_2 \longrightarrow CO_2 + H_2O$. If one reacts 0.3218 moles of $C_4H_8$ with 2.000 moles of $O_2$, how many moles of water should be produced?

a) 1.333 moles
b) 1.609 moles
c) 0.6436 moles
! d) 1.287 moles
e) 2.574 moles

54. Given the chemical reaction, $AsF_3 + C_2Cl_6 \longrightarrow AsCl_3 + C_2Cl_2F_4$, if one reacts 1.3618 moles of $AsF_3$ with 1.000 mole of $C_2Cl_6$, how many moles of $AsCl_3$ should be produced?

a) 0.3618 moles
b) 0.7343 moles
c) 0.7500 moles
! d) 1.3333 moles
e) 1.3618 moles

55. Given the chemical reaction, $AsF_3 + C_2Cl_6 \longrightarrow AsCl_3 + C_2Cl_2F_4$, if one reacts 1.3618 moles of $AsF_3$ with 1.000 moles of $C_2Cl_6$, how many grams of $C_2Cl_2F_4$ should be produced?

a) 128.1 grams
b) 134.1 grams
! c) 170.9 grams
d) 174.6 grams
e) 185.5 grams

56. Thermal decomposition of $KClO_3(s)$ yields $KCl(s)$ and $O_2(g)$. When 4.288 grams of $KClO_3$ (0.003500 moles) undergo this reaction, how many grams of oxygen are produced?

a) 1.120 grams
b) 0.5601 grams
c) 2.240 grams
! d) 1.680 grams
e) 4.288 grams

57. When aluminum metal reacts with $HCl(aq)$, $AlCl_3(aq)$, and hydrogen gas are produced. If 4.288 grams of Al(0.1590 moles) undergo this reaction with an excess of hydrochloric acid, how many grams of hydrogen gas should be produced?

a) 0.1603 grams
b) 0.4770 grams
c) 6.048 grams
d) 1.388 grams
! e) 0.4808 grams

58. Phosphorus tribromide ($PBr_3$, MW = 270.69 g/mole) and water (MW = 18.015 g/mole) react to form phosphorous acid ($H_3PO_3$, MW = 81.996 g/mole) and hydrogen bromide (MW = 80.912 g/mole). If 0.5000 moles of phosphorus tribromide react with 2.000 moles of water and 98.048 grams of hydrogen bromide were obtained, what was the percent yield from the reaction?

a) 72.16 %
b) 97.22 %
c) 78.62 %
d) 85.93 %
! e) 80.79 %

59. $PI_3$ (MW = 411.69 g/mole) and water (MW = 18.015 g/mole) react to form $H_3PO_3$ (MW = 81.996 g/mole) and HI(MW = 127.91 g/mole). If 0.5000 moles of phosphorus triiodide and 2.500 moles of water are used,

a) 63.96 g of hydrogen iodide should be produced.
b) 205.8 g of hydrogen iodide should be produced.
! c) 191.9 g of hydrogen iodide should be produced.
d) 319.8 g of hydrogen iodide should be produced.
e) 383.7 g of hydrogen iodide should be produced.

60. 150.0 grams of $AsF_3$ was reacted with 180.0 g of $CCl_4$ to produce $AsCl_3$ and $CCl_2F_2$. The theoretical yield of $CCl_2F_2$ produced, in moles, should be

a) 0.7802 moles
b) 0.5685 moles
c) 1.274 moles
! d) 1.170 moles
e) 1.705 moles

61. Sodium hydroxide and sulfuric acid react to produce sodium sulfate and water. When 0.0720 moles of sulfuric acid are reacted with 0.0240 moles of sodium hydroxide, how much sodium sulfate ($Na_2SO_4$) is produced?

! a) 0.0120 moles
b) 0.0240 moles
c) 0.0360 moles
d) 0.0480 moles
e) 0.0960 moles

62. 150.0 grams of $AsF_3$ was reacted with 180.0 g of $CCl_4$ to produce $AsCl_3$ and $CCl_2F_2$. The weight, in grams of $CCl_2F_2$ produced was

! a) 141.5 grams
b) 206.2 grams
c) 137.5 grams
d) 104.4 grams
e) 152.6 grams

## Fill in the Blanks

63. The mass, in grams, of 3.00 moles of phosphorus ($P_4$), is ______ grams. (! 372)

64. A sample of $Na_2SO_4$ weighing 7.10 grams contains ______ moles. (! 0.0500)

65. One formula unit of $(NH_4)_2ClO_4$ contains ______ nitrogen atoms. (! 15)

66. The number of moles of iron atoms which are combined with 7.28 moles of oxygen atoms in a sample of the compound, $Fe_3O_4$, is ______. (! 5.46)

67. A sample of ozone, $O_3$, contains $3.011 \times 10^{12}$ atoms ($10^{12}$ represents a trillion). This also represents ______ moles of ozone. (! $1.667 \times 10^{-12}$)

68. The percent, by weight, of sulfur in $(NH_4)_2SO_4$ is ______ ( ! 24.267 %)

## True and False

69. A mole of oxygen, $O_2$, and a mole of phosphorus, $P_4$, contain different a number of molecules. ( !F)

70. The molecular formula for a substance can never contain fewer atoms than the empirical formula for the same substance. ( !T)

71. A mole of naturally occurring neon weighs more than a mole of naturally occurring oxygen. ( !F)

72. NaCl, a compound which can be prepared from a highly toxic gaseous element and a highly poisonous metal, is a very toxic substance. ( !F)

## Critical Thinking

73. In a quantitative analysis study, 4.624 grams of a compound containing carbon, hydrogen and oxygen only yielded 7.210 g of $CO_2$ and 2.656 g of $H_2$ in a combustion analysis apparatus. Determine the empirical formula of the compound.

a) $C_9H_{16}O_8$
! b) $C_{10}H_{18}O_9$
c) $C_{11}H_{20}O_{10}$
d) $C_{12}H_{22}O_{11}$
e) $C_{13}H_{24}O_{12}$

74. A 5.000 g sample of a mixture which contains $MgCO_3$ *and* sand ($SiO_2$) was strongly heated for 2.30 hours until no further reaction occurred. It was then cooled and the residue, which contained MgO and unchanged sand, was weighed. The reaction is: $MgCO_3 \longrightarrow MgO + CO_2(g)$. The residue remaining (minus the $CO_2$ which was completely driven off) weighed 3.397 grams. Calculate the percent, by weight, of $MgCO_3$ in the original sample.

a) 22.59 %
b) 50.00 %
! c) 61.42 %
d) 90.40 %
e) 95.01 %

75. A mixture, weighing 6.000 grams, contains $Na_3PO_4$ (41.50%) and $BaCl_2$ (58.50 % of the mixture, by weight). When dissolved in water, a precipitate forms via the double displacement reaction:

$$BaCl_2 + Na_3PO_4 \longrightarrow NaCl + Ba_3(PO_4)_2(s)$$

If the reaction is quantitative, the yield of the solid barium phosphate should be

! a) 3.382 grams
b) 3.357 grams
c) 4.571 grams
d) 3.510 grams
e) 1.983 grams

76. A mixture containing silver nitrate ($AgNO_3$) and potassium nitrate ($KNO_3$), weighing 5.000 grams, was treated with an **excess** of potassium chloride. A **quantatitive** reaction, the reaction of silver nitrate with potassium chloride occurred, and 1.582 grams of a white precipitate, silver chloride, was obtained. Determine the percent, by weight, of silver nitrate in the original mixture.

a) 32.55 %
! b) 37.50 %
c) 42.50 %
d) 62.50 %
e) 62.74 %

77. Lead sulfide, in ores, can be assayed by the reaction:

$$2\ PbS(s) + 6\ HNO_3(aq) + K_2Cr_2O_7(aq) \longrightarrow 2\ PbCrO_4(s) + 2\ S(s) + 4\ NO_2(g) + 2\ KNO_3(aq) + 2\ H_2O$$

A 6.053 g sample of ore, treated with excess nitric acid and potassium dichromate yielded 6.094 g of $PbCrO_4(s)$. Calculate the percent, by weight, of lead sulfide in the ore.

! a) 74.53%
b) 68.22%
c) 83.11%
d) 77.22%
e) 79.16%

78. A laboratory manual gave precise instructions for carrying out the reaction described in the chemical equation, $C_2H_6O + O_2$ -----> $C_2H_4O_2 + H_2O$. It stated that if you bubble oxygen gas through a solution containing the reactant for 24 hours under a well defined set of conditions, then the yield of $C_2H_4O_2$ would be 7.50% of the theoretical amount. If you desire to produce 700 grams of the $C_2H_4O_2$ in a single batch, how many kg of $C_2H_6O$ should you begin with? ______ ( ! 7.16)

79. Calcium nortonate, $CaNtO_4$, is 54.698% nortonium, by weight. Calculate the percent, by weight, of sodium in sodium nortonate. ______ ( ! 19.513%)

## Chemicals in Use 3: Sodium Chloride

80. Sodium chloride which is used commercially is obtained primarily from

a) The large scale chemical reaction between sodium metal and chlorine gas.
b) The reaction of sodium hydroxide with hydrochloric acid.
c) The neutralization of sodium carbonate with hydrochloric acid.
d) The mining of sodium chloride deposits central Africa.
! e) The evaporation of seawater.

81. Most of the sodium chloride produced in the United States is used for

a) preservation of food
! b) manufacture of other chemicals
c) fabrication of salt shelters for the storage of nuclear waste material
d) in the food trades industry for seasoning food.
e) the aircraft industry, in preparation of airframe metals

## Chapter 4 Chemical Reactions in Solution

### Multiple Choice

1. Which one of the following chemical compounds is not an electrolyte?

a. barium nitrate
b. calcium chloride
c. lithium phosphate
! d. methyl alcohol
e. potassium sulfate

2. Which one of the following chemical compounds is not an electrolyte?

a. ammonium chloride
! b. ethylene glycol
c. nickel sulfate
d. sodium perchlorate
e. zinc nitrate

3. Which one of the following compounds produces 4 ions per formula unit by dissociation when dissolved in water?

a. $(NH_4)_2SO_4$
b. $Hg_2Cl_2$
c. $Ca(NO_3)_2$
! d. $Li_3PO_4$
e. $(NH_4)_4Fe(CN)_6$

4. Which one of the following compounds produces 4 ions per formula unit by dissociation when dissolved in water?

a. $K_2C_2O_4$
! b. $Al(NO_3)_3$
c. $Hg_2(NO_3)_2$
d. $NaBrO_3$
e. $Na_2S_2O_3$

5. Which one of the following compounds produces 3 ions per formula unit by dissociation when dissolved in water?

a. sodium nitrate
b. nickel sulfate
! c. calcium perchlorate
d. aluminum sulfate
e. ammonium bromate

6. Which one of the following compounds produces 3 ions per formula unit by dissociation when dissolved in water?

a. $Hg_2SO_4$
b. $NaClO_2$
c. $LiClO_4$
d. KClO
! e. $(NH_4)_2SO_4$

7. Which statement below is correctly worded and states a fact?

a. Ionic acids are strong electrolytes and ionize completely when dissolved in water.
b. Ionic bases are weak electrolytes and ionize completely when dissolved in water.
c. Ionic bases are strong electrolytes and ionize completely when dissolved in water.
! d. Ionic salts are strong electrolytes and dissociate completely when dissolved in water.
e. Ionic salts are weak electrolytes and ionize partially when dissolved in water.

8. Which statement below is correctly worded and states a fact?

a. All acids are strong electrolytes and dissociate completely when dissolved in water.
! b. Some acids are strong electrolytes and ionize completely when dissolved in water.
c. Some acids are strong electrolytes and dissociate completely when dissolved in water.
d. Some acids are weak electrolytes and ionize completely when dissolved in water.
e. Some acids are weak electrolytes and dissociate partially when dissolved in water.

9. According to the Arrhenius theory of Acids and Bases, acids are substances which

a. exhibit a sour taste
b. react with all metals to release hydrogen gas
c. react with all metals to release carbon dioxide gas
! d. release hydrogen ions when dissolved in water to form a solution
e. turn litmus paper from blue to red

10. The equation for the third step in the stepwise ionization of arsenic acid is

a. $H_2AsO_4^-(aq) + H_2O(l) \longrightarrow HAsO_4^{2-}(aq) + H_3O^+(aq)$
b. $H_2AsO_3^-(aq) + H_2O(l) \longrightarrow HAsO_3^{2-}(aq) + H_3O^+(aq)$
c. $H_2AsO_4^-(aq) + H_2O(l) \longrightarrow HAsO_3^{2-}(aq) + 2\,OH^-(aq)$
! d. $HAsO_4^{2-}(aq) + H_2O(l) \longrightarrow AsO_4^{3-}(aq) + H_3O^+(aq)$
e. $HAsO_4^{2-}(aq) + H_2O(l) \longrightarrow AsO_3^{3-}(aq) + OH^-(aq)$

11. Which one of the following listed species is a molecular base?

a. $NaOH(s)$
! b. $NH_3(g)$
c. $C_2H_5OH(l)$
d. $Li_2O(s)$
e. $NH_4^+(aq)$

12. Which one of the following is the acid anhydride for the acid, HCl(*aq*)?

a. ClO
b. $ClO_2$
c. $Cl_2O_7$
! d. HCl(*aq*) does not have an anhydride
e. $H_2O$

13. Which one of the following is the acid anhydride for the acid, $HClO_4$?

a. ClO
b. $ClO_2$
c. $ClO_3$
d. $ClO_4$
! e. $Cl_2O_7$

14. Which one of the following is a basic anhydride

a. $SO_2(g)$
b. $CO_2(g)$
c. $O_2(g)$
! d. BaO(*s*)
e. $N_2O_5(s)$

15. Which one of the following species is a weak electrolyte?

a. $HClO_4(aq)$
b. HCl(*aq*)
c. NaOH(*aq*)
! d. $NH_3(aq)$
e. LiOH(*aq*)

16. The chemical species present in all solutions which are moderately to strongly acidic, which is responsible for the acidic property is

a. $H_2O(aq)$
b. $OH^-(aq)$
c. $H_2O^+(aq)$
d. $H_3O^+(l)$
! e. $H_3O^+(aq)$

17. The anhydride of orthophosphoric acid, $H_3PO_4$, is

a. $PO_2$
b. $PO_3$
c. $P_2O_3$
! d. $P_4O_{10}$
e. $P_2O$

18. Which one of the following acids is NOT a known strong acid?

a. $HBr(aq)$
b. $HCl(aq)$
c. $HClO_3(aq)$
! d. $HF(aq)$
e. $HI(aq)$

19. Which one of the following bases is NOT a known strong base?

a. $Ca(OH)_2(aq)$
b. $Ba(OH)_2(aq)$
c. $KOH(aq)$
d. $NaOH(aq)$
! e. $NH_3(aq)$

20. Which one of the following acids is NOT a known strong acid?

a. $HBr(aq)$
! b. $HC_2H_3O_2(aq)$
c. $HClO_3(aq)$
d. $HClO_4(aq)$
e. $HNO_3(aq)$

21. Which one of the following acids is a known strong acid?

a. $HBrO(aq)$
b. $HClO_2(aq)$
c. $HF(aq)$
! d. $HI(aq)$
e. $H_3PO_3(aq)$

22. Which one of the following listed solutions is the least acidic (contains the lowest concentration of hydronium ions)?

a. 1.0 molar $HCl(aq)$
b. 1.0 molar $H_2SO_4(aq)$
c. 1.0 molar $HClO_4(aq)$
d. 1.0 molar $HNO_3(aq)$
! e. 1.0 molar $HCHO_2(aq)$

23. Which one of the following substances will ***ionize*** when dissolved in water to form an aqueous solution?

a. $NaClO_4(s)$
! b. $HNO_3(l)$
c. $C_6H_{12}O_6(s)$
d. $(NH_4)_2SO_4(s)$
e. $Ba(OH)_2(s)$

24. Which set below represents a group containing ***only*** weak electrolytes?

a. $HNO_3(aq)$, $H_2SO_4(aq)$, $HCN(aq)$
b. $KOH(aq)$, $H_3PO_4(aq)$, $NaClO_4(aq)$
! c. $NH_3(aq)$, $HC_2H_3O_2(aq)$, $HCN(aq)$
d. $NH_4Cl(aq)$, $HClO_2(aq)$, $HCN(aq)$
e. $NaOH(aq)$, $H_2SO_4(aq)$, $HC_2H_3O_2(aq)$

25. Which set below represents a group containing only weak acids?

! a. $HC_2H_3O_2(aq)$, $HCN(aq)$, $HNO_2(aq)$
b. $HC_2H_3O_2(aq)$, $HCN(aq)$, $HNO_3(aq)$
c. $HC_2H_3O_2(aq)$, $HCl(aq)$, $HNO_2(aq)$
d. $HClO(aq)$, $HCN(aq)$, $HBrO_3(aq)$
e. $HNO_2(aq)$, $HC_2H_3O_2(aq)$, $HI(aq)$

26. A dynamic equilibrium is soon reached when acetic acid is dissolved in water. Which of the species shown below is present in largest amount in the solution?

a. $C_2H_3O_2^-$
b. $H^+$
! c. $HC_2H_3O_2$
d. $H_3O^+$
e. $OH^-$

27. An acid-base neutralization is the reaction of

a. $H_2(g)$ with $O_2(g)$ to form $H_2O(l)$
b. $H_2(aq)$ with $OH^-(aq)$ to form $H_2O(l)$
c. $H^+(aq)$ with $O_2(g)$ to form $H_2O(l)$
! d. $H^+(aq)$ with $OH^-(aq)$ to form $H_2O(l)$
e. $Na^+(aq)$ with $OH^-(aq)$ to form $NaOH(aq)$

28. Which one of the following is ***not*** an example of an acid salt?

a. aluminum bicarbonate
b. barium dihydrogen phosphate
c. nickel(II) bichromate
! d. potassium monohydrogen chloride
e. sodium monohydrogen sulfate

29. Which one of the following is ***not*** an example of an acid salt?

a. ammonium dihydrogen arsenate
b. barium bicarbonate
c. chromium(III) monohydrogen sulfate
! d. cobalt(II) monohydrogen acetate
e. nickel(II) bichromate

30. Which statement below is *not* correct?

! a. Acid salts can be formed by partial neutralization of a monoprotic acid by a monoprotic base.
b. Acid salts can be formed by partial neutralization of a diprotic acid by a monoprotic base.
c. Acid salts can be formed by partial neutralization of a diprotic acid by a diprotic base.
d. Acid salts can be formed by partial neutralization of a triprotic acid by a diprotic base.
e. Acid salts can be formed by partial neutralization of a polyprotic acid by a monoprotic base.

31. In the reaction, $K_2SO_4(aq) + Ba(NO_3)_2(aq) \longrightarrow BaSO_4(s) + 2\ KNO_3(aq)$, which ions are the spectator ions?

a. $Ba^{2+}$ and $SO_4^{2-}$
b. $Ba^{2+}$ and $K^+$
c. $Ba^{2+}$ and $NO_3^-$
d. $K^+$ and $SO_4^{2-}$
! e. $K^+$ and $NO_3^-$

32. In the reaction, $K_2S(aq) + 2\ HCl(aq) \longrightarrow 2\ KCl(aq) + H_2S(g)$, which ions are the spectator ions?

! a. $K^+$ and $Cl^-$
b. $K^+$ and $H^+$
c. $H^+$ and $S^{2-}$
d. $K^+$ and $S^{2-}$
e. $S^{2-}$ and $Cl^-$

33. Which set of compounds below is a set in which all members are considered soluble in water?

a. $BaCO_3$, $NaBrO_3$, $Ca(OH)_2$, and $PbCl_2$
! b. $NaCl$, $BaCl_2$, $NH_4NO_3$, and $LiClO_4$
c. $NiCO_3$, $PbSO_4$, $AgCl$, and $Mg(OH)_2$
d. $NaCl$, $AgBr$, $Na_2CO_3$, and $Hg_2(NO_3)_2$
e. $PbCl_2$, $Pb(NO_3)_2$, $AgClO_4$, and $HgCl_2$

34. Which set of compounds below is a set in which all members are considered insoluble in water?

a. $BaCO_3$, $NaBrO_3$, $Ca(OH)_2$, and $PbCl_2$
b. $NaCl$, $BaCl_2$, $NH_4NO_3$, and $LiClO_4$
c. $NaCl$, $AgBr$, $Na_2CO_3$, and $Hg_2(NO_3)_2$
! d. $NiCO_3$, $PbSO_4$, $AgCl$, and $Mg(OH)_2$
e. $PbCl_2$, $Pb(NO_3)_2$, $AgClO_4$, and $HgCl_2$

35. Which set of compounds below is a set in which all members are considered insoluble in water?

a. $BaSO_4$, $BaCl_2$, $BaCO_3$, and $Ba(OH)_2$
b. $CrSO_4$, $CrCl_2$, $Cr(ClO_4)_2$, and Cr
c. $Na_2SO_4$, NaCl, $Na_2CO_3$, and NaOH
d. $NiSO_4$, $NiCl_2$, $Ni(C_2H_3O_2)_2$, and $NiCO_3$
! e. $PbSO_4$, $PbCl_2$, $PbCO_3$, and $Pb(OH)_2$

36. The reaction, $BaCl_2(aq) + H_2SO_4(aq) \longrightarrow BaSO_4(s) + 2\,HCl(aq)$ is an example of

a. An ionic redox reaction
b. An ionic neutralization reaction
c. A molecular neutralization reaction
! d. A molecular precipitation reaction
e. A molecular redox reaction

37. $ZnCO_3(s)$ is insoluble in water, but dissolves in 6.0 molar $HCl(aq)$. Which set below best fits as a description of what takes place?

a. $ZnCO_3(s) \longrightarrow Zn^{2+}(aq) + CO_3^{2-}(aq)$
b. $ZnCO_3(s) + H_2O(l) \longrightarrow ZnO(aq) + H_2CO_3(aq)$
! c. $ZnCO_3(s) + 2\,H^+(aq) \longrightarrow Zn^{2+}(aq) + H_2O(l) + CO_2(g)$
d. $ZnCO_3(s) + 2\,H_2O(l) \longrightarrow Zn(OH)_2(aq) + H_2CO_3(aq)$
e. $ZnCO_3(s) + 2\,H_2O(l) \longrightarrow Zn(OH)_2(aq) + H_2O(l) + CO_2(g)$

38. Which of the following properties is NOT a characteristic of solutions of strong acids

! a. react with metals to yield $CO_2$ gas.
b. exhibit a sour taste.
c. turn litmus from blue to red.
d. neutralize bases.
e. react with carbonate salts to yield $CO_2$ gas.

39. Which reaction below can be used to produce carbon dioxide gas?

a. react aqueous ammonia with aqueous sodium carbonate
b. react aqueous barium hydroxide solution with solid calcium carbonate
c. react aqueous barium hydroxide solution with aqueous carbonic acid ($H_2CO_3$)
! d. react aqueous sulfuric acid and solid calcium carbonate
e. react sodium metal with aqueous carbonic acid ($H_2CO_3$)

40. Which reaction below can be used to produce sulfur dioxide gas?

a. react aqueous barium hydroxide solution with aqueous sodium sulfite solution
b. react aqueous nitric acid solution with aqueous sodium sulfate solution
c. react aqueous nitric acid solution with concentrated sodium sulfide solution
! d. react aqueous sulfuric acid solution with solid sodium sulfite
e. react solid barium sulfate with aqueous nitric acid solution

41. Which one of the reactions below will yield $NiCr_2O_7$ as one of its products?

! a. nickel(II) hydroxide and dichromic acid
b. nickelic acid and chromium(II) hydroxide
c. nickel(II) hydroxide and chromic acid
d. nickel(II) hydroxide and chromate acid
e. nickel(II) hydroxide and bichromic acid

42. What is the oxidation number of each sulfur atom in the $S_2O_8^{2-}$ ion?

a. -2
b. +1
c. +3
d. +5
! e. +7

43. What is the oxidation number of the chlorine atom in the $HClO_4$ molecule?

a. -1
b. +3
c. +5
! d. +7
e. +9

44. What is the oxidation number of each oxygen atom in the compound, $BaO_2$?

! a. -1
b. -2
c. +1
d. +2
e. +3

45. What is the oxidation number of the sodium atom in the compound, $NaO_2$?

a. -0.5
b. -1
! c. +1
d. +2
e. +4

46. What is the oxidation number of each boron atom in the compound, $Na_2B_4O_7$?

a. -3
b. +1
c. +1.25
! d. +3
e. +6

47. What is the oxidation number of each cobalt atom in the compound, $Co_2(SO_4)_3$?

! a. +3
b. +6
c. +9
d. +15
e. +30

48. What is the oxidation number of each carbon atom in the compound, $K_2C_2O_4$?

a. 0
b. -4
! c. +3
d. +4
e. +6

49. What is the oxidation number of each sulfur atom in the compound, $Rb_2S_2O_4$ molecule?

a. -2
b. +1
! c. +3
d. +5
e. +6

50. What is the oxidation number of the vanadium atom in the compound, $(NH_4)_3VO_4$?

a. +2
b. +3
! c. +5
d. +6
e. +7

51. What is the oxidation number of each nitrogen atom in $(NH_4)_2SO_4$?

! a. -3
b. -2
c. +1
d. +2
e. +3

52. What is the oxidation number of the arsenic atom in the $AsO_4^{3-}$ ion?

a. +1
b. +3
c. +4
! d. +5
e. +6

53. Balance the half reaction, $NO_3^-(aq) \longrightarrow NH_4^+(aq)$ taking place in acidic media. Which answer below describes how many electrons are needed to balance the half reaction?

a. 2 electrons, left side
b. 3 electrons, right side
c. 4 electrons, left side
! d. 8 electrons, left side
e. 8 electrons, right side

54. Balance the half reaction, $H_2S(aq) \longrightarrow S(s)$ taking place in acidic media. Which answer below describes how many electrons are needed to balance the half reaction?

a. 2 electrons, left side
! b. 2 electrons, right side
c. 4 electrons, left side
d. 4 electrons, right side
e. 8 electrons, right side

55. Balance the half reaction, $Cl_2O_7(g) \longrightarrow HClO(aq)$ taking place in acidic media. Which answer below describes how many electrons are needed to balance the half reaction?

a. 2 electrons, left side
b. 3 electrons, right side
! c. 6 electrons, left side
d. 6 electrons, right side
e. 8 electrons, right side

56. Balance the half reaction, $C_5O_5^{2-}(g) \longrightarrow CO_3^{2-}(aq)$ taking place in basic media. Which answer below describes how many hydroxide ions are needed to balance the half reaction?

a. 8 ions, left side
b. 12 ions, right side
c. 12 ions, left side
! d. 20 ions, left side
e. 20 ions, right side

57. Balance the half reaction, $C_5O_5^{2-}(g) \longrightarrow CO_3^{2-}(aq)$ taking place in basic media. Which answer below describes how many electrons are needed to balance the half reaction?

a. 4 electrons, left side
b. 8 electrons, right side
c. 8 electrons, left side
d. 12 electrons, left side
! e. 12 electrons, right side

58. Balance the half reaction, $C_2H_6O(aq) \longrightarrow HC_2H_3O_2(aq)$ taking place in acidic media. How many electrons are needed to balance the charge?

a. 2
b. 3
! c. 4
d. 6
e. 8

59. Balance the half reaction, $C_8H_{10}(l) \longrightarrow C_8H_4O_4^{2-}(aq)$ taking place in basic media. Which answer below describes how many electrons are needed to balance the half reaction?

a. 4 electrons, left side
b. 8 electrons, right side
c. 8 electrons, left side
d. 12 electrons, left side
! e. 12 electrons, right side

60. When you balance the redox equation below, the reducing species is
$C_4H_{10}$ + $Cr_2O_7^{2-}(aq)$ + $H^+(aq) \longrightarrow H_6C_4O_4 + Cr^{3+}(aq) + H_2O$

! a. $C_4H_{10}$
b. $Cr_2O_7^{2-}$
c. $H^+$
d. $H_6C_4O_4$
e. $Cr^{3+}$

61. Complete the balancing of the following half reaction, taking place in basic media, $Br^-(aq) \longrightarrow BrO_3^-(aq)$. Which answer below describes how many electrons are needed to balance the half reaction?

a. 2 electrons, left side
b. 2 electrons, right side
c. 4 electrons, right side
! d. 6 electrons, right side
e. 6 electrons, left side

62. Complete the balancing of the following half reaction, taking place in basic media, $Br^-(aq) \longrightarrow BrO_3^-(aq)$. Which answer below describes how many hydroxide ions are needed to balance the half reaction?

a. 2, on the left side
b. 4, on the left side
c. 4, on the right side
! d. 6, on the left side
e. 6, on the right side

63. Complete the balancing of the following half reaction, taking place in basic media, $Cr(OH)_4^-(aq) \longrightarrow CrO_4^{2-}(aq)$. Which answer below describes how many electrons are needed to balance the half reaction?

a. 1 electron, on the left side
b. 2 electrons, on the right side
! 3. 3 electrons, on the right side
d. 3 electrons, on the left side
e. 4 electrons, on the right side

64. Complete the balancing of the following half reaction, taking place in basic media, $Cr(OH)_4^-(aq) \longrightarrow CrO_4^{2-}(aq)$. Which answer below describes how many hydroxide ions are needed to balance the half reaction?

a. 1, on the left side
b. 3, on the right side
c. 3, on the left side
! d. 4, on the left side
e. 4, on the right side

65. Complete the balancing of the following half reaction, taking place in basic media: $NtO_4^{2-}(aq) \longrightarrow Nt(OH)_2(s)$. Which answer below describes how many electrons are needed to balance the half reaction?

a. 2 electrons, on the left side
b. 3 electrons, on the right side
! c. 4 electrons, on the left side
d. 4 electrons, on the right side
e. 6 electrons, on the right side

66. Complete the balancing of the following half reaction, taking place in basic media: $NtO_4^{2-}(aq) \longrightarrow Nt(OH)_2(s)$. Which answer below describes how many water molecules are needed to balance the half reaction?

a. 2, on the left side
! b. 4, on the left side
c. 4, on the right side
d. 6, on the left side
e. 6, on the right side

67. When the balancing of the equation for the reaction,

$$C_8H_{10}(l) + NtO_4^{2-}(aq) + H^+ \longrightarrow C_8H_4O_4^{2-}(aq) + Nt^{2+}(aq) + H_2O(l),$$

taking place in acidic media is completed, what is the sum of ALL the coefficients in the equation?

a. 6
b. 12
c. 20
! d. 28
e. 32

68. A solution is made by taking 54.62 grams of $K_2CrO_4$ in enough water to make 250.0 ml of solution. The molarity of the solution is therefore

a. 0.0002813 molar
b. 0.001125 molar
c. 0.2813 molar
! d. 1.125 molar
e. 1.409 molar

69. When the balancing of the equation for the reaction,
$Fe^{2+}(aq) + Cr_2O_7^{2-}(aq) + H^+ \longrightarrow Cr^{3+}(aq) + Fe^{3+}(aq) + H_2O(l)$,
taking place in acidic media is properly completed, what is the sum of ALL the coefficients in the equation?

a. 6
b. 22
c. 30
d. 32
! e. 36

70. 66.7 ml of 18.0 molar sulfuric acid solution was dissolved in enough water to make 500 ml of solution. The molarity of the diluted mixture is

! a. 2.40 molar
b. 0.135 molar
c. 36.0 molar
d. 9.00 molar
e. 0.00741 molar

71. 66.7 ml of 18.0 molar sulfuric acid solution was added to 233.3 ml of 1.50 molar sulfuric acid, and enough water added to make 500 ml of solution. The final molarity of the diluted mixture is

a. 0.135 molar
b. 0.318 molar
c. 1.55 molar
d. 2.40 molar
! e. 3.10 molar

72. A solution of $Na_2SO_4$ (M = 142.04 g/mole) has an assay of 54.00 % $Na_2SO_4$ by weight and a density of 1.480 g/ml. The molarity of the solution is therefore:

a. 0.7992 molar
b. 3.802 molar
! c. 5.627 molar
d. 10.42 molar
e. 51.58 molar

73. A mixture was obtained by mixing the following: 50.0 ml of 0.200 molar NaCl, 25.0 ml of 0.300 molar $BaCl_2$ and 125 ml of 0.150 molar $NH_4Cl$. The total number of grams of the element, chlorine, in the mixture is

a. 0.133 grams
b. 0.428 grams
c. 0.665 grams
d. 0.1.47 grams
! e. 1.55 grams

74. How many ml of 3.25 molar $(NH_4)_2SO_4$(*aq*) would be used, if it must contain 8.60 g of $(NH_4)_2SO_4$ which is to be used up in a chemical reaction?

a. 2.65 ml
! b. 20.0 ml
c. 50.0 ml
d. 265 ml
e. 378 ml

75. How many grams of $NiSO_4 \cdot 7H_2O$ would be required to prepare 500 ml of a solution that is 0.300 molar in $NiSO_4$(*aq*)?

a. 0.534 grams
b. 14.3 grams
c. 23.2 grams
! d. 42.1 grams
e. 52.2 grams

76. How many grams of $Fe(NO_3)_3 \cdot 9H_2O$ would be required to prepare 250 ml of a solution that is 0.100 molar in $Fe(NO_3)_3(aq)$?

a. 6.05 grams
! b. 10.1 grams
c. 14.3 grams
d. 20.2 grams
e. 52.2 grams

77. When 25.0 ml of sulfuric acid solution was completely neutralized in a titration with 0.050 molar NaOH solution, it took 18.3 ml of the base to complete the job. What was the molarity of the sulfuric acid solution?

a. 0.0100
b. 0.0148
! c. 0.0183
d. 0.0325
e. 0.0366

78. A student wants to know how many ml of 0.300 molar sulfuric acid is to be added to 50.0 ml of 0.250 molar barium hydroxide solution to neutralize it completely. The answer is

a. 20.8 ml
b. 30.0 ml
! c. 41.7 ml
d. 60.0 ml
e. 110 ml

79. How many ml of 0.446 molar $KMnO_4(aq)$ are required to react with 50.0 ml of 0.200 molar $H_2C_2O_4(aq)$ in the presence of excess $H_2SO_4(aq)$? The reaction is:

$$2\ KMnO_4(aq) + 5\ H_2C_2O_4(aq) + 3\ H_2SO_4(aq) \longrightarrow 2\ MnSO_4(aq) + 10\ CO_2(g) + 8\ H_2O + K_2SO_4(aq)$$

! a. 8.97 ml
b. 17.9 ml
c. 44.8 ml
d. 55.8 ml
e. 112 ml

80. How many ml of 0.200 molar $Na_2SO_4(aq)$ solution are required to completely react with 3.23 grams of $BaCl_2$ (M = 208.2 g/mole) to form products as shown below?

$$BaCl_2(s) + Na_2SO_4(aq) \longrightarrow BaSO_4(s) + 2\ NaCl(aq)$$

a. 0.0155 ml
b. 0.0776 ml
c. 15.5 ml
d. 31.0 ml
! e. 77.6 ml

81. 45.00 ml of an aqueous $H_2SO_4$ solution required 32.0 ml of 0.200 molar NaOH(*aq*) to neutralize it completely. The molarity of the $H_2SO_4$ solution is therefore:

! a. 0.0711 molar
b. 0.142 molar
c. 0.200 molar
d. 0.281 molar
e. 0.284 molar

82. How many grams of lead(II) iodate, $Pb(IO_3)_2$, are precipitated when 32.0 ml of 2.85 M $Pb(NO_3)_2(aq)$ are mixed with 38.6 ml of 5.12 M $NaIO_3(aq)$ solution? The reaction is,

$$Pb(NO_3)_2(aq) + KIO_3(aq) \longrightarrow Pb(IO_3)_2(s) + KNO_3(aq)$$

a. 25.4 g
b. 39.8 g
c. 48.3 g
! d. 50.8 g
e. 51.9 g

83. A 2.710 g sample contains some $CaCl_2$, which is inert to HCl(*aq*) and also CaO, which reacts: $CaO(s) + HCl(aq) \longrightarrow CaCl_2 + H_2O$. It took 32.05 ml of 2.445 molar HCl(*aq*) to react with all the CaO in the sample. The percent, by weight, of CaO in the sample is

a. 35.15 %
b. 61.67 %
c. 77.62 %
! d. 81.08 %
e. 84.17 %

## Fill in the Blanks

84. How many ions per formula unit are produced in solution by dissociation when $(NH_4)_3AsO_4$ dissolves in water? ______ (! 4)

85. How many ions per formula unit are produced in solution by dissociation when $Hg_2SO_4$ dissolves in water? ______ (! 2)

86. All acids share the common property of reacting with water to produce ______. (! hydronium ions)

87. Ionization differs from dissociation in that there are no ______ existing prior to the ionization step. (! ions)

88. Write the equation for the ionization reaction of perchloric acid with water. ____________ (! $HClO_4(l) + H_2O(l) \longrightarrow H_3O(aq) + ClO_4^-(aq)$)

89. A sample from each of the following solutions, 1.00 M $Ba(NO_3)_2(aq)$ 1.00 M $Ca(NO_3)_2(aq)$ 1.00 M $AgNO_3(aq)$ 1.00 M $Pb(NO_3)_2(aq)$ 1.00 M $Zn(NO_3)_2(aq)$ was treated with some 1.00 M $H_2SO_4(aq)$. How many of the solutions will show a net reaction? ______ (! 3)

90. Write the net ionic equation for the reaction which occurs when $HClO_4(aq)$ reacts with $Ba(OH)_2(aq)$: ________________ ( ! $H^+(aq) + OH^-(aq) \longrightarrow H_2O(l)$ )

91. A sample from each of the following solutions: 1.00 M $Na_2SO_3(aq)$ 1.00 M $NH_4NO_3(aq)$ 1.00 M $K_2S(aq)$ 1.00 M $Li_3PO_4(aq)$ 1.00 M $Ni(ClO_4)_2(aq)$ was treated with some 1.00 M $H_2SO_4(aq)$. How many solutions will show a reaction? ______ ( ! 2)

92. A sample from each of the following solutions: 1.00 M $H_2SO_4(aq)$ 1.00 M $Ni(NO_3)_2(aq)$ 1.00 M $BaCl_2(aq)$ 1.00 M $AgNO_3(aq)$ 1.00 M $K_2SO_4(aq)$ was treated with some 0.500 M $Na_2CO_3(aq)$. How many solutions will show a reaction? ______ (! 4)

93. Is the process, $Cr_2O_7^{2-}(aq) \longrightarrow Cr^{3+}(aq)$ an oxidation or a reduction? ______
( ! reduction)

94. Is the process, $NO_3^-(aq) \longrightarrow NO(g)$ an oxidation or a reduction? ______
( ! reduction)

95. Is the process, $S_2O_3^{2-}(aq) \longrightarrow S_4O_6^{2-}(aq)$ an oxidation or a reduction? ______
( ! oxidation)

96. Is the process, $C_2O_4^{2-}(aq) \longrightarrow CO_3^{2-}(aq)$ an oxidation or a reduction? ______
( ! oxidation)

97. A 50.0 ml sample of aqueous sodium chloride solution was treated with excess silver nitrate solution. The precipitate formed was collected, dried, and weighed. It had a mass of 2.973 grams. What is the molarity of the NaCl solution? ______ ( ! 0.415 molar)

98. An aqueous solution of ammonium nortonate (M = 159.83 g/mole) contains 26.37 grams of solute. What is the volume of the solution, in ml? ______ ( ! 750.0 ml )

99. In the course of a titration of barium ion in a sample by sulfate ion in a titrant solution, 83.5 ml of 0.200 molar $Na_2SO_4(aq)$ was required to precipitate all the barium ion. How many grams will the dried precipitate weigh? ______ ( ! 3.90 grams)

## True and False

100. When solid calcium nitrate dissolves in water, the process can be represented by the equation, $Ca(NO_3)_2(s) \longrightarrow Ca^{2+}(l) + 2\ NO_3^-(l)$ ( !F)

101. When solid nickel sulfate dissolves in water, the process can be represented by the equation, $NiSO_4(s) \longrightarrow Ni^{2+}(aq) + S^{2-}(aq) + 2\ O_2(g)$ ( !F)

102. The symbol, $H^+(aq)$, is often used as a substitute for $H_2(aq)$. ( !F)

103. A 0.540 molar aqueous solution of sodium tetrafluoroborate (M = 109.79 g/mole) contains 14.82 grams of solute in 250 ml of solution. ( !T)

## Critical Thinking

104. When the balancing of the half reaction, $C_{10}H_8(s) \longrightarrow C_8H_6O_4(s) + CO_2(g)$, taking place in acidic media is completed, the number of electrons showing in the half reaction is ______. ( ! 18)

105. What is the oxidation number of the sulfur in the compound, $FeS_2$? In order to think about this problem, consider the existence of $H_2O_2$, $BaO_2$, $NaO_2$, $KO_2$, among others, and the periodic relationships between these and the species under consideration. What is your best reasoned response? ______ (! -1)

106. A finely powdered well mixed sample is a mixture containing NaCl and KCl. A 4.624 gram sample of the mixture was dissolved in 80.00 ml of water and treated with 0.264 molar $AgNO_3(aq)$. It required 277.0 ml of the $AgNO_3(aq)$ solution to combine with all of the chloride ion present. The percent, by weight, of NaCl in the mixture is therefore ______ % (! 64.9 — 65.0)

107. A high school friend who is working part time as a technician in a chemical manufacturing plant needs help with this problem. He has some 6.00 molar aqueous sulfuric acid and some 0.500 molar aqueous sulfuric acid. The water supply has been shut off, and he wants to make up 2.000 liters of 1.20 molar aqueous sulfuric acid from just the two reagents above. How many ml of each should he take to make up the 2.000 liters? ______________ (! 345 ml of 6.00 molar and 1655 ml of 0.500 molar)

108. A concentrated aqueous solution of potassium nitrate contains 24.00 % by weight of the solute and has a density of 1.1623 g/ml. How many ml of this solution should be used, if you want to dilute some of it to prepare 2.500 liters of 0.2000 molar potassium nitrate solution from it? ______ (! 181.2 ml)

109. 50.0 ml of an aqueous solution of sodium chromate was treated with an excess amount of lead chloride solution. The precipitate, which was obtained in quantitative yield (theoretical yield) was carefully dried and weighed. A mass of 4.04 grams was obtained. What is the molarity of the sodium chromate solution? ______ (! 0.250 molar)

110. Hydrochloric acid solution can be neutralized by using lime (CaO). The reaction is: $HCl(aq) + CaO(s) \longrightarrow CaCl_2(aq) + H_2O(l)$. How many kg of CaO (MW=56.077) are required for neutralization of 35.50 gallons of 2.40M HCl(*aq*)? ______ (! 9.04 kg)

## Special Topics

111. A chemical reaction by which washing soda "softens" hard water is

a. $H^+(aq) + CO_3^{2-}(aq) \longrightarrow HCO_3^-(aq)$
! b. $Ca^{2+}(aq) + CO_3^{2-}(aq) \longrightarrow CaCO_3(s)$
c. $Ca^{2+}(aq) + 2\,HCO_3^-(aq) \longrightarrow Ca(HCO_3)_2(s)$
d. $2\,H^+(aq) + CO_3^{2-}(aq) \longrightarrow H_2O(l) + CO_2(g)$
e. $CaCO_3(s) + 2\,H^+(aq) \longrightarrow Ca^{2+}(aq) + H_2O(l) + CO_2(g)$

112. The active chemical reagent in ordinary chlorine bleach is

a. $Cl_2O_7$
b. $NaClO_3$
! c. NaClO
d. $H_2O_2$
e. $CCl_2F_2$

# Chapter 5 Energy and Thermochemistry

## Multiple Choice

1. Chemical energy is

   a. the kinetic energy resulting from violent decomposition of energetic chemicals
   b. the heat energy associated with combustion reactions
   c. the electrical energy produced by fuel cells
   ! d. the potential energy which resides in chemical bonds
   e. the energy living plants receive from solar radiation

2. Which one of the following statements is true?

   a. Molecules in gases possess kinetic energy since they are in constant motion, while molecules in liquids and solids are not in constant motion and hence possess no kinetic energy.
   b. Molecules in gases and liquids possess kinetic energy since they are in constant motion, while molecules in solids are not in constant motion and hence possess no kinetic energy.
   ! c. Molecules in gases, liquids and solids possess kinetic energy since they are in constant motion.
   d. Polyatomic molecules possess kinetic energy in the liquid and gaseous states since the atoms can move about in the molecule even if the molecule cannot move.
   e. Since solids are rigid, their molecules do not possess kinetic energy unless the solid is melted.

3. A chemical reaction has just occurred in an insulated isolated system which caused an overall increase in the potential energy of the system. Which statement below is true?

   a. heat was taken in from the surroundings by the system
   b. heat was given off to the surroundings by the system
   c. the temperature of the system increased
   ! d. the temperature of the system decreased
   e. the total energy of the system increased

4. A chemical reaction has just occurred in an insulated isolated system which caused an overall decrease in the potential energy of the system. Which statement below is true?

   a. heat was taken in from the surroundings by the system
   b. heat was given off to the surroundings by the system
   ! c. the temperature of the system increased
   d. the temperature of the system decreased
   e. the total energy of the system decreased

5. An exothermic chemical reaction has just occurred in an insulated isolated system. Which statement below is true?

   a. heat was taken in from the surroundings by the system
   b. heat was given off to the surroundings by the system
   c. the potential energy of the system increased
   ! d. the potential energy of the system decreased
   e. the total energy of the system increased

6. A 500.0 gram sample of aluminum is initially at 25.0 °C. It absorbs 32.60 kJ of heat from its surroundings. What is its final temperature, in °C? (specific heat = 0.9930 J/g-degree for aluminum)

a. 40.4 °C
b. 64.7 °C
c. 65.7 °C
d. 89.7 °C
! e. 90.7 °C

7. A 350.0 gram sample of copper is initially at 25.0 °C. It absorbs 12.50 kJ of heat from its surroundings. What is its final temperature, to the nearest tenth of a degree? (specific heat = 0.3874 J/g-degree for copper)

a. 38.8 °C
b. 67.2 °C
c. 92.2 °C
! d. 117.2 °C
e. 156.7 °C

8. A calorimeter consists of metal parts with a heat capacity of 850.0 J/degree and 1100 grams of oil with a specific heat of 2.184 J/g-degree. What is the heat capacity, in joules per degree, of the ***entire*** assembly?

a. 1354 J/deg
b. 1952 J/deg
c. 2956 J/deg
! d. 3252 J/deg
e. 4259 J/deg

9. A calorimeter consists of metal parts with a heat capacity of 925.0 J/degree and 1100 grams of oil with a specific heat of 2.814 J/g-degree. What is the heat capacity, in joules per degree, of the ***entire*** assembly?

a. 1321 J/deg
b. 2028 J/deg
c. 3703 J/deg
! d. 4020 J/deg
e. 5698 J/deg

10. A certain oil used in industrial transformers has a density of 1.086 g/ml and a specific heat of 1.826 J/g-degree. Calculate the heat capacity of one gallon of this oil. (1 gallon = 3.785 liters)

a. 0.4442 kJ/degree
b. 0.5239 kJ/degree
c. 2.251 kJ/degree
d. 6.364 kJ/degree
! e. 7.506 kJ/degree

11. A certain oil used in industrial transformers has a density of 1.068 g/ml and a specific heat of 1.628 J/g-degree. Calculate the heat capacity of one gallon of this oil. (1 gallon = 3.785 liters)

a. 0.3747 kJ/degree
b. 0.4027 kJ/degree
c. 2.483 kJ/degree
d. 5.770 kJ/degree
! e. 6.581 kJ/degree

12. A calorimeter consists of metal parts with a heat capacity of 950.0 J/degree and 850 grams of oil with a specific heat of 2.418 J/g-degree. Calculate the amount of heat energy required, in kJ, to raise its temperature from 25.00 °C to 31.60 °C.

a. 4.91 kJ
b. 11.9 kJ
! c. 19.8kJ
d. 20.8 kJ
e. 28.7 kJ

13. A calorimeter consists of metal parts with a heat capacity of 925.0 J/degree and 1100 grams of oil with a specific heat of 2.184 J/g-degree. Calculate the heat required, in kJ, to raise its temperature from 24.40 °C to 29.75 °C.

a. 0.827 kJ
b. 7.64 kJ
! c. 17.8 kJ
d. 23.7 kJ
e. 99.0 kJ

14. A coffee cup calorimeter contains 480.0 grams of water at 25.00 °C. To it are added:

380.0 grams of water at 53.5 °C
525 grams of water at 65.5 °C.

Consider the heat absorbed by the styrofoam as negligible, and calculate the expected final temperature.

a. 38.2 °C
! b. 48.2 °C
c. 67.6 °C
d. 88.7 °C
e. 94.4 °C

15. A coffee cup calorimeter contains 525.0 grams of water at 25.0 °C. To it are added:

350.0 grams of water at 48.3 °C
480 grams of water at 63.8 °C.

Neglect the heat absorbed by the styrofoam, and calculate the final temperature. The specific heat of water is 4.184 J/g-degree.

a. 39.6 °C
b. 45.7 °C
! c. 44.8 °C
d. 66.7 °C
e. 92.4 °C

16. A calorimeter consists of metal parts with a heat capacity of 850.0 J/degree and 1050 grams of oil with a specific heat of 2.148 J/g-degree. Both are at 24.50 °C. A 500 g copper slug, at 220.0 °C is added. What is the final temp? (specific heat of copper = 0.3874 J/g-degree)

a. 33.4 °C
! b. 36.0 °C
c. 36.8 °C
d. 89.7 °C
e. 120.5 °C

17. A calorimeter has metal parts (heat capacity of 850.0 J/degree) and 1100 grams of oil (specific heat = 2.184 J/g-degree), both at 24.50 °C. Adding a 460 g slug, at 240.0 °C caused the temperature to rise to 32.5 °C. Find the specific heat of the metal!

a. 0.236 J/g-degree
! b. 0.273 J/g-degree
c. 0.309 J/g-degree
d. 0.357 J/g-degree
e. 2.28 J/g-degree

18. A calorimeter has metal parts (heat capacity of 925.0 J/degree) and 1100 grams of oil (specific heat = 2.824 J/g-degree), both at 25.40 °C. Adding a 550 g slug, at 220.0 °C caused the temperature to rise to 35.2 °C. Find the specific heat of the metal!

a. 0.365 J/g-degree
! b. 0.389 J/g-degree
c. 0.395 J/g-degree
d. 0.551 J/g-degree
e. 1.20 J/g-degree

19. Which statement below is true?

a. A state function is one whose value in a system depends on the method of preparation of the reactants and products.
b. Hess' law states that the enthalpy of the products is always greater than the enthalpy of the reactants.
c. The molar heat capacity of a substance is the number of joules in one mole of the substance
d. A state function is one whose value in a system is determined by the temperature of the system, and not on the composition of the system.
! e. A state function is one whose value in a system is determined by the composition of the system, the volume, the temperature and the pressure.

20. Standard reference conditions in a thermochemical situation has the values

a. temperature: 0.00 K; pressure: 1.000 standard atmosphere
b. temperature: 0.00 °C; pressure: 1.000 standard atmosphere
c. temperature: 273.15 K; pressure: 1.000 Pascal
! d. temperature: 298.15 K; pressure: 1.000 standard atmosphere
e. temperature: 298.15 K; pressure: 1.000 Pascal

21. The thermochemical equation which is associated with the standard enthalpy of formation for HCl($g$) is

a. $H(g) + Cl(g) \longrightarrow HCl(g)$
b. $H_2(g) + Cl_2(g)\ 2\,HCl(g)$
! c. $\frac{1}{2}H_2(g) + \frac{1}{2}Cl_2(g) \longrightarrow HCl(g)$
d. $H_2(g) + Cl_2(l) \longrightarrow 2\,HCl(g)$
e. $\frac{1}{2}H_2(g) + \frac{1}{2}Cl_2(l) \longrightarrow HCl(g)$

22. The thermochemical equation which is associated with the standard enthalpy of formation for urea, $CO(NH_2)_2(s)$, is

a. $CO(g) + 2\,NH_3(g) \longrightarrow CO(NH_2)_2(s) + H_2(g)$
b. $CO(g) + 2\,H_2(g) + N_2(g) \longrightarrow CO(NH_2)_2(s)$
c. $C(s) + O(g) + N_2(g) + 2\,H_2(g) \longrightarrow CO(NH_2)_2(s)$
! d. $C(s) + \frac{1}{2}O_2(g) + N_2(g) + 2\,H_2(g) \longrightarrow CO(NH_2)_2(s)$
e. $C(s) + \frac{1}{2}O_2(g) + 2\,NH_2(g) \longrightarrow CO(NH_2)_2(s)$

23. Given the reaction, $4B + 3A \longrightarrow 4C + 7D$, and some standard enthalpies of formation:
A: 15.7 kJ/mole B: -86.4 kJ/mole C: -52.7 kJ/mole D: -71.6 kJ/mole
What is the standard enthalpy of reaction, in kJ for the reaction shown?

a. -53.6 kJ
! b. - 413.5 kJ
c. -515.6 kJ
d. -853.6 kJ
e. -908.4 kJ

24. Given the reaction, $3B + 5A \longrightarrow 7C + 3D$, and some standard enthalpies of formation: A: 15.7 kJ/mole B: -86.4 kJ/mole C: -52.7 kJ/mole D: -71.6 kJ/mole
What is the standard enthalpy of reaction, in kJ for the reaction shown?

a. +26.6 kJ
b. -53.6 kJ
c. -198.8 kJ
! d. -403.0 kJ
e. -413.5 kJ

25. A chemical compound has a molecular weight of 89.05 g/mole. 1.400 grams of this compound underwent complete combustion under constant pressure conditions in a calorimeter with a heat capacity of 2980 J/degree (metal shell, water substitute--a special oil, thermocouple, *etc*). The temperature went up by 11.95 degrees. Calculate the molar heat of combustion of the compound.

a. 35.6 kJ/mole
b. 686.2 kJ/mole
c. 1681 kJ/mole
d. 1886 kJ/mole
! e. 2265 kJ/mole

26. The standard enthalpy of combustion for xylene, $C_8H_{10}(l)$, is -3908 kJ/mole. Using this data and the standard enthalpies of formation: $H_2O(l)$ = -285.9 kJ/mole; $CO_2(g)$ = -393.5 kJ/mole, calculate the standard enthalpy of formation of $C_8H_{10}(l)$, in kJ per mole

! a. -669.5 kJ
b. 3228.6 kJ
c. -3228.6 kJ
d. 4587.4 kJ
e. 8485.5 kJ

27. The standard enthalpy of combustion for naphthalene, $C_{10}H_8(s)$, is -5156.8 kJ/mole. Using this data and the standard enthalpies of formation: $H_2O(l)$ = -285.9 kJ/mole; $CO_2(g)$ = -393.5 kJ/mole, calculate the standard enthalpy of formation of $C_{10}H_8(s)$, in kJ/mole

! a. 78.2 kJ
b. 935.9 kJ
c. -1065.4 kJ
d. 3619.7 kJ
e. -10235.4 kJ

28. When aluminum metal reacts with iron(III) oxide to form aluminum oxide and iron metal, 429.6 kJ of heat are given off for each mole of aluminum metal consumed, under constant pressure and standard conditions. What is the correct value for the standard enthalpy of reaction in the thermochemical equation below?

$$2\,Al(s) + Fe_2O_3(s) \longrightarrow 2\,Fe(s) + Al_2O_3(s)$$

a. + 429.6 kJ
b. - 429.6 kJ
c. + 859.2 kJ
! d. - 859.2 kJ
e. - 1289 kJ

29. Complete combustion of hydrocarbons, or compounds with C,H, and O as the only elements give $CO_2$ and $H_2O$ as the only products. If carried out under standard conditions, the $CO_2$ is a gas while the $H_2O$ is a liquid. Given these standard enthalpies of combustion: $C_2H_4(g)$ = -1411.08 kJ/mole, $C_2H_2(g)$ = -1299.65 kJ per mole, $H_2(g)$ = -285.90 kJ per mole, $C(s)$ = -393.50 kJ per mole. Calculate the standard enthalpy of reaction for the process, $C_2H_2(g) + H_2(g) \longrightarrow C_2H_4(g)$

! a. -174.47 kJ
b. 397.33 kJ
c. -961.47 kJ
d. -2424.83 kJ
e. -2996.63 kJ

30. For the reaction, $A_2(g) + 2\,BX(g) \longrightarrow B_2(s) + 2\,AX(g)$, the enthalpy of the products is lower than that of the reactants. Which one of the following is true for the system?

a. the kinetic energy of the system decreases as the reactants are converted to products
! b. the potential energy of the system decreases as the reactants are converted to products
c. the total energy of the system decreases as the reactants are converted to products
d. the total mass of the system decreases as the reactants are converted to products

31. Complete combustion of hydrocarbons, or compounds with C,H, and O as the only elements give $CO_2$ and $H_2O$ as the only products. If carried out under standard conditions, the $CO_2$ is a gas while the $H_2O$ is a liquid. Given these standard enthalpies of combustion: $C_6H_{12}(l)$ = -3919.86 kJ/mole, $C_6H_6(l)$ = -3267.80 kJ per mole, $H_2(g)$ = -285.90 kJ per mole, C(s) = -393.50 kJ per mole. Calculate the standard enthalpy of reaction for the process, $C_6H_6(l) + 3\,H_2(g) \longrightarrow C_6H_{12}(l)$

! a. -205.64 kJ
b. 366.16 kJ
c. 759.66 kJ
d. 2155.36 kJ
e. 5684.36 kJ

32. For the reaction, $D_2(s) + 2\,AX(g) \longrightarrow A_2(g) + 2\,DX(g)$, the enthalpy of the reactants is lower than that of the products. Which one of the following is true for the system?

! a. the kinetic energy of the system decreases as the reactants are converted to products
b. the potential energy of the system decreases as the reactants are converted to products
c. the total energy of the system decreases as the reactants are converted to products
d. the total mass of the system decreases as the reactants are converted to products

## Fill in the Blanks

33. A 500.0 gram sample of water is initially at 25.0 °C. It absorbs 50.0 kJ of heat from its surroundings. What is its final temperature, in °C? (specific heat = 4.184 J/g-degree for water)
______ (! 48.9)

34. A calorimeter consists of metal parts with a heat capacity of 925.0 j/degree and 975 grams of oil with a specific heat of 2.214 J/g-degree. Both are at 25.40 °C. A 550 g iron slug, at 240.0 °C is added. What is the final temperature? (specific heat of iron = 0.4498 J/g-degree)
______ (! 37.0 °C)

35. A sample of zinc weighing 425 grams was initially at a temperature of 25.40 °C. It required 1360 joules of heat energy to increase the temperature to 33.70 °C. What is the molar heat capacity of the zinc? ______ (! 25.2 J/mole-degree)

36. A sample of nickel weighing 425 grams was initially at a temperature of 26.20 °C. It required 975 joules of heat energy to increase the temperature to 31.55 °C. What is the specific heat of the nickel? ______ (! 0.429 J/g-degree)

37. Complete combustion of hydrocarbons, or compounds with C,H, and O as the only elements give $CO_2$ and $H_2O$ as the only products. If carried out under standard conditions, the $CO_2$ is a gas while the $H_2O$ is a liquid. Given these standard enthalpies of combustion: $C_2H_4(g)$ = -1411.08 kJ/mole, $C_6H_{12}(l)$ = -3919.86 kJ per mole, $H_2(g)$ = -285.90 kJ per mole, C(s) = -393.50 kJ per mole. Calculate the standard enthalpy of reaction for the process, $3\,C_2H_4(g) \longrightarrow C_6H_{12}(l)$
______ (! -313.38 kJ)

38. A sample of chromium weighing 254 g was initially at a temperature of 25.88 °C. It required 843 joules of heat energy to increase the temperature to 32.75 °C. What is the molar heat capacity of the chromium? ______ (! 25.1 J/mole-degree)

39. Complete combustion of hydrocarbons, or compounds with C,H, and O as the only elements give $CO_2$ and $H_2O$ as the only products. If carried out under standard conditions, the $CO_2$ is a gas while the $H_2O$ is a liquid. Given these standard enthalpies of combustion: $CH_3CHO(l)$ = -1166.37 kJ/mole, $C_6H_{12}O_3(l)$ = -3340.34 kJ per mole, $H_2(g)$ = -285.90 kJ per mole, $C(s)$ = -393.50 kJ per mole. Calculate the standard enthalpy of reaction for the process, $3\ CH_3CHO(l) \longrightarrow C_6H_{12}O_3(l)$
______ ( ! -158.77 kJ)

40. Complete combustion of hydrocarbons, or compounds with C,H, and O as the only elements give $CO_2$ and $H_2O$ as the only products. If carried out under standard conditions, the $CO_2$ is a gas while the $H_2O$ is a liquid. Given these standard enthalpies of combustion: $C_2H_2(g)$ = -1299.65 kJ/mole, $C_6H_6(l)$ = -3267.80 kJ per mole, $H_2(g)$ = -285.90 kJ per mole, $C(s)$ = -393.50 kJ per mole. Calculate the standard enthalpy of reaction for the process, $3\ C_2H_2(g) \longrightarrow C_6H_6(l)$
______ ( ! -631.15 kJ)

41. Use these reactions, standard enthalpies

$2\ ZbO(s) + \frac{1}{2}\ O_2(g) \longrightarrow Zb_2O_3(s)$ -128.0 kJ
$ZbO(s) + \frac{1}{2}\ O_2(g) \longrightarrow ZbO_2(s)$ -380.0 kJ
$2\ ZbO(s) + \frac{3}{2}\ O_2(g) \longrightarrow Zb_2O_5(s)$ -344.5 kJ

find the value for

$Zb_2O_3(s) + O_2(g) \longrightarrow Zb_2O_5(s)$ ______ ( ! -216.5 kJ)

42. Use these reactions, standard enthalpies

$2\ ZbO(s) + \frac{1}{2}\ O_2(g) \longrightarrow Zb_2O_3(s)$ -128.0 kJ
$ZbO(s) + \frac{1}{2}\ O_2(g) \longrightarrow ZbO_2(s)$ -380.0 kJ
$2\ ZbO(s) + \frac{3}{2}\ O_2(g) \longrightarrow Zb_2O_5(s)$ -344.5 kJ

find the value for

$Zb_2O_3(s) + Zb_2O_5(s) \longrightarrow 4\ ZbO(s)$ ______ ( ! +472.5 kJ)

## True and False

43. The statement, "the total energy of the universe is constant", is a logical extension of the law of conservation of energy. ( ! T)

44. Heat energy can always be quantitatively converted into various other forms of energy. ( ! F)

## Critical Thinking

45. A calorimeter has metal parts (heat capacity of 850.0 J/degree) and 1020 grams of oil (specific heat = 2.248 J/g-degree), both at 24.50 °C. Two metal slugs, one a 460.0 g piece of cobalt and the other a 360.0 g piece of cadmium were removed from an oven which was maintained at 240.0 °C and added to the calorimeter. If no heat was lost to the surroundings, what would be the final temperature in the calorimeter? (cadmium, *molar* heat capacity = 25.34 J/mole-degree; cobalt, *molar* heat capacity = 25.12 J/mole-degree) ______ ( ! 42.0 °C)

46. Use these reactions, and the standard enthalpies as shown

$M(s) + CO_2(g) \longrightarrow CO(g) + MO(s)$ -140.0 kJ
$3\ MO(s) + CO_2(g) \longrightarrow CO(g) + M_3O_4(s)$ 1.20 kJ
$2\ M_3O_4(s) + CO_2(g) \longrightarrow CO(g) + 3\ M_2O_3(s)$ -380.0 kJ
$M_2O_3(s) + CO_2(g) \longrightarrow CO(g) + 2\ MO_2(s)$ -121.8 kJ
$2\ MO_2(s) + CO_2(g) \longrightarrow CO(g) + M_2O_5(s)$ +344.5 kJ

to find the value for

$3\ M_2O_5(s) + 7\ CO(g) \longrightarrow 7\ CO_2(g) + 2\ M_3O_4(s)$ _____ (! -288.1 kJ)

47. Use these reactions, and the standard enthalpies as shown

$M(s) + CO_2(g) \longrightarrow CO(g) + MO(s)$ -140.0 kJ
$3\ MO(s) + CO_2(g) \longrightarrow CO(g) + M_3O_4(s)$ 1.20 kJ
$2\ M_3O_4(s) + CO_2(g) \longrightarrow CO(g) + 3\ M_2O_3(s)$ -380.0 kJ
$M_2O_3(s) + CO_2(g) \longrightarrow CO(g) + 2\ MO_2(s)$ -121.8 kJ
$2\ MO_2(s) + CO_2(g) \longrightarrow CO(g) + M_2O_5(s)$ +344.5 kJ

to find the value for

$M_2O_3(s) + 3\ CO(g) \longrightarrow 3\ CO_2(g) + 2\ M(s)$ _____ (! +405.9 kJ)

48. Use these reactions, and the standard enthalpies as shown

$M(s) + CO_2(g) \longrightarrow CO(g) + MO(s)$ -140.0 kJ
$3\ MO(s) + CO_2(g) \longrightarrow CO(g) + M_3O_4(s)$ 1.20 kJ
$2\ M_3O_4(s) + CO_2(g) \longrightarrow CO(g) + 3\ M_2O_3(s)$ -380.0 kJ
$M_2O_3(s) + CO_2(g) \longrightarrow CO(g) + 2\ MO_2(s)$ -121.8 kJ
$2\ MO_2(s) + CO_2(g) \longrightarrow CO(g) + M_2O_5(s)$ +344.5 kJ

to find the value for

$MO_2(s) + CO(g) \longrightarrow CO_2(g) + MO(s)$ _____ (! 123.8 kJ)

49. Given the thermochemical equation, $3\ M(s) + 3\ O_2(g) \longrightarrow 3\ MO_2(s)$ with a standard enthalpy of reaction = -1443 kJ, calculate a value for the standard enthalpy of reaction in the thermochemical reaction, $MO_2(s) \longrightarrow M(s) + O_2(g)$. _____ (! +481 kJ)

50. Given the thermochemical equation, $2\ M_2O_5(s) \longrightarrow 4\ MO_2(s) + O_2(g)$ with a standard enthalpy of reaction = +74.2 kJ, calculate a value for the standard enthalpy of reaction in the thermochemical reaction, $2\ MO_2(s) + \frac{1}{2}\ O_2(g) \longrightarrow M_2O_5(s)$ _____ (! +37.1 kJ)

51. Given the reaction, 7A + 5B $\longrightarrow$ 3C + 4D, and some standard enthalpies of formation: A: 15.7 kJ/mole B: -86.4 kJ/mole C: -52.7 kJ/mole D: -71.6 kJ/mole
What is the standard enthalpy of reaction, in kJ for the reaction shown? _____ (! -122.4 kJ)

52. Using the standard enthalpies of formation: $H_2O(l)$ = -285.9 kJ/mole; $C_2H_4(g)$ = 52.284 kJ/mole; $C_2H_5OH(l)$ = -277.63 kJ/mole, calculate the standard enthalpy of reaction for

$C_2H_4(g) + H_2O(l) \longrightarrow C_2H_5OH(l)$ _____ (! -44.0 kJ)

53. Using the standard enthalpies of formation: $NO(g)$ = 90.37 kJ/mole; $NO_2(g)$ = 33.8 kJ/mole; $SO_2(g)$ = -296.9 kJ/mole; $SO_3(g)$ = -395.2 kJ/mole, calculate the standard enthalpy of reaction for

$SO_2(g) + NO_2(g) \longrightarrow SO_3(g) + NO(g)$ _____ (! -41.7 kJ)

54. Using the standard enthalpies of formation: $CO(g)$ = -110.5 kJ/mole; $CO_2(g)$ = -393.5 kJ/mole; $SO_2(g)$ = -296.9 kJ/mole; $SO_3(g)$ = -395.2 kJ/mole, calculate the standard enthalpy of reaction for

$$CO_2(g) + SO_2(g) \longrightarrow SO_3(g) + CO(g)$$ _____ (! 184.7 kJ)

55. Using the standard enthalpies of formation: $CO(g)$ = -110.5 kJ/mole; $CO(NH_2)_2(s)$ = -333.19 kJ/mole; $NH_3(g)$ = -46.19 kJ/mole, calculate the standard enthalpy of reaction for

$$CO(NH_2)_2(s) + H_2(g) \longrightarrow 2NH_3(g) + CO(g)$$ _____ (! 130.3 kJ)

56. The standard enthalpy of combustion for oxalic acid, $H_2C_2O_4(s)$, is -251.9 kJ/mole. Using this data and the standard enthalpies of formation: $H_2O(l)$ = -285.9 kJ/mole; $CO_2(g)$ = -393.5 kJ/mole, calculate the standard enthalpy of formation of $H_2C_2O_4(s)$, in kJ per mole. _____ (! -821.0 kJ)

57. The standard enthalpy of combustion for ethylene glycol, $C_2H_6O_2(l)$, is -1179.5 kJ/mole. Using this data and the standard enthalpies of formation: $H_2O(l)$ = -285.9 kJ/mole; $CO_2(g)$ = -393.5 kJ/mole, calculate the standard enthalpy of formation of $C_2H_6O_2(l)$, in kJ per mole. _____ (! -465.2 kJ)

58. 500.0 ml of 0.220 molar HCl(*aq*) was added to a high quality insulated calorimeter containing 500.0 ml of 0.200 molar NaOH(*aq*). Both solutions have a density of 1.000 g/ml and a specific heat of 4.184 J/g-degree. The calorimeter had a heat capacity of 850.0 J/degree. The temperature of the ***entire*** system rose from 25.60 °C to 26.70 °C. Calculate the heat of reaction, in kJ, per mole of HCl(*aq*). _____ (! 55.4 kJ)

59. 600.0 ml of 0.240 molar perchloric acid, $HCl_4(aq)$ was added to a high quality insulated calorimeter containing 400.0 ml of 0.300 molar KOH(*aq*). Both solutions have a density of 1.000 g/ml and a specific heat of 4.184 J/g-degree. The calorimeter had a heat capacity of 950.0 J/degree. The temperature of the ***entire*** system rose from 25.30 °C to 26.59 °C. Calculate the heat of reaction, in kJ, per mole of $HClO_4(aq)$. _____ (! 55.2 kJ)

## Chemicals In Use 4: The Greenhouse Effect

60. Which one of the following "greenhouse" gases, found in the Earth's atmosphere, is not produced by chemical reactions which take place in nature?

a. $CO_2$
b. $CH_4$
c. $N_2O$
! d. $CCl_2F_2$
e. $O_2$

61. Gases which trap heat contribute to the greenhouse effect by

a. preventing $CO_2$ from escaping into outer space from the earth's atmosphere
b. preventing absorption of heat energy from the infrared solar radiation
! c. preventing the radiation of heat from the earth's surface back into space
d. facilitating the chemical reaction between $CH_4$ and $O_2$ in the upper atmosphere which produces extra heat
e. preventing the reaction between $N_2$ and $O_2$ in the upper atmosphere to produce $N_2O$ which consume energy and produce a cooling trend

# Chapter 6 Atomic and Electronic Structure

## Multiple Choice

1. The frequency of an electromagnetic wave is

   a. the number of complete oscillations or cycles over a distance of one meter
   ! b. the number of complete oscillations or cycles in a one second time interval
   c. the distance between successive maxima in the wave in one complete cycle
   d. the number of complete oscillations or cycles over a distance of one centimeter
   e. the distance between successive nodes in the wave

2. The wavelength of an electromagnetic wave is

   a. the number of complete oscillations or cycles over a distance of one meter
   b. the number of complete oscillations or cycles in a one second time interval
   ! c. the distance between successive maxima in the wave
   d. the number of complete oscillations or cycles over a distance of one centimeter
   e. the distance between a minimum and the nearest maximum in the oscillation

3. What is the wavelength of electromagnetic radiation which has a frequency of $4.464 \times 10^{14}\ s^{-1}$?

   a. $1.338 \times 10^{23}$ m
   b. $1.489 \times 10^{-6}$ m
   ! c. $6.71 \times 10^{-7}$ nm
   d. 671.6 nm
   e. $7.472 \times 10^{-15}$ nm

4. What is the wavelength of electromagnetic radiation which has a frequency of $3.818 \times 10^{14}\ s^{-1}$?

   a. 1145 nm
   b. $1.274 \times 10^{-1}$ nm
   c. $1.274 \times 10^{-7}$ m
   d. $7.852 \times 10^{-7}$ nm
   ! e. $7.852 \times 10^{-7}$ m

5. Calculate the frequency of visible light having a wavelength of 464.1 nm.

   a. $139.1\ s^{-1}$
   b. $1.548 \times 10^{-6}\ s^{-1}$
   c. $1.548 \times 10^{-15}\ s^{-1}$
   ! d. $6.460 \times 10^{14}\ s^{-1}$
   e. $6.460 \times 10^{5}\ s^{-1}$

6. A police radar unit is operating on a frequency of 9.527 Gigahertz. What is the wavelength of the radiation being employed?

   a. 314.7 nm
   b. 314.7 m
   ! c. 3.147 cm
   d. 314.7 cm
   e. 31.78 m

7. Calculate the frequency of visible light having a wavelength of 568.8 nm.

a. 170.5 $s^{-1}$
b. 1.897 x $10^{6}$ $s^{-1}$
c. 1.897 x $10^{15}$ $s^{-1}$
d. 5.271 x $10^{-9}$ $s^{-1}$
! e. 5.271 x $10^{14}$ $s^{-1}$

8. Which one of the following types of radiation has the lowest frequency?

! a. FM radio waves
b. infrared waves
c. microwaves
d. x-rays
e. ultraviolet rays

9. Which one of the following types of radiation has the highest frequency?

! a. blue visible light
b. FM radio
c. infrared waves
d. microwave radiation
e. short wave radio waves

10. Which one of the following types of radiation has the shortest wavelength?

a. FM radio waves
b. infrared waves
c. microwaves
d. ultraviolet rays
! e. x-rays

11. Which one of the following types of radiation has the longest wavelength?

a. gamma rays
b. green visible light
! c. red visible light
d. ultraviolet rays
e. x-rays

12. What is the energy, in joules, of one photon of microwave radiation with a wavelength of 0.158 m?

! a. 1.26 x $10^{-24}$ J
b. 3.14 x $10^{-26}$ J
c. 3.19 x $10^{25}$ J
d. 3.49 x $10^{-43}$ J
e. 7.15 x $10^{40}$ J

13. What is the energy, in joules, of one photon of visible radiation with a wavelength of 464.1 nm?

a. $1.026 \times 10^{-48}$ J
b. $2.100 \times 10^{35}$ J
c. $2.341 \times 10^{11}$ J
! d. $4.280 \times 10^{-19}$ J
e. $4.280 \times 10^{-12}$ J

14. What is the energy, in joules, of one mole of photons associated with visible light having a wavelength of 486.1 nm?

a. 12.41 kJ
b. $2.461 \times 10^{-4}$ J
! c. $2.461 \times 10^{5}$ J
d. $6.167 \times 10^{14}$ J
e. $8.776.15 \times 10^{25}$ J

15. What is the energy, in joules, of one mole of photons associated with visible light having a wavelength of $4.89 \times 10^{2}$ nm?

a. $1.48 \times 10^{42}$ J
b. $1.95 \times 10^{-16}$ J
! c. $2.45 \times 10^{5}$ J
d. $3.24 \times 10^{-40}$ J
e. $4.06 \times 10^{-19}$ J

16. What is the energy, in joules, of one mole of photons associated with radiation which has a frequency of $6.336 \times 10^{15}$ Hz?

! a. $2.528 \times 10^{6}$ J
b. $3.882 \times 10^{14}$ J
c. $3.955 \times 10^{-7}$ J
d. $4.198 \times 10^{-18}$ J
e. $6.298 \times 10^{-26}$ J

17. What is the energy, in joules, of one mole of photons associated with radiation which has a frequency of $3.818 \times 10^{15}$ Hz?

a. $1.045 \times 10^{-25}$ J
! b. $1.524 \times 10^{6}$ J
c. $2.530 \times 10^{-18}$ J
d. $6.564 \times 10^{-7}$ J
e. $9.568 \times 10^{24}$ J

18. Which statement among the ones presented below is true?

a. The line spectra of elements are the same provided they belong to the same family.
b. The line spectra of elements are the same provided they belong to the same family *and* are combined with oxygen.
c. The line spectra of elements are the same provided they belong to the same family *and* are in the same physical state.
d. The line spectra of elements can be used for separation of elements from mixtures.
! e. The line spectra of elements can be used for identification purposes.

19. Which statement among the ones presented below is true?

a. The line spectra of atoms consists of a series of white lines superimposed on a colored background.
b. The line spectra of atoms consists of a series of white lines superimposed on a dark background.
! c. The line spectra of atoms consists of a series of colored lines superimposed on a dark background.
d. The line spectra of atoms consists of a series of dark lines superimposed on a white background.
e. The line spectra of atoms consists of a series of dark lines superimposed on a colored background.

20. Which statement among the ones presented below is true?

a. The spectrum of sunlight consists of a series of white lines superimposed on a colored background.
b. The spectrum of sunlight consists of a series of white lines superimposed on a dark background.
c. The spectrum of sunlight consists of a series of colored lines superimposed on a dark background.
! d. The spectrum of sunlight consists of a series of dark lines superimposed on a colored background.
e. The spectrum of sunlight consists of a series of dark lines superimposed on a white background.

21. Which statement below is true with regard to Bohr's model of the atom?

a. The model accounted for the absorption spectra of atoms but not for the emission spectra.
! b. The model could account for the emission spectrum of hydrogen and for the Rydberg equation.
c. The model was based on the wave properties of the electron.
d. The model accounted for the emission spectra of atoms, but not for the absorption spectra.
e. The model was generally successful for all atoms to which it was applied.

22. What is the energy, in joules, of one mole of photons whose wavelength is $5.461 \times 10^{2}$ nm?

a. $2.191 \times 10^{-4}$ J
b. $2.437 \times 10^{-12}$ J
! c. $2.191 \times 10^{5}$ J
d. $1.376 \times 10^{6}$ J
e. $4.06 \times 10^{-19}$ J

23. Which statement below is true about the spectrum of hydrogen obtained in gas discharge tubes experiments?

a. A photon is absorbed as the electron goes from a state with a ***higher*** energy to one with a ***lower*** energy.
b. A photon is absorbed as the electron goes from a state with a ***lower*** energy to one with a ***higher*** energy.
! c. A photon is emitted as the electron goes from a state with a ***higher*** energy to one with a ***lower*** energy.
d. A photon is emitted as the electron goes from a state with a ***lower*** energy to one with a ***higher*** energy.
e. An electron is emitted as the photon goes from a state with a ***higher*** energy to one with a ***lower*** energy.

24. The definite energies associated with specific wavelengths in the emission spectrum of atomic hydrogen suggest that

a. electrons have a smaller rest mass than photons
b. photons have a smaller rest mass than electrons
! c. energy states in the hydrogen atom are quantized
d. atomic hydrogen is more stable and has a lower potential energy than molecular hydrogen
e. the potential energy of electrons in the atom can have any arbitrary value over a period of time, but the kinetic energy may only have certain specific values

25. Calculate the energy required to excite a hydrogen atom by causing an electronic transition from the energy level with n = 1 to the level with n = 4. Recall that the energies of the levels in the hydrogen atom are given by: $E_n = (21.79 \times 10^{-19})(1/n^2)$ joules.

a. $2.017 \times 10^{-29}$ J
! b. $2.043 \times 10^{-18}$ J
c. $2.192 \times 10^{5}$ J
d. $2.254 \times 10^{-18}$ J
e. $3.275 \times 10^{-17}$J

26. Calculate the frequency of the light emitted by a hydrogen atom during a transition of its electron from the energy level with n = 4 to the level with n = 1. Recall that the energies of the levels in the hydrogen atom are given by: $E_n = (21.79 \times 10^{-19})(1/n^2)$ joules.

a. $1.028 \times 10^{7}\ s^{-1}$
b. $1.215 \times 10^{-7}\ s^{-1}$
c. $2.467 \times 10^{15}\ s^{-1}$
! d. $3.083 \times 10^{15}\ s^{-1}$
e. $8.228 \times 10^{6}\ s^{-1}$

27. Calculate the wavelength of the light emitted by a hydrogen atom during a transition of its electron from the energy level with n = 4 to the level with n = 1. Recall that the energies of the levels in the hydrogen atom are given by: $E_n = (21.79 \times 10^{-19})(1/n^2)$ joules.

a. 0.6126 nm
b. $6.856 \times 10^{-18}$ nm
c. 29.16 nm
! d. 97.23 nm
e. 121.5 nm

28. Calculate the frequency of the light emitted by a hydrogen atom during a transition of its electron from the energy level with n = 6 to the level with n = 3. Recall that the energies of the levels in the hydrogen atom are given by: $E_n = (21.79 \times 10^{-19})(1/n^2)$ joules.

a. $1.665 \times 10^{-26}\ s^{-1}$
b. $1.824 \times 10^{-15}\ s^{-1}$
! c. $2.741 \times 10^{14}\ s^{-1}$
d. $3.649 \times 10^{-15}\ s^{-1}$
e. $9.132 \times 10^{13}\ s^{-1}$

29. Calculate the energy required to excite a hydrogen atom by causing an electronic transition from the energy level with n = 3 to the level with n = 6. Recall that the energies of the levels in the hydrogen atom are given by: $E_n = (21.79 \times 10^{-19})(1/n^2)$ joules.

a. $1.211 \times 10^{-19}$ J
! b. $1.816 \times 10^{-19}$ J
c. $2.421 \times 10^{-19}$ J
d. $3.632 \times 10^{-19}$ J
e. $6.658 \times 10^{-19}$ J

30. Calculate the wavelength of the light emitted by a hydrogen atom during a transition of its electron from the energy level with n = 6 to the level with n = 3. Recall that the energies of the levels in the hydrogen atom are given by: $E_n = (21.79 \times 10^{-19})(1/n^2)$ joules.

a. $2.954 \times 10^{-5}$ m
! b. 1094 nm
c. 547 nm
d. 821 nm
e. 1640 nm

31. In an electron microscope, electrons are accelerated to high speeds. Using the de Broglie relation and ignoring relativistic effects, calculate the wavelength of an electron traveling with a speed of $6.550 \times 10^4$ meters per second. The electron has a mass of $5.485712 \times 10^{-4}$ a.m.u.

! a. $1.111 \times 10^{-8}$ m
b. $1.111 \times 10^{-11}$ m
c. $1.844 \times 10^{-35}$ m
d. $2.426 \times 10^{-12}$ m
e. $5.528 \times 10^{-27}$ m

32. Using the de Broglie relation and ignoring relativistic effects, calculate the wavelength of a $^{19}F^{+}$ ion which is moving with a speed of $4.255 \times 10^{5}$ m/s. The mass of F-19 is 18.9984 a.m.u.

a. $4.936 \times 10^{-17}$ m
! b. $4.936 \times 10^{-13}$ m
c. $1.484 \times 10^{-29}$ m
d. $1.484 \times 10^{-26}$ m
e. $4.936 \times 10^{-20}$ m

33. Using the de Broglie relation, calculate the wavelength of a neutron which is moving with a speed of $4.505 \times 10^{4}$ m/s. The mass of the neutron is 1.008665 a.m.u.

a. $1.458 \times 10^{-38}$ m
b. $2.632 \times 10^{-3}$ m
c. $4.372 \times 10^{-30}$ m
d. $8.781 \times 10^{-15}$ m
! e. $8.781 \times 10^{-12}$ m

34. The first full presentation describing the behavior of the electron in the hydrogen atom by application of the wave mechanics approach was presented by

a. Louis de Broglie
b. Werner von Heisenberg
c. Wolfgang Pauli
d. Ernest Rutherford
! e. Erwin Schrödinger

35. "No two electrons in the same atom can have all its quantum numbers the same." This statement is based on the work of

a. Louis de Broglie
b. Werner von Heisenberg
c. Jesse McKinley Nicholson
! d. Wolfgang Pauli
e. Erwin Schrödinger

36. Which statement below best summarizes the known facts about the quantum numbers which characterize the equations that describe the behavior of electrons in atoms?

a. all of the quantum numbers are allowed to have values which are not integers
b. only the principal quantum number is allowed to have values which are not integers
! c. only the spin quantum numbers are allowed to have values which are not integers
d. only the azimuthal quantum number is allowed to have values which are not integers
e. no quantum numbers are allowed to have values which are not integers

37. The wave functions which are solutions to the wave functions which describe the behavior of the electron in the hydrogen atom are described by how many quantum numbers?

a. 1
b. 2
! c. 3
d. 4
e. 5

38. The letter designation for the subshell is based on

! a. the value of the azimuthal quantum number
b. the value of the principal quantum number
c. the value of the magnetic quantum number, $m_l$
d. the value of the spin quantum number, $m_s$
e. the transverse polarization of the optical emission from the H atom

39. The three quantum numbers which characterize the solutions to the wave equation describing the behavior of the electron in the H atom are usually designated as

a. 1s 2s 2p
b. $n$ $l$ $m_s$
c. $m_l$ $m_s$ $m_p$
! d. $n$ $l$ $m_l$
e. $l$ $m_l$ $m_s$

40. All orbitals with the same value of the principal quantum number are said

! a. to belong to the same shell
b. to belong to the same subshell
c. to belong to the same group
d. to belong to the same period
e. to belong to the same class

41. The spectroscopic notation (number + letter designation) for the subshell with $n = 5$ and $l = 3$ is

a. 5d subshell
b. 5p subshell
! c. 5f subshell
d. 5g subshell
e. 5s subshell

42. The spectroscopic notation (number + letter designation) for the subshell with $n = 4$ and $l = 2$ is

! a. 4d subshell
b. 4p subshell
c. 4f subshell
d. 4s subshell
e. there is no subshell fitting this description

43. The maximum number of electrons that would be required to fill all the energy levels having a value of 4 for $n$, the principal quantum number, is

a. $n$
b. $n+1$
c. $2n$
d. $n^2$
! e. $2n^2$

44. The spectroscopic notation (number + letter designation) for the subshell with $n = 3$ and $l = 3$ is

a. 3d subshell
b. 3f subshell
c. 3p subshell
d. 3s subshell
! e. There is no subshell fitting this description.

45. Given the following sets of quantum numbers for $n, l, m_\ell, m_s$, which one of these sets is not a possible set for an electron in an atom?

a. 3 2 2 -½
b. 3 1 -1 ½
c. 4 3 2 ½
d. 4 3 -2 -½
! e. 5 2 3 ½

46. Given the following sets of quantum numbers for $n, l, m_\ell, m_s$, which one of these sets is not a possible set for an electron in an atom?

! a. 3 1 -1 0
b. 3 2 2 -½
c. 4 3 2 ½
d. 4 3 -2 -½
e. 5 3 2 ½

47. Given the following sets of quantum numbers for $n, l, m_\ell, m_s$, which one of these sets is not a possible set for an electron in an atom?

a. 3 1 -1 ½
! b. 3 3 2 -½
c. 4 3 2 ½
d. 4 3 -2 -½
e. 5 3 2 ½

48. The statement that the ground state configuration of an atom is generated by filling in levels from the lowest (energy-wise) to the highest with electrons observing the maximum for each of these levels is

! a. the Aufbau principle
b. Bustamente's principle
c. Hund's Rule
d. Murphy's rule
e. the Pauli Principle

49. There are several possible arrangements of electrons when you try to place 7 electrons in a 3d subshell. To determine the correct distribution for the ground state we are guided by

a. the Aufbau principle
! b. Hund's Rule
c. the Pauli Principle
d. the wave particle duality principle
e. the uncertainty principle

50. A correct description for the ground state configuration of the vanadium atom is

a. [Ar] $4s^1$ $3d^4$, paramagnetic
! b. [Ar] $4s^2$ $3d^3$, paramagnetic
c. [Ar] $4s^3$ $3d^2$, paramagnetic
d. [Ar] $3d^5$, paramagnetic
e. [Ar] $3s^2$ $3d^3$, paramagnetic

51. A correct description for the ground state configuration of the vanadium atom is

a. [Ar] $4s^1$ $3d^4$, paramagnetic
! b. [Ar] $4s^2$ $3d^3$, paramagnetic
c. [Ar] $4s^3$ $3d^2$, paramagnetic
d. [Ar] $3d^5$, paramagnetic
e. [Ar] $3s^2$ $3d^3$, paramagnetic

52. A correct description for the ground state configuration of the chromium atom is

! a. [Ar] $4s^1$ $3d^5$, paramagnetic
b. [Ar] $4s^2$ $3d^4$, paramagnetic
c. [Ar] $4s^3$ $3d^3$, paramagnetic
d. [Ar] $3d^6$, paramagnetic
e. [Ar] $3s^2$ $3d^4$, paramagnetic

53. A correct description for the ground state configuration of the selenium atom is

a. [Ar] $4s^1$ $3d^{10}$ $4p^5$, paramagnetic
! b. [Ar] $4s^2$ $3d^{10}$ $4p^4$, paramagnetic
c. [Ar] $4s^2$ $3d^8$ $4p^6$, paramagnetic
d. [Ar] $3d^{10}$,$4p^6$ diamagnetic
e. [Ar] $4s^1$ $3d^9 4p^6$, paramagnetic

54. Which one of the following configurations represents an excited state configuration?

a. [Ar] $4s^1$ $3d^5$
! b. [Ar] $4s^2$ $3d^4$
c. [Xe] $5s^2$ $5p^1$
d. [Xe] $6s^2$ $4f^7$
e. [Rn] $7s^2$

55. Which one of the following configurations represent a situation that can not exist for ground state or for excited state?

a. [Ar] $4s^2$ $3d^{10}$ $4p^5$
! b. [Xe] $5s^2$ $4d^{10}$ $5p^4$
c. [Rn] $7s^2$ $4f^8$
d. [Kr] $5s^1$,$4d^6$
e. [Ne] $6s^1$

56. An atom is described as having the ground state electronic configuration, [Ar] $4s^2\ 3d^5\ 4f^2$. Which element fits the description?

a. one of the chalcogen family
b. one of the halogen family
c. one of the alkaline earth family
d. one of the lanthanide elements
! e. there is no element which fits the description listed above

57. Using X to indicate a filled inner core and x for a valence shell electron, indicate which of the following choices is the correct ground state electronic configuration for the sulfur atom.

| | [Ne] | 3s | 3p | | |
|---|---|---|---|---|---|
| a. | X | xx | x | x | x |
| ! b. | X | xx | xx | x | x |
| c. | X | x | xx | x | xx |
| d. | X | xx | xx | xx | |
| e. | X | xx | xx | xx | |

58. Based on the Aufbau principle and other applicable guiding principles, what ground state electronic configuration would expect to find for technetium (Z = 43)?

a. [Kr] $4s^2\ 3d^5$
b. [Kr] $4s^2\ 4d^5$
c. [Kr] $4d^7$
! d. [Kr] $5s^2\ 4d^5$
e. [Kr] $5s^2\ 5d^5$

59. Which one of the following configurations represents a non-existent state?

a. [Ar] $4s^1\ 3d^5$
b. [Ar] $4s^2\ 3d^4$
! c. [Xe] $5s^2\ 5p^1$
d. [Xe] $6s^2\ 4f^7$
e. [Rn] $7s^2$

60. Which one of the following configurations represents an alkaline earth element?

a. [Ar] $4s^1\ 3d^5$
b. [Ar] $4s^2\ 3d^4$
c. [Xe] $5s^2\ 5p^1$
d. [Xe] $6s^2\ 4f^7$
! e. [Rn] $7s^2$

61. Which one of the following configurations represents a halogen element?

a. [Ar] $4s^1\ 3d^5$
b. [Ar] $4s^2\ 3d^4$
! c. [Kr] $5s^2\ 5p^5\ 4d^{10}$
d. [Xe] $6s^2\ 4f^7$
e. [Xe] $6s^2\ 4f^{14}\ 5d^{10} 6p^4$

62. A possible set of quantum numbers for an electron in the partially filled subshell in the gallium atom in its ground state configuration would be

| | $n$ | $l$ | $m_l$ | $m_s$ |
|---|---|---|---|---|
| a. | 3 | 1 | 0 | -½ |
| b. | 3 | 1 | 1 | ½ |
| c. | 4 | 0 | 0 | -½ |
| ! d. | 4 | 1 | 0 | ½ |
| e. | 4 | 2 | 1 | ½ |

63. A possible set of quantum numbers for an electron in the partially filled subshell in the vanadium atom in its ground state configuration would be

| | $n$ | $l$ | $m_l$ | $m_s$ |
|---|---|---|---|---|
| a. | 3 | 1 | 0 | -½ |
| ! b. | 3 | 2 | 1 | ½ |
| c. | 4 | 0 | 0 | -½ |
| d. | 4 | 1 | 0 | ½ |
| e. | 4 | 2 | 1 | ½ |

64. A possible set of quantum numbers for an electron in the partially filled subshell in the technetium atom in its ground state configuration would be

| | $n$ | $l$ | $m_l$ | $m_s$ |
|---|---|---|---|---|
| a. | 3 | 1 | 0 | -½ |
| b. | 3 | 2 | 1 | ½ |
| c. | 4 | 0 | 0 | -½ |
| d. | 4 | 1 | 0 | ½ |
| ! e. | 4 | 2 | 1 | ½ |

65. Which one of the following situations is not ever encountered in the ground state configuration of an atom OR an ion?

a. a . . . $3d^4$ with 4 unpaired electrons
! b. a . . . $3d^7$ with 7 unpaired electrons
c. a . . . $3p^1$ with 1 unpaired electron
d. a . . . $3d^9$ with 1 unpaired electron
e. a . . . $4f^7$ with 7 unpaired electrons

66. Which one of the following statements is true?

a. a 3f orbital is larger (extends farther) than a 3 p orbital in a particular atom
b. a 3p orbital has four lobes, one in each of the four quadrants
c. a $3f_{xyz}$ has eight lobes, one is each octant
d. all 3d orbitals have the same shape, just different orientations
! e. the 7s orbital can be represented by sphere

67. Based on the expected ground state electronic configuration, element #120

a. should be an inert gas element
b. should have two unpaired electrons
c. should be a chalcogen element
! d. should be an alkaline earth element
e. should have four unpaired electrons

68. Based on the order in which the subshells are filled in the ground state electronic configuration of atoms, element #111

! a. should be a coinage metal
b. should be an alkali metal
c. should be a halogen element
d. should be an actinide element
e. should be an inert gas element

69. Which one of the species below should have the smallest radius?

a. Ar
b. $Ca^{2+}$
c. $K^{+}$
! d. $Mg^{2+}$
e. $Na^{+}$

70. Which one of the species below should have the smallest radius?

a. $Cl^{-}$
! b. $F^{-}$
c. $O^{2-}$
d. $S^{2-}$
e. $Se^{2-}$

71. Which one of the species below should have the largest radius?

! a. Ar
b. $Ca^{2+}$
c. $K^{+}$
d. $Mg^{2+}$
e. $Na^{+}$

72. Which one of the atoms listed below has the largest value for its first ionization energy?

! a. Al
b. Sr
c. Ga
d. Cr
e. Fr

## Fill In The Blanks

73. The number of orbitals in a shell with $n = 3$ is ______. (! 9)

74. The number of orbitals in a subshell with $l = 3$ is ______. (! 7)

75. A particular energy level in a multielectron atom has a value of 3 for the secondary quantum number. What is the maximum number of electrons that can occupy this energy level? ______ (! 14)

76 The behavior of an electron in an atom can be described in terms of solutions to the wave equation. There are variable quantum numbers in these solutions (functions). How many variable quantum numbers are there for describing ONE electron in the atom? ______ ( ! 4)

77. Even though the _______ might suggest that the ground state configuration of the beryllium atom is $1s^4$, the __________ says otherwise. ( ! Aufbau method, Pauli principle)

78. How many completely filled subshells are there in the ground state configuration for a calcium atom? ______ ( ! 6)

79. How many unpaired electrons are there in the ground state electronic configuration for an iron atom? ______ ( ! 4)

80. Which one of the atoms in the set: Y, Cr, Mg, N, Ba, Se, Sn has the largest first ionization energy? ______ ( ! N)

81. Which one of the atoms in the set: Sr, Fr, Hg, Ga, Cr, Sn has the smallest first ionization energy? ______ ( ! Fr)

82. Which one of the atoms in the set: O, F, Ne, Ar, Cl, K, Ga has the greatest affinity for electrons? ______ ( ! F)

83. Which one of the atoms in the set: O, F, Ne, Ar, Cl, K, Ga has the largest first ionization energy? ______ ( ! Ne)

84. Which one of all the atoms has the smallest atomic radius? ______ ( ! He)

85. Which one of the atoms in the set: Ba, Cr, N, Sn, Mg, Sn, Se would you expect to have the largest atomic radius? ______ ( ! Ba)

86. Which one of the atoms in the set: Sr, Hf, Hg, Ga, Cr, Fr, Sn would you expect to have the largest atomic radius? ______ ( ! Fr)

87. Which one of the atoms in the set: Ba, Cr, N, Sn, Mg, Sn, Se would you expect to have the largest atomic radius? ______ ( ! Ba)

## True and False

88. Even though the fluoride ion is larger than the fluorine atom, and the chloride ion is larger than the chlorine atom, the oxide ion is smaller than the oxygen atom. ( !F)

89. The uncertainty principle was proposed to explain the fact that certain atoms, for instance silver, have different electronic configurations in their ground states than predicted by the Aufbau method. ( !F)

90. In the tellurium atom, the valence shell electronic configuration is $ns^2\ np^6$. ( !F)

91. There are more unpaired electrons in the ground state electronic configuration for a chromium atom than there are for a manganese atom. ( !T)

92. Generally speaking, anions and cations have larger radii than their parent atoms. ( !F)

93. One of the ways in which very very small particles display wavelike behavior is in their ability to be diffracted when they are in particle beams. ( !T)

94. The periodic structure of the periodic table can be correlated with periodic behavior of the electronic stuctural configuration of the elements. ( !T)

## Critical Thinking

95. A phototube in a photoelectric device has a photosensitive element. When this element is beamed with radiation having a wavelength of 374.2 nm, electrons are emitted with a maximum kinetic energy of $1.488 \times 10^{-19}$ J/electron. What is the wavelength of a laser which will turn the photocell on and cause electrons to be ejected which have absolutely no excess energy? ______ ( ! 520 nm)

96. A phototube in a photoelectric device has a photosensitive element. When this element is beamed with radiation having a wavelength of 354.2 nm, electrons are emitted with a maximum kinetic energy of $1.988 \times 10^{-19}$ J/electron. What is the wavelength of a laser which will turn the photocell on and cause electrons to be ejected which have absolutely no excess energy? ______ ( ! 549 nm)

## Special Topics

97. A device which utilizes the particle properties of light beams is

a. the electron microscope
b. the magnetic resonance imaging spectrometer (MRI)
! c. the photoelectric cell
d. the radiotelescope
e. the visible light laser

98. A device which utilizes the wave properties of accelerated electron beams is

! a. the electron microscope
b. the infrared laser
c. the magnetic resonance imaging spectrometer (MRI)
d. the phase constrast optical microscope
e. the ROM BIOS

## Chemicals in Use 5: Lasers in Chemistry

99. The term, laser, is an acronym for:

a. light amplification stimulated by equivalents of radiation
b. light amplification stimulated by equivalents of radiation
c. limited amplification stimulated by emission of radiation
d. limited amplification by stoichiometric equivalents of radiation
! e. light amplification by stimulated emission of radiation

100. Lasers produce light which is coherent. This means

! a. in phase, and hence can be easily focused to a tiny spot
b. the laser beam is composed of photons which are all of the same wavelength, hence possess one energy
c. the beam clings to the surface against which it is projected
d. when used to read CD's, the words recorded are highly intelligible
e. a laser beam is composed of standing waves only

101. In which portion of the electromagnetic spectrum do we find laser emission from the carbon dioxide laser?

! a. infrared
b. radio frequency
c. ultraviolet
d. visible
e. x-ray

# Chapter 7 Chemical Bonding I

## Multiple Choice

1. Sodium tends to form ions which have the electronic configuration of a noble gas. What is the electronic configuration of the noble gas which the sodium ion mimics?

a. $1s^2$
b. $1s^2\ 2p^6$
! c. $1s^2\ 2s^2\ 2p^6$
d. $1s^2\ 2s^2\ 2p^6\ 3s^2$
e. $1s^2\ 2s^2\ 2p^6\ 3s^2\ 3p^6$

2. Bromine tends to form simple ions which have the electronic configuration of a noble gas. What is the electronic configuration of the noble gas which the bromide ion mimics?

a. $1s^2\ 2s^2\ 2p^6\ 3s^2\ 3p^6\ 3d^{10}\ 4p^6$
b. $1s^2\ 2s^2\ 2p^6\ 3s^2\ 3p^6\ 4p^6\ 4d^{10}$
c. $1s^2\ 2s^2\ 2p^6\ 3s^2\ 3p^6\ 4s^2\ 4p^6$
! d. $1s^2\ 2s^2\ 2p^6\ 3s^2\ 3p^6\ 3d^{10}\ 4s^2\ 4p^6$
e. $1s^2\ 2s^2\ 2p^6\ 3s^2\ 3p^6\ 3d^{10}\ 4s^2\ 4p^6\ 4d^{10}$

3. Which one of the following reactions is exothermic to a substantial degree?

! a. $F(g) + e^- \longrightarrow F^-(g)$
b. $F^-(g) + e^- \longrightarrow F^{2-}(g)$
c. $F^{2-}(g) + e^- \longrightarrow F^{3-}(g)$
d. $F^{3-}(g) + e^- \longrightarrow F^{4-}(g)$
e. $F^{4-}(g) + e^- \longrightarrow F^{5-}(g)$

4. Which one of the following reactions is exothermic to a substantial degree?

! a. $Cl(g) + e^- \longrightarrow Cl^-(g)$
b. $He(g) + e^- \longrightarrow He^-(g)$
c. $Ar(g) + e^- \longrightarrow Ar^-(g)$
d. $O^-(g) + e^- \longrightarrow O^{2-}(g)$
e. $N^{2-}(g) + e^- \longrightarrow N^{3-}(g)$

5. If the octet rule were firmly obeyed by arsenic atoms, which one of the following ions of arsenic would be found in binary arsenic compounds?

a. $As^-$
b. $As^{2-}$
! c. $As^{3-}$
d. $As^{4-}$
e. $As^{5-}$

6. The octet rule appears to hold true for cations of which of the following sets?

a. post transition metals and transition metals
! b. Group Ia and Group IIa metals
c. Group IVa elements
d. lanthanide elements
e. actinide elements

7. Which of the following possible states of the element uranium is the one which obeys the octet rule?

a. $U^{2+}$
b. $U^{3+}$
c. $U^{4+}$
! d. $U^{6+}$
e. $U^{2-}$

8. The ions below do not follow the octet rule, with the exception of

a. $Fe^{3+}$
b. $Sn^{2+}$
c. $Ni^{2+}$
! d. $Ti^{4+}$
e. $Cr^{3+}$

9. The ions below do not follow the octet rule, with the exception of

a. $Fe^{3+}$
b. $Ni^{2+}$
! c. $Sc^{3+}$
d. $V^{3+}$
e. $Cr^{3+}$

10. The valence electrons in the potassium ions present in potassium nitrate are

a. the 2p electrons only
b. the 3s electrons only
c. the 3s and 3p electrons only
d. the 3p electrons only
! e. There are no valence electrons in the potassium ions in the compound above.

11. The valence electrons in the chloride ions present in potassium chloride are

a. the 2p electrons only
b. the 3s electrons only
! c. the 3s and 3p electrons only
d. the 3p electrons only
e. There are no valence electrons in the chloride ions in the compound above.

12. Which one of the elements below has 5 valence electrons in its Lewis structure?

a. actinium
b. americium
c. argon
! d. arsenic
e. astatine

13. Which one of the elements below has 3 valence electrons in its Lewis structure?

! a. gallium
b. fluorine
c. iron
d. nickel
e. sulfur

14. Which one of the species below has 8 valence electrons in its Lewis structure?

a. $Ar^+$
b. $F^+$
c. $Mg^+$
! d. $S^{2-}$
e. Si

15. Which one of the elements below has no valence electrons in its Lewis structure?

a. $Ar^+$
b. $Ga^+$
! c. $Mg^{2+}$
d. $S^{2-}$
e. $F^-$

16. Covalent bonds formed by the sharing of electrons are most likely to be formed between

a. an atom with a high electron affinity and an atom with a low ionization energy
b. an atom with a low electron affinity and an atom with a high ionization energy
! c. two atoms with high electron affinities and high ionization energies
d. two atoms with low electron affinities and low ionization energies
e. an atom with a low electronegativity and an atom with a high ionization energy

17. The atoms in the nitrogen molecule, $N_2$, are held together by

a. a single covalent bond
b. a double covalent bond
! c. a triple covalent bond
d. an ionic bond
e. a magnetic dipole bond

18. The atoms in the oxygen molecule, $O_2$, are held together by

a. a single covalent bond
! b. a double covalent bond
c. a triple covalent bond
d. an ionic bond
e. a magnetic dipole bond

19. The atoms in the hydrogen fluoride molecule are held together by

! a. a single covalent bond
b. a double covalent bond
c. a triple covalent bond
d. an ionic bond
e. a magnetic dipole bond

20. The Lewis structure for the carbon atom shows __ valence electrons. The number of bonds which carbon usually forms in order to complete its valence shell and obey the octet rule is ___

a. 4, 1
b. 4, 2
c. 2, 4
d. 4, 3
! e. 4, 4

21. The Lewis structure for the nitrogen atom shows __ valence electrons. The number of bonds which nitrogen usually forms in order to complete its valence shell and obey the octet rule is ___

a. 5, 1
b. 5, 2
c. 3, 4
! d. 5, 3
e. 5, 4

22. An element, whose chemical symbol is illegible, had 6 electrons surrounding it in the Lewis structure. Which element among the following could it possibly be?

a. Cr
b. Ni
c. Ba
! d. Se
e. As

23. The structural formula for one of the heptadiene isomers can be written in abbreviated form as

$$CH_3CHCHCH_2CHCHCH_3$$

How many bonds should there be in this molecule?

a. 16
b. 18
c. 19
! d. 20
e. 22

24. The structural formula for a certain alcohol can be written in abbreviated form as

$$CH_2CHCH_2OH$$

How many bonds should there be in this molecule?

a. 8
b. 9
! c. 10
d. 11
e. 12

25. The structural formula for a certain alcohol can be written in abbreviated form as

$$CH_2CHCH_2CH_3OH$$

How many bonds should there be in this molecule?

a. 13
! b. 14
c. 15
d. 16
e. 17

26. The compound shown immediately below is an example of

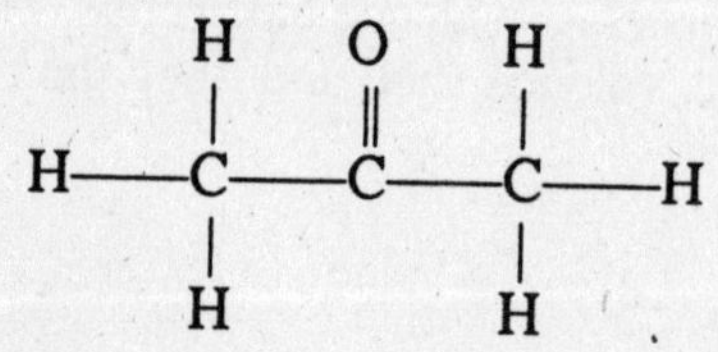

a. an alcohol
! b. a ketone
c. an acid
d. a hydrocarbon
e. a base

27. The compound shown immediately below is an example of

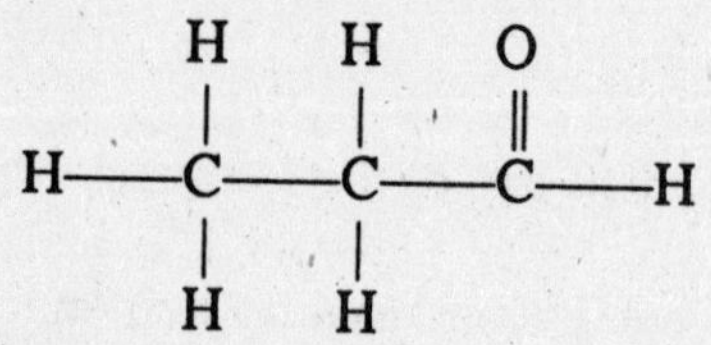

a. an alcohol
b. a ketone
! c. an aldehyde
d. an acid
e. a base

28. The compound shown immediately below is an example of

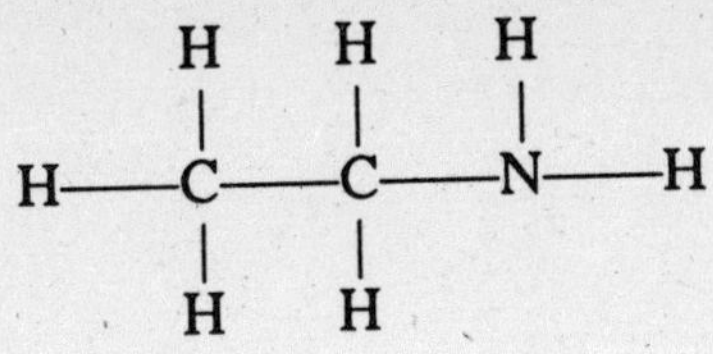

a. an alcohol
b. a ketone
c. an aldehyde
d. an acid
! e. a base

29. The structural formula for a hydrocarbon can be written in abbreviated form as

$$CH_3CHCHCH_2CHCH_2$$

How many bonds should there be in this molecule?

a. 15
b. 16
! c. 17
d. 18
e. 19

30. Which one of the following bonds is the most polar one of the set?

a. H—Br
b. H—Cl
! c. H—F
d. H—I
e. H—N

31. Which one of the following bonds is the most polar one of the set?

a. H—C
! b. H—Cl
c. H—P
d. H—S
e. H—Se

32. Based on the position in the periodic table, which one of the folowing atoms would you expect to be the most electronegative?

a. Ba
b. Ga
c. Mn
! d. N
e. Si

33. Based on the position in the periodic table, which one of the folowing atoms would you expect to be the most electronegative?

! a. Cl
b. Ge
c. P
d. Se
e. Sn

34. Which one of the following is the least electronegative element of the set presented?

a. F
b. N
c. C
d. O
! e. H

35. Which one of the following is the least electronegative element of the set presented?

a. N
b. S
! c. C
d. P
e. 0

36. Some possible skeletons for the Lewis structure of sulfurous acid, $H_2SO_3$ are shown below. Which one of the following is the correct skeleton?

a.
```
            O
H     H     S     O
            O
```

b.
```
      O
H     S     O
      O     H
```

! c.
```
            O
H     O     S     O     H
```

d.
```
H     O     S     O     O     H
```

e.
```
            H
H     O     S     O
            O
```

37. Complete the Lewis structure for $HClO_2$ from the shell presented below by filling in the bonds and the remaining valence electrons (those which are not in the bonds). If the valence shells are filled to the *usual* limit, how many of these non-bonding valence electrons are there in the molecule?

H O Cl O

! a. 14
b. 16
c. 18
d. 20
e. 26

38. Which one of the following is the least electronegative element of the set presented?

a. N
b. O
c. Cl
d. Br
! e. I

39. Complete the Lewis structure for $H_2SeO_3$ from the shell presented below by filling in the bonds and the remaining valence electrons (those which are not in the bonds). If the valence shells are filled to the *usual* limit, how many of these non-bonding valence electrons are there in the molecule?

O

H O Se O H

a. 14
! b. 16
c. 18
d. 20
e. 28

40. Complete the Lewis structure for $HClO_3$ from the shell presented below by filling in the bonds and the remaining valence electrons (those which are not in the bonds). If the valence shells are filled to the *usual* limit, what is the formal charge on the chlorine atom?

O

H O Cl O

a. -1
b. 0
! c. +1
d. +2
e. +3

41. Complete the Lewis structure for $HClO_3$ from the shell presented below by filling in the bonds and the remaining valence electrons (those which are not in the bonds). If the valence shells are filled to the *usual* limit, what is the sum of the absolute values of all the formal charges in the molecule?

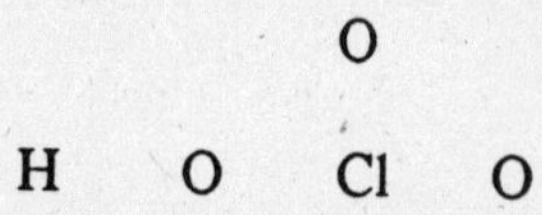


a. 0
b. 1
! c. 2
d. 3
e. 4

42. Complete the Lewis structure for $H_2SeO_4$ from the shell presented below by filling in the bonds and the remaining valence electrons (those which are not in the bonds). If the valence shells are filled to the *usual* limit, what is the sum of the absolute values of all the formal charges in the molecule?

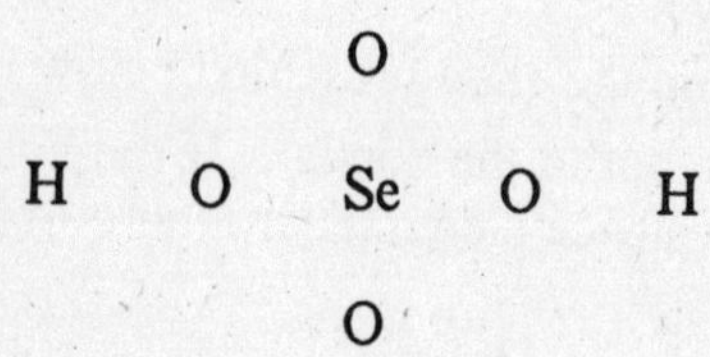


a. 0
b. 1
c. 2
d. 3
! e. 4

43. A student drew four possible Lewis structures for $HBrO_4$

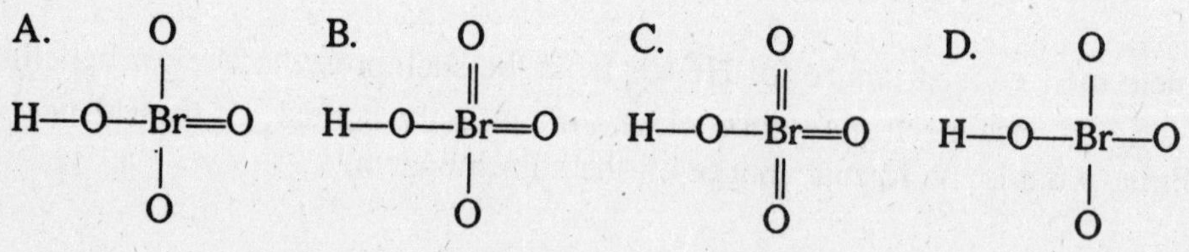


Complete these Lewis structures presented above by filling in the remaining valence electrons (those which are not in the bonds). Based on these structures, the preferred structure would be the structure shown as ____ in which the sum of the absolute values of the formal charges on all the atoms is ____.

a. A, 4
b. B, 2
c. C, 0
d. D, 6
e. D, 0

44. The Lewis stucture for $SF_5^-$ ion when completed, should include ___ bonds and ___ non-bonding valence electrons (the valence electrons which are not in any bond).

a. 5, 30
b. 6, 28
! c. 5, 32
d. 5, 42
e. 5, 31

45. The chlorite ion has a Lewis structure which is based on the skeleton shown below.

O Cl O

Complete the Lewis structure presented above by filling in the remaining valence electrons (those which are not in the bonds) and minimizing the formal charges present in the structure. Based on this "tuned" structure, the chlorite ion has

a. two single bonds, sum of absolute values of formal charges = 3, and no resonance hybrids.
b. one single and one double bond, sum of absolute values of formal charges = 5, and there are two contributing resonance hybrids.
c. one single and one double bond, sum of absolute values of formal charges = 3, and there are two contributing resonance hybrids.
d. two double bonds, sum of absolute values of formal charges = 1, and no resonance hybrids.
e. one single and one double bond, sum of absolute values of formal charges = 1, and there are two contributing resonance hybrids.

46. Draw a correct Lewis structure for $CH_2Cl_2$. Based on this Lewis structure, the calculated value for the formal charge on the carbon atom is

! a. 0
b. +4
c. +2
d. -2
e. -4

47. Draw a correct Lewis structure for $H_3C—NH_2$. Based on this Lewis structure, the calculated value for the formal charge on the nitrogen atom is

a. -2
b. +3
c. -3
d. +2
! e. 0

48. Draw a Lewis structure for $H_3C—NO_2$. Both oxygen atoms are directly attached to the nitrogen atom. What is the formal charge on the nitrogen atom?

a. -3
! b. +1
c. +3
d. -1
e. 0

49. When the fluoride ion reacts with a $BF_3$ molecule (a molecule in which there are no multiple bonds) an ion is formed in which the boron atom in the central atom. The bond between the boron trifluoride and the fluoride ion is

a. an ionic bond
b. a regular covalent bond, where both species contribute 1 electron to the bond.
! c. a coordinate covalent bond
d. a resonance hybrid bond
e. a bond where two atoms share one electron instead of two

50. Complete the Lewis structures for $COCl_2$ and $SOCl_2$ using the skeletons shown below, being sure to follow the procedure for minimizing the sum of the absolute values for the formal charges, where valence shell size restrictions permit. Based on the structures you have completed, which statement below is true?

```
      O                 O

Cl    C    Cl     Cl    S    Cl
```

a. The $COCl_2$ exhibits residual formal charges but no resonance hybrids, while the $SOCl_2$ exhibits both residual formal charges and resonance hybrids.
b. The $COCl_2$ exhibits both residual formal charges and resonance hybrids, while the $SOCl_2$ exhibits residual formal charges but no resonance hybrids.
c. The $COCl_2$ exhibits both residual formal charges and resonance hybrids, while the $SOCl_2$ exhibits residual formal charges but no resonance hybrids.
d. The $COCl_2$ exhibits residual both formal charges and resonance hybrids, while the $SOCl_2$ exhibits resonance hybrids but no residual formal charges.
e. The $COCl_2$ exhibits residual formal charges but no resonance hybrids, while the $SOCl_2$ exhibits resonance hybrids but no residual formal charges.

51. The carbonate ion has the skeletal structure as shown below. Complete the Lewis structure by filling in the bonds and the remaining valence electrons which are not involved in bonds. Which assertion made about the carbonate ion below is true?

```
     O

O    C    O
```

Based on the structure the carbonate ion should have ___ resonance hybrids and ___ atoms with residual formal charges on them in its structure.

a. 2, 2
b. 3, 1
c. 3, 2
d. 2, 1
e. no, 2

52. When the fluoride ion reacts with a $BF_3$ molecule (a molecule in which there are no multiple bonds) an ion is formed in which the boron atom is the central atom. In this reaction,

! a. the fluoride ion behaves as a Lewis base and the $BF_3$ behaves as a Lewis acid.
b. the fluoride ion behaves as a Lewis acid and the $BF_3$ behaves as a Lewis base.
c. the fluoride ion behaves as an Arrhenius base and the $BF_3$ behaves as an Arrhenius acid.
d. the fluoride ion behaves as a Brönsted acid and the $BF_3$ behaves as a Brönsted base.
e. the fluoride ion behaves as an Arrhenius acid and the $BF_3$ behaves as an Arrhenius base.

53. The skeleton for the Lewis structure of nitromethane, $CH_3NO_2$, is shown below. Complete the Lewis structure by filling in the bonds and the remaining valence electrons which are not involved in bonds. Which assertion made about the nitromethane carbonate ion below is true?

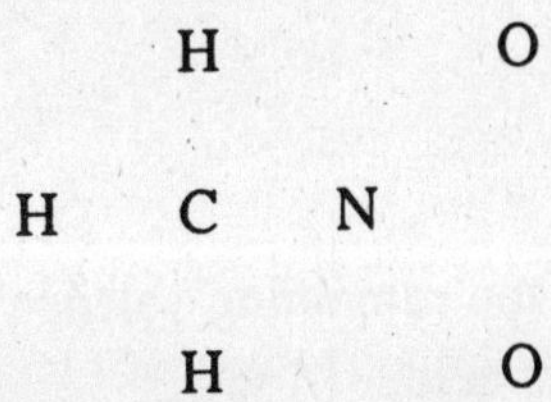


Based on the structure of the nitromethane molecule, it should have ___ resonance hybrids and ___ atoms with residual formal charges on them.

a. no, 2
b. no, 3
c. 2, 2
d. 2, 3
e. 3, 2

54. The sulfide ion can react with the sulfur trioxide molecule to produce a thiosulfate ion, as shown below.

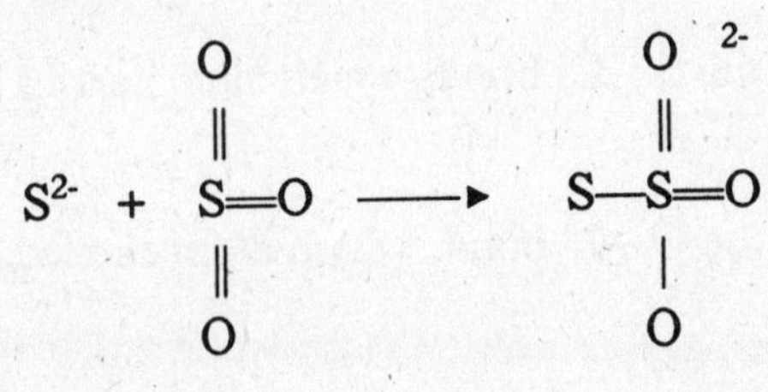


Complete the Lewis structures by filling in the bonds and the remaining valence electrons which are not involved in bonds. Which reagent is the Lewis base, if any? ( ! $S^{2-}$)

55. Selenium atoms can react with the sulfite ion to produce a selenosulfate ion, as shown below

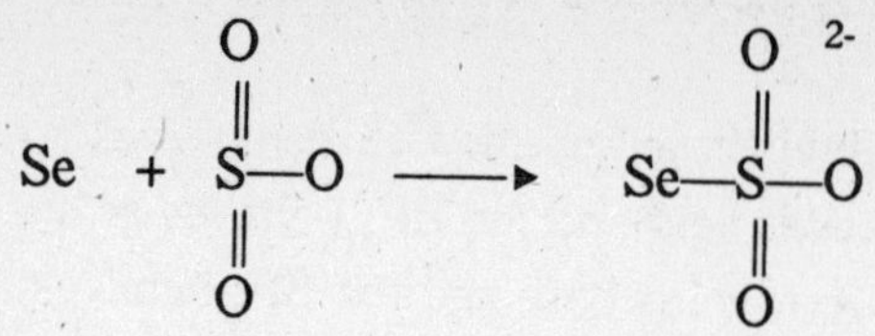


Complete the Lewis structures by filling in the bonds and the remaining valence electrons which are not involved in bonds. Which reagent is the Lewis base, if any? ( ! $SO_3^{2-}$)

56. The oxide ion can react with the carbon dioxide molecule to produce a carbonate ion, as shown below.

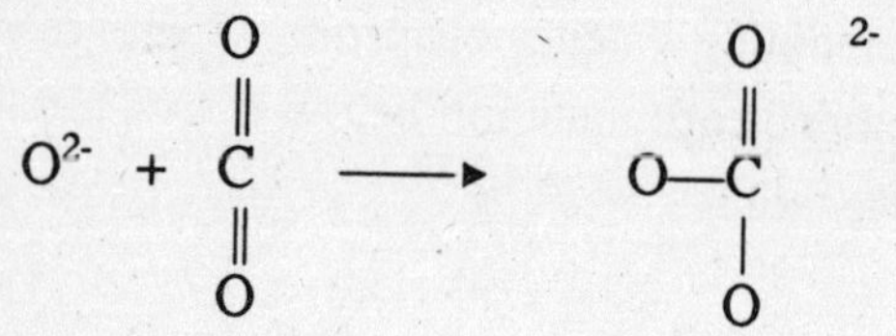


Complete the Lewis structures by filling in the bonds and the remaining valence electrons which are not involved in bonds. Which reagent, if any, is the Lewis acid? ( ! $CO_2$)

## Fill in the Blanks

57. Lattice energies of ionic compounds are proportional (to a significant degree) to the product of the charge on the ions in the lattice. This means that a CaO sample in which there are $Ca^{3+}$ and $O^{3-}$ ions should have much greater lattice energies than the case where the ions are $Ca^{2+}$ and $O^{2-}$. Why is the lattice with the $Ca^{2+}$ and $O^{2-}$ the one which is actually found in nature?
( ! The ionization energies and electron affinity for removal or acquisition of a third electron are so high that the increased lattice energy still does not make it energetically favorable)

58. What is the electronic configuration of the metal ion in chromium(III) sulfate? ________
( ! [Ar] $3d^3$)

59. The total number of bonds in the nitrogen trifluoride molecule is ______. ( ! 3)

60. In the covalent bond between the atoms in the S—Cl bond, which atom carries the partial negative charge? ( ! Cl)

61. In the covalent bond between the atoms in the O—H bond, which atom carried the partial negative charge? ( ! O)

## True and False

62. The energy of the bond holding two covalently bound atoms together is independent of the distance between the two atoms. ( !F)

## Critical Thinking

63. Draw the Lewis structures of the following species: $CO_2$, $CH_3\underset{\underset{O}{\|}}{C}{-}O^-$, CO, $CO_3^{2-}$, $H{-}\underset{\underset{O}{\|}}{C}{-}H$

Which one of the species has the longest carbon oxygen bond(s)?

a. $CO_2$
b. HCHO
! c. $CO_3^{2-}$
d. $CH_3COO^-$
e. CO

64. Draw the Lewis structures of the following species: $CO_2$, $CH_3\underset{\underset{O}{\|}}{C}{-}O^-$, CO, $CO_3^{2-}$, $H{-}\underset{\underset{O}{\|}}{C}{-}H$

Which of the species has (have) the weakest carbon oxygen bond(s)?

a. $CO_2$
b. HCHO
! c. $CO_3^{2-}$
d. $CH_3COO^-$
e. CO

65. Draw the Lewis structures of the following species: $CO_2$, $CH_3\underset{\underset{O}{\|}}{C}{-}O^-$, CO, $CO_3^{2-}$, $H{-}\underset{\underset{O}{\|}}{C}{-}H$

Which of the species has (have) the strongest carbon oxygen bond(s)?

a. $CO_2$
b. HCHO
c. $CO_3^{2-}$
d. $CH_3COO^-$
! e. CO

## Chemicals in Use 6: Molecular Structure and the Computer

66. The technique of choice in studying the spatial relationships between the various parts of large and complex molecules being used currently in research is ______. (! molecular modeling)

67. Molecular modeling is most conveniently and convincingly done using

a. ball and stick models
b. space filling models
c. Biospatial stereomodels
! d. computer generated 3-dimensional molecular models
e. flexible frame models

## Chapter 8 Chemical Bonding II

### Multiple Choice

1. All of the geometries listed below are examples of the five basic geometries for molecules with more than 3 atoms except

   a. planar triangular
   b. octahedral
   c. tetrahedral
   ! d. trihedral
   e. trigonal bipyramidal

2. The concept that electron pairs located in the valence shell of an atom bonded to other atoms tend to stay as far apart as possible so as to minimize repulsions between them is incorporated in the

   a. Pauli principle
   b. Heisenberg uncertainty principle
   ! c. valence shell electron pair repulsion theory
   d. electronegativity and polar bonds theory
   e. Aufbau principle

3. Which one of the following arrangements would best accomodate three electron pairs in the valence shell of a covalently bonded atom?

   ! a. planar triangular
   b. octahedral
   c. tetrahedral
   d. trihedral
   e. trigonal bipyramidal

4. Three different values can be observed for the bond angles in which of the following basic molecular structures for simple molecules?

   a. linear
   b. planar triangular
   c. tetrahedral
   ! d. trigonal bipyramidal
   e. octahedral

5. Application of the concepts of VSEPR theory leads us to conclude that the shape of the $SO_3$ molecule is

   a. trigonal pyramidal
   b. square planar
   c. regular tetrahedral
   ! d. triangular planar
   e. sawhorse shaped

6. The geometric structure of the $SF_4$ molecule is best described as

a. trigonal pyramidal
b. square planar
c. regular tetrahedral
d. triangular planar
! e. sawhorse shaped

7. Application of the concepts of VSEPR theory leads to the prediction that the shape of the $PCl_3$ molecule is

a. bent
b. linear
c. tetrahedral
d. triangular planar
! e. trigonal pyramidal

8. Application of the concepts of VSEPR theory leads to the prediction that the shape of the $PH_3$ molecule is

a. bent
b. linear
c. tetrahedral
d. triangular planar
! e. trigonal pyramidal

9. The geometry of the $CS_2$ molecule is best described as

a. bent
! b. linear
c. tetrahedral
d. triangular planar
e. trigonal pyramidal

10. The geometry of the $ClF_3$ molecule is best described as

a. sawhorse shaped
b. tetrahedral
! c. T-shaped
d. trigonal pyramidal
e. triangular planar

11. Application of the concepts of the VSEPR theory suggest that the geometric arrangement of the atoms in the carbonate ion, $CO_3^{2-}$ is

a. octahedral
b. square planar
c. tetrahedral
! d. triangular planar
d. trigonal pyramidal

12. Based on conclusions from application of the VSEPR theory, which one of the following species is linear?

a. $BF_3$
! b. HCN
c. $H_2CO$
d. $H_2S$
e. $SO_2$

13. Based on conclusions from application of the VSEPR theory, which one of the following molecules is trigonal bipyramidal?

! a. $AsF_5$
b. $NF_3$
c. $SF_4$
d. $SF_6$
e. $XeF_4$

14. Based on conclusions from application of the VSEPR theory, which one of the following molecules is bent (nonlinear)?

a. $CO_2$
b. $CS_2$
c. $KrF_2$
d. $C_2H_2$
! e. $SO_2$

15. Based on conclusions from application of the VSEPR theory, which one of the following molecules or ions is bent (nonlinear)?

! a. $Cl_2O$
b. $CO_2$
c. HCN
d. CO
e. $NO_2^+$

16. Based on conclusions from application of the VSEPR theory, which one of the following molecules or ions is tetrahedral?

a. $BF_3$
! a. $CF_4$
c. $NH_3$
d. $SF_4$
e. $XeF_4$

17. The Se—C—Se molecule is a non-polar molecule because

a. the bonds in the molecule are all non-polar.
b. the bonds in the molecule are polar but their effect on the overall polarity is canceled by the effect of lone pairs in the valence shell of the carbon atom.
! c. the bonds in the molecule are polar but their effect on the overall polarity is canceled by the fact that they are equal in magnitude and oppositely directed.
d. the bonds in the molecule are polar but the polar effect is canceled by the resonance hybrids which distribute the charge evenly.
e. the bonds in the molecule are polar but only slightly so and this is not enough to affect the polarity of the molecule as a whole.

18. The bond angle in $Cl_2O$ is expected to be approximately:

a. 90 degrees
! b. 109.5 degrees
c. 120 degrees
d. 145 degrees
e. 180 degrees

19. The smallest F—S—F bond angle in $SF_6$, in degrees, is:

! a. 90 degrees
b. 109.5 degrees
c. 120 degrees
d. 145 degrees
e. 180 degrees

20. The F—Cl—F bond angles in $ClF_3$, in degrees, are:

a. 90, only
b. 109.5, only
c. 120, only
d. 90, 120 and 180
! e. 90 and 180

21. Based on conclusions from application of the VSEPR theory, the expected bond angles in the $AsF_4^-$ ion should be:

a. 90, only
b. 109.5, only
c. 120, only
! d. 90, 120 and 180
e. 90 and 180

22. Based on observed periodic trends, arrange the following species, HBr, HCl, HF, HI, in order of increasing dipole moment.

a. HF < HCl < HBr < HI
b. HBr < HCl < HF < HI
c. HI < HF < HCl < HBr
! d. HI < HBr < HCl < HF
e. HCl < HBr < HI < HF

23. Which of the following molecules $CO_2$, $CS_2$, $NO_2$, COS is/are polar?

a. $NO_2$ only
b. $CS_2$, $NO_2$, and COS
c. $CO_2$ only
! d. COS and $NO_2$
e. COS only

24. Based on conclusions from application of the VSEPR theory, which one of the following molecules is should be nonpolar?

a. $CH_3Cl$
! b. $CSe_2$
c. $H_2O$
d. $NH_3$
e. $OF_2$

25. Which one of the molecules below is a polar molecule?

a. $Br_2$
b. $BF_3$
c. $CO_2$
d. $CS_2$
! e. IBr

26. Which one of the molecules below has no permanent dipole moment?

! a. $SO_3$
b. $SO_2$
c. CO
d. $NH_3$
e. $CH_2Cl_2$

27. Which one of the following is a polar molecule?

a. $XeF_2$
! b. $BrF_5$
c. $XeF_4$
d. $CCl_4$
e. $PBr_5$

28. Determine the expected molecular geometry and polarity of the $SO_2$ molecule by applying VSEPR theory.

a. linear, nonpolar
b. linear, polar
c. bent, $\approx 109.5°$ angle, polar
d. bent, $\approx 120°$ angle, polar
e. bent, $\approx 109.5°$ angle, polar

29. Predict the molecular geometry and polarity of the O—C—S molecule.

a. linear, nonpolar
b. bent, nonpolar
! c. linear, polar
d. tetrahedral, nonpolar
e. bent, polar

30. Bonding in the hydrogen chloride molecule can be explained by the valence bond theory in terms of an overlap between

a. the 1s orbital of the hydrogen and the 1s orbital of the chlorine.
b. the 1s orbital of the hydrogen and the 2s orbital of the chlorine.
c. the 1s orbital of the hydrogen and the 2p orbital of the chlorine.
d. the 1s orbital of the hydrogen and the 3s orbital of the chlorine.
! e. the 1s orbital of the hydrogen and the 3p orbital of the chlorine.

31. Bonding in the chlorine molecule can be explained by the valence bond theory in terms of an overlap between

a. the 1s orbital of a chlorine atom and the 1s orbital of the other chlorine atom.
b. the 2s orbital of a chlorine atom and the 2s orbital of the other chlorine atom.
c. the 2p orbital of a chlorine atom and the 2p orbital of the other chlorine atom.
d. the 3s orbital of a chlorine atom and the 3s orbital of the other chlorine atom.
! e. the 3p orbital of a chlorine atom and the 3p orbital of the other chlorine atom.

32. Bonding in the water molecule can be explained by the valence bond theory in terms of an overlap between

a. the 1s orbital of the hydrogen and the 1s orbital of the oxygen.
b. the 1s orbital of the hydrogen and the 2s orbital of the oxygen.
! c. the 1s orbital of the hydrogen and the 2p orbital of the oxygen.
d. the 1s orbital of the hydrogen and the 3s orbital of the oxygen.
e. the 1s orbital of the hydrogen and the 3p orbital of the oxygen.

33. Bonding in the nitrogen trifluoride molecule can be explained by the valence bond theory in terms of an overlap between

a. the 2s orbital of the nitrogen and the 2s orbital of the fluorine.
b. the 2s orbital of the nitrogen and the 2p orbital of the fluorine.
! c. the 2p orbital of the nitrogen and the 2p orbital of the fluorine.
d. the 1s orbital of the nitrogen and the 2p orbital of the fluorine.
e. the 2p orbital of the nitrogen and the 3p orbital of the fluorine.

34. Draw a Lewis structure for $H_3C—NO_2$. Both oxygen atoms are directly attached to the nitrogen atom. What is the hybrid orbital set used by the nitrogen atom for bonding?

a. $sp^3d^2$
b. sp
c. $sp^3d$
d. $sp^3$
! e. $sp^2$

35. The central atom in $BrF_5$ uses $sp^3d^2$ hybridization. It must therefore have _____ $\sigma$-bonds and _____ lone pair(s) of electrons in the valence shell of the central atom of the molecule.

a. 1, 5
b. 3, 3
c. 5, 0
! d. 5, 1
e. 6, 0

36. The central atom in $SF_4$ has _____ $\sigma$-bonding pair(s) and _____ lone pair(s) of electrons in the valence shell of the central atom.

a. 3, 2
b. 4, 0
! c. 4, 1
d. 5, 0
e. 5, 1

37. Which one of the following hybrid orbital sets is used by the central atom for $\sigma$-bonding in the $PCl_3$ molecule?

a. sp
b. $sp^2$
! c. $sp^3$
d. $sp^3d$
e. $sp^3d^2$

38. Which one of the following hybrid orbital sets is used by the central atom for $\sigma$-bonding in the $PCl_4^-$ ion?

a. sp
b. $sp^2$
! c. $sp^3$
d. $sp^3d$
e. $sp^3d^2$

39. Which one of the following hybrid orbital sets is used by the central atom for $\sigma$-bonding in the $SF_6$ molecule?

a. sp
b. $sp^2$
c. $sp^3$
d. $sp^3d$
! e. $sp^3d^2$

40. Which one of the following hybrid orbital sets is used by the central atom for $\sigma$-bonding in the $AsF_5$ molecule?

a. sp
b. $sp^2$
! c. $sp^3$
d. $sp^3d$
e. $sp^3d^2$

41. Which one of the following hybrid orbital sets is used by the central atom for $\sigma$-bonding in the $NO_3^-$ ion?

a. sp
! b. $sp^2$
c. $sp^3$
d. $sp^3d$
e. $sp^3d^2$

42. Which one of the following molecules uses an $sp^2$ hybrid orbital set on the central atom for $\sigma$-bonding?

a. $CS_2$
b. $N_2O$
! c. $SO_2$
d. $NF_3$
e. $PF_5$

43. Which one of the following hybrid orbital sets is used by the As atom for $\sigma$-bonding in the $AsF_4^-$ ion?

a. sp
b. $sp^2$
c. $sp^3$
! d. $sp^3d$
e. $sp^3d^2$

44. Which one of the following hybrid orbital sets is used by the central atom for $\sigma$-bonding in the $ClO_3^-$ ion?

a. sp
b. $sp^2$
c. $sp^3$
! d. $sp^3d$
e. $sp^3d^2$

45. The skeleton for the Lewis structure of thionyl chloride, $SOCl_2$, is shown below. Complete the Lewis structure by filling in the bonds and the remaining valence electrons which are not involved in bonds.

Cl

O S

Cl

After adjusting for formal charges, what type of orbitals are predicted to lie around the sulfur atom by the valence bond theory?

a. 5 $sp^3d$ hybrid orbitals and no $\pi$ orbitals
! b. 4 $sp^3$ hybrid orbitals and one localized $\pi$ orbital
c. 4 $sp^3$ hybrid orbitals and one delocalized $\pi$ orbital
d. 3 $sp^3d$ hybrid orbitals and one localized $\pi$ orbital
e. 3 $sp^3d$ hybrid orbitals and one delocalized $\pi$ orbital

46. Draw a Lewis structure for $H_3C—NO_2$. Both oxygen atoms are directly attached to the nitrogen atom. How many sigma bonds are there in the molecule?

a. 7
! b. 6
c. 1
d. 5
e. 2

47. How many $\sigma$ bonds and $\pi$ bonds, respectively, are there in a $CO_2$ molecule?

a. 1 $\sigma$ and 2 $\pi$
b. 2 $\sigma$ and 0 $\pi$
! c. 2 $\sigma$ and 2 $\pi$
d. 2 $\sigma$ and 4 $\pi$
e. 4 $\sigma$ and 0 $\pi$

48. How many $\sigma$ bonds and $\pi$ bonds, respectively, are there in a $N_2H_2$ molecule?

a. 2 $\sigma$ bonds and 2 $\pi$ bonds
b. 4 $\sigma$ bonds and 0 $\pi$ bonds
c. 3 $\sigma$ bonds and 0 $\pi$ bond
! d. 3 $\sigma$ bonds and 1 $\pi$ bond
e. 2 $\sigma$ bonds and 1 $\pi$ bond

49. Which one of the following statements about the ammonia molecule is true?

a. The N—H bonds are polar, but the molecule is non-polar.
b. The molecule has both $\sigma$ and $\pi$ bonds in it.
! c. The nitrogen atom can be said to utilize $sp^3$ hybrid orbitals for $\sigma$-bonding.
d. The H—N—H bond angles are greater than 109.5 degrees.
e. The molecular geometry is planar triangular.

50. How many $\pi$ bonds are there in the $CN^-$ ion?

a. 0
b. 1
c. 1.5
! d. 2
e. 3

51. Ozone ($O_3$) is an allotropic form of oxygen. Based on VSEPR theory, which statement below would you expect to be true for this molecule?

a. The molecule is linear, and has no lone pairs in the valence shell on the center atom.
b. The molecule is linear, and has two $\pi$ bonds.
c. The molecule is T-shaped.
d. The molecule is bent, and has two lone pairs in the valence shell on the center atom.
! e. The molecule is bent, and has one $\pi$ bond.

52. What is the bond order for the N—O bonds in the nitrite ion?

a. 1.0
b. 1.333
! c. 1.5
d. 2.0
e. 2.5

53. When we compare the carbon monoxide molecule with the carbon dioxide molecule, from VSEPR theory we find

a. $CO_2$ is polar and CO is non-polar.
b. the bond in CO is weaker than the ones in $CO_2$.
c. $CO_2$ has shorter bonds than CO.
! d. the bond order in CO is higher than the bond order in $CO_2$.
e. Both CO and $CO_2$ are non-polar.

54. What is the bond order for the carbon-carbon bond in the $C_2H_2$ molecule?

a. 1.0
b. 1.5
c. 2.0
d. 2.5
! e. 3.0

55. What is the bond order for the carbon-carbon bond in the $C_2F_4$ molecule?

a. 1.0
b. 1.5
! c. 2.0
d. 2.5
e. 3.0

56. Which statement below is true for the $BrF_5$ molecule?

a. There are two lone pairs in the valence shell of the bromine atom.
b. The molecule is non-polar.
! c. Some of the F—Br—F bond angles are close to 90 degrees.
d. At least one of the bond angles, F—Br—F, is about 109.5 degrees.
e. Some of the F—Br—F bond angles are close to 120 degrees.

57. A comparison of the two molecules, $SbF_5$ and $IF_5$, using VSEPR theory reveals that

! a. the $IF_5$ is polar and the $SbF_5$ is non-polar.
b. both molecules are non-polar.
c. the $SbF_5$ is polar and the $IF_5$ is non-polar.
d. both molecules are polar.
e. both molecules have lone pairs in the valence shell of the central atom .

58. Application of the bonding theories and other concepts to the $NCl_3$ molecule leads to the prediction that

! a. the molecule should have trigonal pyramidal geometry.
b. the molecule should have delocalized valence electrons.
c. the molecule should be non-polar.
d. the formal charge on the center atom is +3.
e. the molecule should have a planar triangular geometry.

59. Which statement below is true about the $BF_3$ molecule?

a. The molecule can function as a Lewis base.
b. The molecule has one $\pi$ bond and three $\sigma$ bonds.
c. The B—F bond order is 1.33.
! d. The boron atom is electron deficient in its valence shell.
e. The boron atom uses $sp^3$ hybridization.

60. The restrictions placed on the assignment of electrons to molecular orbitals in the ground state electronic configuration of multinuclear species

a. must obey the Pauli principle but not Hund's rule or the Aufbau principle
b. must obey the Pauli principle and the Aufbau principle but not Hund's rule
c. must obey the Aufbau principle but the Pauli principle or Hund's rule
d. are not governed by either the Aufbau principle, Hund's rule, or the Pauli principle
! e. must obey the Aufbau principle, Hund's rule, and the Pauli principle

61. The energy level scheme for the valence orbitals of 2nd period homonuclear diatomic molecules shows the molecular orbitals in the following order of increasing energy

$$\sigma_{2s} < \sigma^*_{2s} < \sigma_{2p(x)} < \pi_{2p(y)}, \pi_{2p(z)} < \pi^*_{2p(y)}, \pi^*_{2p(z)} < \sigma^*_{2p(x)}$$

Based on this energy level scheme, the bond order for the bond in the $N_2^+$ ion is

a. 1.0
b. 1.5
c. 2.0
! d. 2.5
e. 3.0

62. The energy level scheme for the valence orbitals of 2nd period homonuclear diatomic molecules shows the molecular orbitals in the following order of increasing energy

$$\sigma_{2s} < \sigma^*_{2s} < \sigma_{2p(x)} < \pi_{2p(y)}, \pi_{2p(z)} < \pi^*_{2p(y)}, \pi^*_{2p(z)} < \sigma^*_{2p(x)}$$

Based on this energy level scheme, the bond order for the bond in the $O_2^+$ ion in its ground state is

a. 1.0
b. 1.5
c. 2.0
! d. 2.5
e. 3.0

63. The energy level scheme for the valence orbitals of 2nd period homonuclear diatomic molecules shows the molecular orbitals in the following order of increasing energy

$$\sigma_{2s} < \sigma^*_{2s} < \sigma_{2p(x)} < \pi_{2p(y)}, \pi_{2p(z)} < \pi^*_{2p(y)}, \pi^*_{2p(z)} < \sigma^*_{2p(x)}$$

Based on this energy level scheme, the bond order for the bond in the $O_2^+$ ion in its ground state is

a. 1.0
! b. 1.5
c. 2.0
d. 2.5
e. 3.0

64. The energy level scheme for the valence orbitals of 2nd period homonuclear diatomic molecules shows the molecular orbitals in the following order of increasing energy

$$\sigma_{2s} < \sigma^*_{2s} < \sigma_{2p(x)} < \pi_{2p(y)}, \pi_{2p(z)} < \pi^*_{2p(y)}, \pi^*_{2p(z)} < \sigma^*_{2p(x)}$$

Based on this energy level scheme, how many electrons are there in all of the antibonding molecular orbitals of the $O_2$ molecule in its ground state?

a. 1
b. 2
c. 3
! d. 4
e. 5

65. The energy level scheme for the valence orbitals of 2nd period homonuclear diatomic molecules shows the molecular orbitals in the following order of increasing energy

$$\sigma_{2s} < \sigma^*_{2s} < \sigma_{2p(x)} < \pi_{2p(y)}, \pi_{2p(z)} < \pi^*_{2p(y)}, \pi^*_{2p(z)} < \sigma^*_{2p(x)}$$

Based on this energy level scheme, how many electrons are there in all of the antibonding molecular orbitals of the $F_2^+$ ion in its ground state?

a. 1
b. 2
c. 3
d. 4
! e. 5

66. The energy level scheme for the valence orbitals of 2nd period homonuclear diatomic molecules shows the molecular orbitals in the following order of increasing energy

$$\sigma_{2s} < \sigma^*_{2s} < \sigma_{2p(x)} < \pi_{2p(y)}, \pi_{2p(z)} < \pi^*_{2p(y)}, \pi^*_{2p(z)} < \sigma^*_{2p(x)}$$

Based on this energy level scheme, how many unpaired electrons are there in all of the antibonding molecular orbitals of the $O_2$ molecule in its ground state?

a. 1
! b. 2
c. 3
d. 4
e. 5

67. The energy level scheme for the valence orbitals of 2nd period homonuclear diatomic molecules shows the molecular orbitals in the following order of increasing energy

$$\sigma_{2s} < \sigma^*_{2s} < \sigma_{2p(x)} < \pi_{2p(y)}, \pi_{2p(z)} < \pi^*_{2p(y)}, \pi^*_{2p(z)} < \sigma^*_{2p(x)}$$

Based on this energy level scheme, how many unpaired electrons are there in all of the antibonding molecular orbitals of the $O_2^-$ ion in its ground state?

! a. 1
b. 2
c. 3
d. 4
e. 5

68. The energy level scheme for the valence orbitals of 2nd period homonuclear diatomic molecules shows the molecular orbitals in the following order of increasing energy

$$\sigma_{2s} < \sigma^*_{2s} < \sigma_{2p(x)} < \pi_{2p(y)}, \pi_{2p(z)} < \pi^*_{2p(y)}, \pi^*_{2p(z)} < \sigma^*_{2p(x)}$$

Based on this energy level scheme, which one of the following species has the *largest* number of unpaired electrons in its ground state?

a. $N_2^+$
b. $O_2^+$
c. $O_2^-$
! d. $O_2$
e. $F_2^+$

69. The energy level scheme for the valence orbitals of 2nd period homonuclear diatomic molecules shows the molecular orbitals in the following order of increasing energy

$$\sigma_{2s} < \sigma^*_{2s} < \sigma_{2p(x)} < \pi_{2p(y)}, \pi_{2p(z)} < \pi^*_{2p(y)}, \pi^*_{2p(z)} < \sigma^*_{2p(x)}$$

Based on this energy level scheme, how many unpaired electrons are there in all of the antibonding molecular orbitals of the $O_2^{2-}$ ion in its ground state?

! a. 0
b. 1
c. 2
d. 3
e. 4

70. The energy level scheme for the valence orbitals of 2nd period homonuclear diatomic molecules shows the molecular orbitals in the following order of increasing energy

$$\sigma_{2s} < \sigma^*_{2s} < \sigma_{2p(x)} < \pi_{2p(y)}, \pi_{2p(z)} < \pi^*_{2p(y)}, \pi^*_{2p(z)} < \sigma^*_{2p(x)}$$

Based on this energy level scheme, which statement below is true about the bonds in the set, $O_2$, $O_2^+$, $O_2^-$, $O_2^{2-}$?

a. $O_2^-$ has the shortest bond.
b. $O_2^{2-}$ has the shortest bond.
c. $O_2$ has the shortest bond.
d. $O_2^+$ has the shortest bond.
e. Actually, the bonds in all these species are of equal length.

71. The $H_2^-$ molecular ion, in its ground state electronic configuration, should have a bond order of

a. 0
! b. 0.5
c. 1.0
d. 1.5
e. 2.0

72. Which one of the sets below includes all those from the list, $NO_3^-$, $NO_2^-$, $O_3$, that have delocalized molecular orbitals?

a. $NO_3^-$ and $NO_2^-$
b. $NO_2^-$ and $O_3$
! c. $NO_3^-$, $NO_2^-$ and $O_3$
d. $NO_3^-$ only
e. $NO_3^-$ and $O_3$

73. What is the total number of $\pi$ bonds in the benzene molecule?

a. 0
b. 1
c. 2
! d. 3
e. 4

74. Metallic magnesium is a good conductor of electricity. The conduction is explained by

a. the presence of a delocalized 3s band which extends throughout the solid and is fully populated by electrons which can move freely in this 3s band.
b. the presence of a localized 3s band which extends throughout the solid and is fully populated by electrons which can move freely in this 3s band.
c. the presence of a delocalized 3p band which extends throughout the solid and is fully populated by electrons which can move freely in this 3p band.
d. the presence of a delocalized 3p band which extends throughout the solid and is partially populated (half filled) by electrons which can move freely in this 3p band.
! e. the presence of a delocalized 3p band which extends throughout the solid, is empty, and overlaps the fully delocalized 3s valence band which is fully populated by electrons. This provides easy access for electrons to move freely throughout the metal.

75. Metallic sodium is a good conductor of electricity. The conduction is explained by

a. the presence of a delocalized 3s band which extends throughout the solid and is fully populated by electrons which can move freely in this 3s band.
b. the presence of a localized 3s band which extends throughout the solid and is partially populated by electrons which can move freely in this 3s band.
! c. the presence of a delocalized 3s band which extends throughout the solid and is partially populated by electrons which can move freely in this 3s band.
d. the presence of a delocalized 3p band which extends throughout the solid and is partially populated (half filled) by electrons which can move freely in this 3p band.
e. the presence of a delocalized 3p band which extends throughout the solid, is empty, and overlaps the fully delocalized 3s valence band which is fully populated by electrons. This provides easy access for electrons to move freely throughout the metal.

76. In which one of the elements below (when in the solid state) is overlap between the 3s and 3p bands which are fully delocalized throughout the solid most important in the mechanism which allows high electrical conductivity?

a. Al
! b. Mg
c. Na
d. P
e. S

## Fill in the Blanks

77. The $PCl_5$ molecule has (polar)(non-polar) bonds and is a (polar)(non-polar) molecule. ______, ______ ( ! polar, non-polar)

78. The shape of dichlorophosphine, $PHCl_2$, would be best described as ______. ( ! trigonal pyramidal)

79. If we use the same molecular orbital energy level scheme for the valence orbitals in the CO molelcule as we use for homonuclear diatomic molecules,

$$\sigma_{2s} < \sigma^*_{2s} < \sigma_{2p(x)} < \pi_{2p(y)}, \pi_{2p(z)} < \pi^*_{2p(y)}, \pi^*_{2p(z)} < \sigma^*_{2p(x)}$$

how many valence electrons are there in the antibonding orbitals of the system? ______ ( !2)

80. If we use the same molecular orbital energy level scheme for the valence orbitals in the $NO^-$ ion as we use for homonuclear diatomic molecules,

$$\sigma_{2s} < \sigma^*_{2s} < \sigma_{2p(x)} < \pi_{2p(y)}, \pi_{2p(z)} < \pi^*_{2p(y)}, \pi^*_{2p(z)} < \sigma^*_{2p(x)}$$

how many unpaired electrons are there in the antibonding orbitals of the system? ______ ( !2)

## True and False

81. A covalent bond is formed when two atoms position themselves so that maximum overlap of the atomic orbitals occurs with an increase in the potential energy of the system. ( !F)

82. When two hydrogen interact so that their valence orbitals form a bonding molecular orbital and an antibonding molecular orbital, the formation of these molecular orbitals lowers the potential energy of the system. ( ! T)

83. The numerical restrictions placed on the number of electrons that can occupy an orbital by the Pauli principle applies to atomic orbitals but not to molecular orbitals. ( ! F)

84. The molecular orbital energy level scheme for the $He_2$ molecule (generated by using the valence orbitals) employs two atomic valence orbitals to form 4 molecular orbitals. ( ! F)

85. The ground state electronic configuration for the $He_2^+$ molecular ion is $(\sigma_{1s})^2 (\sigma^*_{1s})^1$. ( ! T)

86. The conductivity of some metalloids can be modified. The modified materials derive their conductivity from impurities that affect the population of the valence band (which would be ordinarily be filled if the material were pure and thus be considerably lower in conductivity). ( !T)

## Critical Thinking

87. Consider these species: $ClO_2^-$, $NO_2^-$, $ClO_2$, $NO_2$, $ClO_2^+$, $NO_2^+$. Determine the molecular geometry using VSEPR theory, then list them in order of *increasing* bond angle.

(! $ClO_2^- < ClO_2 < ClO_2^+ \approx NO_2^- < NO_2 < NO_2^+$)

---

88. Two adjacent atoms, a and b, can form bonds by overlap of their atomic orbitals. Using the axis system immediately below, if the internuclear axis is labeled as the x axis, which pair of orbitals will overlap correctly to form a σ-bond?

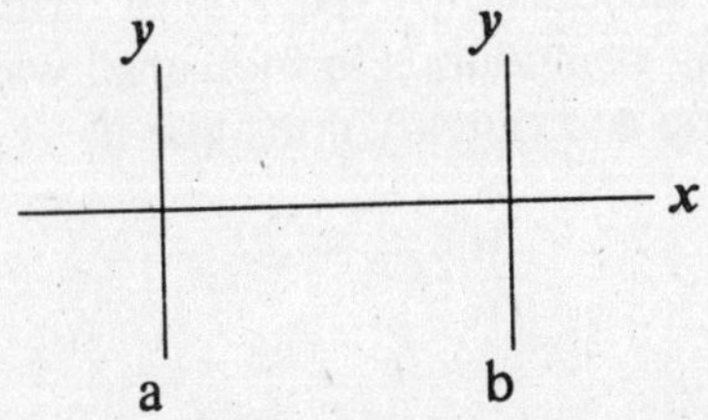


a. the $d_{xz}$ on atom *a* with the $d_{xz}$ on atom *b*
b. the $d_{xy}$ on atom *a* with the $d_{xy}$ on atom *b*
c. the $p_z$ on atom *a* with the $p_z$ on atom *b*
d. the $p_y$ on atom *a* with the $p_y$ on atom *b*
! e. the $p_x$ on atom *a* with the $p_x$ on atom *b*

89. Two adjacent atoms, a and b, can form bonds by overlap of their atomic orbitals. Using the axis system immediately below, if the internuclear axis is labeled as the x axis, which pair of orbitals will overlap correctly to form a π-bond?

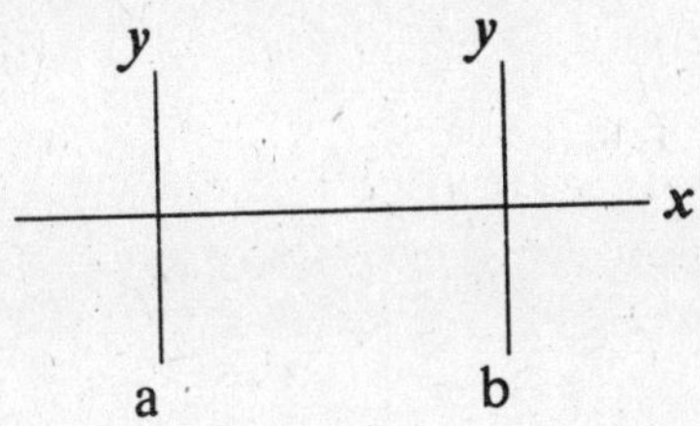


a. the $d_{xz}$ on atom *a* with the $p_x$ on atom *b*
b. the $d_{yz}$ on atom *a* with the $d_{xz}$ on atom *b*
c. the $p_x$ on atom *a* with the $p_y$ on atom *b*
! d. the $p_y$ on atom *a* with the $d_{xy}$ on atom *b*
e. the $p_x$ on atom *a* with the $d_{x^2-y^2}$ on atom *b*

## Chemicals in Use 7: Ozone in the Stratosphere

90. The reaction which Cl atoms undergo with ozone molecules, $Cl + O_3 \longrightarrow ClO_2 + O$, followed by $O + O \longrightarrow O_2$ clearly shows that each chlorine atom consumes only one ozone molecule. This still causes serious depletion because the ozone concentration is not very high in unpolluted atmospheres. ( !F)

91. The reaction sequence of chlorine atoms with ozone molecules is such that one chlorine atoms is capable of consuming hundreds of ozone molecules before some reaction occurs to terminate its reactive state. ( !T)

92. Atmospheric ozone is consumed by

a. the bromine atoms produced by chlorine induced displacement reactions from bromine containing compounds in industrial waste gases present in the atmosphere
b. the iodine atoms produced by chlorine induced displacement reactions from iodine containing compounds in industrial waste gases present in the atmosphere
c. the chlorine atoms produced by action of ultraviolet light on chlorine containing compounds in industrial waste gases present in the atmosphere.
d. the fluorine atoms produced by action of ultraviolet light on fluorine containing compounds in industrial waste gases present in the atmosphere.
e. the nitrogen chlorides, like $N_2Cl$, produced by reaction of irridiated nitrogen molecules with chlorine containing compounds in industrial waster gases. These nitrogen chlorides react with ozone and quickly consume it.

# Chapter 9 Chemical Reactions: Periodic Correlations

## Multiple Choice

1. In the chemical reaction, $HCl(g) + H_2O(l) \longrightarrow H_3O^+(aq) + Cl^-(aq)$ ıe conjugate base is

   a. $HCl(aq)$
   b. $H_2O(l)$
   c. $H_3O^+(aq)$
   ! d. $Cl^-(aq)$
   e. $OH^-(aq)$

2. In the chemical reaction, $H_2PO_4^-(aq) + H_2O(l) \longrightarrow H_3O^+(aq) + HPO_4^{2-}(a$ the conjugate base is

   a. $H_2PO_4^-(aq)$
   b. $HPO_3^{2-}(aq)$
   c. $H_3O^+(aq)$
   d. $OH^-(aq)$
   ! e. $HPO_4^{2-}(aq)$

3. In the chemical reaction, $NH_3(g) + H_2O(l) \longrightarrow NH_4^+(aq) + OH^-(aq)$ the conjug acid is

   a. $H_3O^+(aq)$
   b. $H^+(aq)$
   ! c. $NH_4^+(aq)$
   d. $H_2O(l)$
   e. $NH_3^+(aq)$

4. In the chemical reaction, $C_2H_3O_2^-(aq) + H_2O(l) \longrightarrow HC_2H_3O_2(aq) + OH^-(aq)$ the conjugate acid is

   a. $H_2O(l)$
   ! b. $HC_2H_3O_2(aq)$
   c. $H_3O^+(aq)$
   d. $H_2C_2H_3O_2^+(aq)$
   e. $C_2H_3O_2^-(aq)$

5. The conjugate acids to the species, $H_2PO_4^-(aq)$, $NH_3(aq)$ and $SO_4^{2-}(aq)$ are

   a. $H_3PO_3(aq)$, $NH_4^+(aq)$, and $H_2SO_4(aq)$
   b. $H_3PO_4(aq)$, $NH_4^+(aq)$, and $H_2SO_4(aq)$
   c. $H_3PO_4(aq)$, $NH_3^+(aq)$, and $H_2SO_4(aq)$
   d. $HPO_4^{2-}(aq)$, $NH_4^+(aq)$, and $HSO_4^-(aq)$
   ! e. $H_3PO_4(aq)$, $NH_4^+(aq)$, and $HSO_4^-(aq)$

6. Which pair among the following represents a conjugate acid — base pair?

a. $HPO_4^{2-}$, $H_3PO_4$
b. $HClO_2$, $HClO_3$
c. $H_2SO_4$, $HSO_3^-$
d. $H_3O^+$, $OH^-$
! e. $NH_3$, $NH_2^-$

7. When bisulfate ion reacts with monohydrogen phosphate ion, a reaction occurs in which the bisulfate ion functions as a proton donor and the other reactant behaves as a proton acceptor. The conjugate acid formed by the forward reaction described is

a. $H_2SO_4$
b. $H_3O^+$
c. $H_3PO_4$
! d. $H_2PO_4^-$
e. $H_4PO_4^+$

8. Which one of the following species is amphoteric as judged by its frequent chemical behavior?

a. $Cl^-(aq)$
b. $SO_4^{2-}(aq)$
! c. $H_2O(l)$
d. $Cl_2(aq)$
e. $Na^+(aq)$

9. Which one of the following species is amphoteric as judged by its frequent chemical behavior?

! a. $HCO_3^-(aq)$
b. $SO_4^{2-}(aq)$
c. $H_2SO_4(aq)$
d. $Cl^-(aq)$
e. $H_3O^+(aq)$

10. The conjugate base to the $HPO_4^{2-}$ ion is

a. $OH^-$
b. $H_2PO_4^-$
c. $PO_3^{3-}$
d. $PO_3^-$
! e. $PO_4^{3-}$

11. Which one of the following listed acids ionizes to the smallest extent if they are all 1.00 molar?

! a. $HC_2H_3O_2(aq)$
b. $HNO_3(aq)$
c. $HClO_4(aq)$
d. $H_2SO_4(aq)$
e. $HCl(aq)$

12. The strongest conjugate base among the following listed species is

a. $NO_3^-(aq)$
! b. $F^-(aq)$
c. $Cl^-(aq)$
d. $B^-(aq)$
e. $I^-(aq)$

13. Which one of the following compounds should be the most acidic?

a. $BeH_2$
b. $CH_4$
c. LiH
d. $NH_3$
! e. $OH_2$

14. Which one of the following compounds should be the most acidic?

a. $MgH_2$
b. NaH
c. $PH_3$
! d. $SH_2$
e. $SiH_4$

15. Which one of the following oxides should be the most strongly basic?

a. FeO
b. MgO
! c. $Na_2O$
d. $SO_2$
e. $SiO_2$

16. Which one of the following hydrates should be the most strongly acidic?

! a. $Al(H_2O)_6^{3+}(aq)$
b. $Zn(H_2O)_6^{2+}(aq)$
c. $Fe(H_2O)_6^{3+}(aq)$
d. $Ni(H_2O)_6^{2+}(aq)$
e. $Cd(H_2O)_6^{2+}(aq)$

17. List the species, $H_2O$, $H_3O^+(aq)$ and $OH^-(aq)$ in order of increasing acid strength

a. $H_2O < H_3O^+(aq) < OH^-(aq)$
b. $H_2O < OH^-(aq) < H_3O^+(aq)$
! c. $OH^-(aq) < H_2O < H_3O^+(aq)$
d. $H_3O^+(aq) < H_2O < OH^-(aq)$
e. $OH^-(aq) < H_3O^+(aq) < H_2O$

18. Judicious research has shown that the following is the correct listing for the acids included here in order of increasing strength.

$$HCN(aq) < HCO_3^-(aq) < HC_2H_3O_2(aq) < HF(aq) < H_3PO_4(aq) < HCl(aq)$$

The weakest conjugate base, of those listed below, is

a. $H_2CO_3(aq)$
b. $CN^-(aq)$
c. $HPO_4^{2-}(aq)$
d. $C_2H_3O_2^-(aq)$
! e. $H_2PO_4^-(aq)$

19. The anhydride of the acid, $H_3PO_4$, is

a. $HPO_3$
b. $P_2O_3$
c. $P_4O_6$
! d. $P_4O_{10}$
e. $PO_2$

20. Which one of the following listed acids does not have an anhydride?

! a. HCN(aq)
b. $H_3PO_3(aq)$
c. $H_2SO_4(aq)$
d. $HNO_3(aq)$
e. $HClO_3(aq)$

21. Which one of the following listed acids does not have an anhydride?

! a. HCl(aq)
b. HClO(aq)
c. $H_2SO_3(aq)$
d. $H_3PO_4(aq)$
e. $HClO_4(aq)$

22. The anhydride of the compound, NaOH, is

a. NaO
! b. $Na_2O$
c. $Na_2O_2$
d. $NaO_2$
e. $NaOH_2^+$

23. All of the following compounds are basic anhydrides except

a. CaO
b. MgO
c. $Na_2O$
d. NiO
! e. $SeO_3$

24. The acid anhydride of the compound, dichromic acid, is

a. $Cr_2O_3$
b. $Cr_2O_7^{2-}$
c. $Cr_2O_7$
! d. $CrO_3$
e. $CrO_4^{2-}$

25. In a chemical reaction, zinc metal reacts with perchloric acid solution to produce zinc perchlorate (in solution) and hydrogen gas, which escapes. The species being oxidized in this reaction is

a. $HClO_4$(aq)
b. $H_2$(g)
c. $Zn^{2+}$ (aq)
! d. Zn(s)
e. $Zn(ClO_4)_2$(aq)

26. In a chemical reaction, zinc metal reacts with nitric acid solution to produce zinc nitrate (in solution) ammonium nitrate (in solution) and water. The species being oxidized in this reaction is

a. $HNO_3$(aq)
b. $NH_4^+$ (aq)
c. $Zn^{2+}$ (aq)
! d. Zn(s)
e. $Zn(NO_3)_2$(aq)

27. In a chemical reaction, magnesium metal reacts with aqueous sulfuric acid solution to produce magnesium sulfate (in solution) and hydrogen gas. In the course of the reaction, which element undergoes an increase in oxidation number?

a. hydrogen, only
! b. magnesium, only
c. oxygen, only
d. sulfur, only
e. magnesium and hydrogen

28. In a series of chemical test reactions the following observations were made when some metallic element samples were treated with 3.00 molar hydrochloric acid solution. Zn: metal dissolves, hydrogen gas is emitted. Cu: no reaction, Ag: no reaction Mg: metal dissolves, hydrogen gas is emitted. Mn: metal dissolves, hydrogen gas is emitted. Which statement below is true?

a. Mg, Mn, Zn are more reactive than Ag, Cu but less reactive than $H_2$.
b. Mg, Mn, Zn are less reactive than Ag, Cu but more reactive than $H_2$.
! c. Mg, Mn, Zn are more reactive than Ag, Cu and more reactive than $H_2$.
d. Mg, Mn, Zn are less reactive than Ag, Cu and less reactive than $H_2$.
e. These observations are insufficient to form any judgment about relative reactivity of any of the elements listed above.

29. In which situation described below can the product of the chemical reaction be predicted with a high degree of confidence?

! a. an active metal reacting with a non-oxidizing monoprotic acid
b. an active metal reacting with an oxidizing monoprotic acid
c. an active metal reacting with a non-oxidizing diprotic acid
d. an active metal reacting with an oxidizing diprotic acid
e. an active metal reacting with a non-oxidizing triprotic acid

30. Three metallic elements, copper, gold and zinc, can be distinguished from one another on the basis of how they react with two strong acids, $HNO_3$(aq) and $H_2SO_4$(aq) Which set below, using the abbreviations R (for reaction occurs) and NR (for no reaction) correctly describes what occurs.

a. Au: HCl( R), $HNO_3$(NR) Cu: HCl( R), $HNO_3$(NR) Zn: HCl( R), $HNO_3$(NR)
b. Au: HCl(NR), $HNO_3$(NR) Cu: HCl( R), $HNO_3$( R) Zn: HCl( R), $HNO_3$(NR)
! c. Au: HCl(NR), $HNO_3$(NR) Cu: HCl(NR), $HNO_3$( R) Zn: HCl( R), $HNO_3$( R)
d. Au: HCl(NR), $HNO_3$(NR) Cu: HCl( R), $HNO_3$( R) Zn: HCl(NR), $HNO_3$( R)
e. Au: HCl(NR), $HNO_3$(NR) Cu: HCl( R), $HNO_3$(NR) Zn: HCl( R), $HNO_3$( R)

31. Three metallic elements, copper, magnesium and silver, can be distinguished from one another on the basis of how they react with two strong acids, $HNO_3$(aq) and $H_2SO_4$(aq) Which set below, using the abbreviations R (for reaction occurs) and NR (no reaction) correctly describes what occurs.

a. Ag: HCl( R), $HNO_3$(NR) Cu: HCl( R), $HNO_3$(NR) Mg: HCl( R), $HNO_3$(NR)
b. Ag: HCl(NR), $HNO_3$(NR) Cu: HCl( R), $HNO_3$( R) Mg: HCl( R), $HNO_3$(NR)
c. Ag: HCl(NR), $HNO_3$(NR) Cu: HCl( R), $HNO_3$( R) Mg: HCl(NR), $HNO_3$( R)
d. Ag: HCl(NR), $HNO_3$(NR) Cu: HCl( R), $HNO_3$(NR) Mg: HCl( R), $HNO_3$( R)
! e. Ag: HCl(NR), $HNO_3$(NR) Cu: HCl(NR), $HNO_3$( R) Mg: HCl( R), $HNO_3$( R)

32. In terms of activity, the series in decreasing for metals is found to be,

Au < Ag < Cu < Sn < Cd < Zn < Al < Mg < Na < Cs

Which reaction below ***does not occur spontaneously*** upon mixing the reagents shown.

! a. $Cd(s) + Al^{3+}(aq) \longrightarrow Cd^{2+}(aq) + Al(s)$
b. $Cd(s) + Cu^{2+}(aq) \longrightarrow Cd^{2+}(aq) + Cu(s)$
c. $Zn(s) + Cu^{2+}(aq) \longrightarrow Zn^{2+}(aq) + Cu(s)$
d. $Al(s) + Ag^{+}(aq) \longrightarrow Al^{3+}(aq) + Ag(s)$
e. $Cu(s) + Au^{3+}(aq) \longrightarrow Cu^{2+}(aq) + Au(s)$

33. In terms of activity, the series in decreasing for metals is found to be,

Au < Ag < Cu < Sn < Cd < Zn < Al < Mg < Na < Cs

Which reaction below ***occurs spontaneously*** upon mixing the reagents shown.

a. $Sn(s) + Zn^{2+}(aq) \longrightarrow Sn^{2+}(aq) + Zn(s)$
b. $Ag(s) + Mg^{2+}(aq) \longrightarrow Ag^{+}(aq) + Mg(s)$
! c. $Zn(s) + Au^{3+}(aq) \longrightarrow Zn^{2+}(aq) + Au(s)$
d. $Ag(s) + Mn^{2+}(aq) \longrightarrow Ag^{+}(aq) + Mn(s)$
e. $Sn(s) + Al^{3+}(aq) \longrightarrow Sn^{2+}(aq) + Al(s)$

34. Which one of the following elements would you expect to react most rapidly, if at all, with water?

a. Ag
b. Al
c. Cu
! d. Na
e. Zn

35. Which one of the following elements would you expect to react most rapidly, if at all, with water?

a. Al
b. Cd
! c. Cs
d. Mg
e. Zn

36. Which one of the following reagents is the strongest oxidizing agent?

a. $Br_2$(l)
b. $Cl_2$(g)
! c. $F_2$(g)
d. $I_2$(s)
e. Ar(g)

37. When the hydrocarbon, $C_8H_{16}$, undergoes complete combustion, a specific set of products are formed. If you write the equation for the reaction and balance it, the sum of the coefficients for the reagents in the balanced equation will be

a. 17
b. 19
c. 21
d. 26
! e. 29

38. When the carbohydrate, $C_{12}H_{22}O_{11}$, undergoes complete combustion, a specific set of products are formed. If you write the equation for the reaction and balance it, the sum of the coefficients for the reagents in the balanced equation will be

a. 24
b. 35
! c. 36
d. 47
e. 83

## Fill in the Blanks

39. The conjugate base of hydrocyanic acid, HCN(aq), is ______. (! $CN^-$)

40. The anhydride of $HClO_4$ is ______. ( ! $Cl_2O_7$)

41. The conjugate acid of $H_2PO_4^-$ is ______. ( ! $HPO_4^{2-}$)

42. The anhydride of $La(OH)_3$ is ______. ( ! $La_2O_3$)

## True and False

43. The conjugate acid of $H_2SO_4$ is $H_3O^+$. ( !F)

44. The conjugate base of $NH_3$ is $OH^-$. ( !F)

45. The reaction, $HCl(aq) + Na_2CO_3(aq) \longrightarrow NaCl(aq) + H_2O(l) + CO_2(g)$ involves changes in oxidation number and is therefore classified as a redox reaction. ( !F)

46. The reaction, $Al(s) + HClO_4(aq) \longrightarrow Al(ClO_4)_3(aq) + H_2(g)$ involves changes in oxidation number and is therefore classified as a redox reaction. ( !T)

47. The anhydride of $CuSO_4$ is CuO. ( !F)

48. The reaction, $Cl_2(g) + NaBr(aq) \longrightarrow NaCl(aq) + Br_2(l)$, involves changes in oxidation number and is therefore classified as a redox reaction. ( !F)

49. The reaction, $NiS(s) + O_2(g) \longrightarrow NiO(s) + SO_2(g)$, involves changes in oxidation number and is therefore classified as a redox reaction. ( !T)

50. The reaction, $AgNO_3(aq) + NH_4Br(aq) \longrightarrow AgCl(s) + NH_4NO_3(aq)$, involves changes in oxidation number and is therefore classified as a redox reaction. ( !F)

# Chapter 10 Properties of Gases

## Multiple Choice

1. An open end manometer constructed from a U-shaped tube was operated using an oil which has a density of 1.088 g/ml. In a particular measurement, the level in the end connected to the gas manifold on which the experiment was being conducted measured 76.2 cm above the U-neck, while the level in the open end (to the atmosphere) was 23.8 cm above the U-neck. The outside air pressure in the laboratory was measured as 754 torr. What is the pressure in the gas manifold? For reference, the density of mercury is 13.533 g/ml at room temperature.

a. 42.0 torr
b. 230 torr
c. 272 torr
d. 516 torr
! e. 712 torr

2. An open end manometer constructed from a U-shaped tube was operated using an oil which has a density of 1.164 g/ml. In a particular measurement, the level in the end connected to the gas manifold on which the experiment was being conducted measured 82.8 cm above the U-neck, while the level in the open end (to the atmosphere) was 17.2 cm above the U-neck. The outside air pressure in the laboratory was measured as 764 torr. What is the pressure in the gas manifold? For reference, the density of mercury is 13.533 g/ml at room temperature.

a. 200 torr
b. 563 torr
c. 706 torr
! d. 708 torr
e. 738 torr

3. A closed end manometer which utilizes an oil having a density of 0.820 g/ml is being used to measure gas pressure in a manifold designed for studies on the gaseous state. In a particular experiment, the oil in the closed end of this manometer was 62.2 cm above the U-neck, while the level in the open end (connected to the manifold) was measured as 25.2 cm above the U-neck. By way of comparison, the density of mercury at room temperature is 13.533 g/ml. What is the pressure, in atmospheres, in the gas collection manifold?

a. $2.66 \times 10^{-2}$ atm
! b. $2.95 \times 10^{-2}$ atm
c. 0.345 atm
d. 0.784 atm
e. 1.16 atm

4. A closed end manometer was constructed from a U-shaped glass tube which was 80.5 cm high. It was loaded with mercury so that the closed side was filled all the way to the top (80.0 cm above the neck) while the open end was at a level 18.0 cm above the neck. It was taken into a chamber used for training astronauts. What is the highest pressure, in torr, that can be read on this manometer?

a. 62.0 torr
b. 62.5 torr
c. 98.0 torr
d. 98.5 torr
! e. 620 torr

5. A properly designed Torricelli barometer (the simplest type of barometer) should be at least

a. 50 cm tall
! b. 80 cm tall
c. 2.50 meters tall
d. 100 in tall
e. 76 mm tall

6. A sample of a gas in a cylindrical chamber with a movable piston occupied a volume of 6.414 liters when the pressure was 850 torr and the temperature was 27.2 °C. The pressure was readjusted to 5.82 atm by moving the piston. What was the volume occupied by the sample under the later conditions if the temperature remained constant throughout?

a. 0.837 liters
b. 0.937 liters
! c. 1.23 liters
d. 1.53 liters
e. 3.34 liters

7. A sample of a gas in a cylindrical chamber with a movable piston occupied a volume of 2.86 cubic meters when the pressure was $9.85 \times 10^4$ Pa and the temperature was 25.8 °C. The pressure was readjusted to $1.08 \times 10^5$ Pa by moving the piston. What was the volume occupied by the sample under the later conditions if the temperature remained constant throughout?

a. 0.383 $m^3$
! b. 2.61 $m^3$
c. 3.14 $m^3$
d. 8.23 $m^3$
e. 26.1 $m^3$

8. A sample of a gas in a cylindrical chamber with a movable piston occupied a volume of 1.40 liters when the pressure was 762 torr and the temperature was 26.9 °C. The volume of the system was readjusted to 0.150 liters by moving the piston. What was the pressure exerted on the surface of the piston, in atmospheres, by the gas if the temperature of the system remained constant?

a. 0.107 atm
b. 2.89 atm
c. 9.30 atm
! d. 9.36 atm
e. 9.38 atm

9. A sample of a gas in a cylindrical chamber with a movable piston occupied a volume of 8.50 liters when the pressure was $9.85 \times 10^4$ Pa and the temperature was 24.9 °C. The volume of the system was readjusted to 11.6 liters by moving the piston. What was the pressure exerted on the surface of the piston, in pascals, by the gas if the temperature of the system remained constant?

a. 7.22 Pa
b. $2.66 \times 10^4$ Pa
! c. $7.22 \times 10^4$ Pa
d. $1.34 \times 10^5$ Pa
e. $3.90 \times 10^5$ Pa

10. The standard reference conditions for studies of the gaseous state of matter has the values

a. temperature: 0.00 K; pressure: 1.000 standard atmosphere
! b. temperature: 0.00 °C; pressure: 1.000 standard atmosphere
c. temperature: 273.15 K; pressure: 1.000 Pascal
d. temperature: 298.15 K; pressure: 1.000 standard atmosphere
e. temperature: 298.15 K; pressure: 1.000 Pascal

11. A sample of a gas in a cylindrical chamber with a movable piston occupied a volume of 6.414 liters when the pressure was 850 torr and the temperature was 27.2 °C. The temperature was readjusted to 65.5 °C while the load on the piston was kept constant to keep the pressure constant in the system. What was the volume occupied by the sample at the new temperature?

a. 2.66 liters
b. 5.689 liters
c. 7.21 liters
! d. 7.232 liters
e. 15.4 liters

12. A sample of a gas in a cylindrical chamber with a movable piston occupied a volume of 2.86 cubic meters when the pressure was $9.85 \times 10^4$ Pa and the temperature was 25.8 °C. The temperature was readjusted to -20.5 °C while the load on the piston was kept constant to keep the pressure constant in the system. What was the volume occupied by the sample under the later conditions?

a. 2.27 $m^3$
! b. 2.42 $m^3$
c. 3.38 $m^3$
d. 3.60 $m^3$
e. 15.2 $m^3$

13. A sample of a gas in a cylindrical chamber with a movable piston occupied a volume of 1.40 liters when the pressure was 768 torr and the temperature was 26.9 °C. The volume of the system was readjusted to 2.16 liters by changing the temperature while the load on the piston was kept constant to keep the pressure in the system constant. What was the temperature in the system at this point?

a. 41.5 °C
b. 41.9 °C
! c. 189.8 °C
d. 194.7 °C
e. 288.6 °C

14. A sample of a gas in a cylindrical chamber with a movable piston occupied a volume of 0.00256 cubic meters when the pressure was $1.20 \times 10^5$ Pa and the temperature was 77.8 °C. The volume of the system was readjusted to 1.88 liters by changing the temperature while the load on the piston was kept constant to keep the pressure in the system constant. What was the temperature in the system at this point?

! a. -15.4°C
b. 36.1 °C
c. 57.1 °C
d. 204.7 °C
e. 571 °C

15. A sample of an unknown gas was isolated in a gas containment bulb on a manifold used in this type work. The volume of the bulb was 1.425 liters. The temperature was 25.40 °C, and the manifold pressure was 583.0 torr. What volume, in liters, would this gas sample occupy at STP?

! a. 1.000 liters
b. 1.195 liters
c. 1.700 liters
d. 2.030 liters
e. 11.76 liters

16. A sample of a gas was isolated in a gas containment bulb on a manifold used in this type work. The volume of the bulb was 1.524 liters. The temperature was 28.40 °C, and the manifold pressure was 85.00 kPa. What volume, in liters, would this gas sample occupy at STP?

a. 1.069 liters
! b. 1.158 liters
c. 1.412 liters
d. 1.645 liters
e. 2.006 liters

17. A student prepared hydrogen gas by the reaction,

$$Zn(s) + 2\,HCl(aq) \longrightarrow ZnCl_2(aq) + H_2(g)$$

The gas evolved was collected over water by downward displacement in a large gas collection tube. The volume of the saturated gas collected was 135.5 ml and the pressure measured 752.5 torr. The temperature of the collection system was measured as 25.8 °C, at which temperature the saturation vapor pressure of water is 24.9 torr. If a large excess of the HCl solution was used, how many milli-grams of zinc were there in the sample?

a. 4.60 mg
! b. 346 mg
c. 358 mg
d. 359 mg
e. 377 mg

18. The reaction, $MnO_2(s) + 4\,HCl(aq) \longrightarrow MnCl_2(aq) + Cl_2(g) + H_2O(l)$ was being studied. 12.2 g of $MnO_2$ were reacted with 70.0 ml of 2.50 M HCl solution. The chlorine produced was collected in a vessel. The pressure measured 0.971 atm at a temperature of 78.2 °C. Determine the volume of chlorine produced if the theoretical yield was obtained.

a. 0.289 l
b. 1.22 l
! c. 1.30 l
d. 4.17 l
e. 5.20 l

19. A sample of a gas was isolated in a gas containment chamber used in this type work. The volume of the chamber was 1.245 cubic meters. The temperature was 22.80 °C, and the pressure in the chamber was 88.26 kPa. What volume, in cubic meters, would this gas sample occupy at STP?

! a. 1.001 $m^3$
b. 1.007 $m^3$
c. 1.175 $m^3$
d. 1.319 $m^3$
e. 1.549 $m^3$

20. A sample of a gas occupies a volume of 1.820 liters at STP. What pressure, in atmospheres, would it exert if transferred to a 1.425 liter vessel in which the temperature is maintained at 25.2 °C?

a. 0.7168 atm
b. 0.8552 atm
c. 1.169 atm
d. 1.278 atm
! e. 1.395 atm

21. A sample of a gas occupies a volume of 1.664 liters at STP. What pressure, in kilopascals, would it exert if it is transferred to a 1.524 liter vessel if the temperature in the vessel is maintained at 29.5 °C?

a. 83.75 kPa
b. 99.85 kPa
c. 102.8 kPa
d. 112.3 kPa
! e. 122.6 kPa

22. A sample of a gas occupies a volume of 1.566 cubic meters at STP. What pressure, in kilopascals, would it exert if it is transferred to a gas chamber with a volume of 1.824 cubic meters if the temperature in the chamber is maintained at 22.5 °C?

a. 80.37 kPa
b. 86.26 kPa
! c. 94.16 kPa
d. 109.0 kPa
e. 127.7 kPa

23. A sample of a gas occupies a volume of 185.5 ml at STP. What pressure, in torr, would it exert if it is transferred to a gas bulb with a volume of 255.5 ml in which the temperature is maintained at 34.5 °C?

a. 489.9 torr
b. 569.4 torr
! c. 621.5 torr
d. 929.4 torr
e. 1179 torr

24. A sample of a gas occupies a volume of 1.462 liters at STP. It was placed in a different vessel in which the pressure was measured as 722.5 torr when the temperature was 25.20 °C. What was the volume of this new vessel, in liters?

a. 1.272 liters
b. 1.408 liters
c. 1.518 liters
d. 1.539 liters
! e. 1.680 liters

25. A sample of a gas occupies a volume of 1.524 cubic meters at STP. It was placed in a different gas chamber in which the pressure was measured as 88.95 kPa when the temperature was 29.60 °C. What was the volume of this new gas chamber, in cubic meters?

a. 1.207 $m^3$
b. 1.483 $m^3$
c. 1.566 $m^3$
d. 1.763 $m^3$
! e. 1.924 $m^3$

26. A sample of a gas occupies a volume of 1.624 liters at STP. It was placed in a different vessel in which the pressure was measured as 106.3 kPa when the temperature was 28.6 °C. What was the volume of this new vessel, in liters?

a. 1.401 liters
b. 1.542 liters
c. 1.567 liters
! d. 1.710 liters
e. 1.882 liters

27. A sample of a gas occupies a volume of 1.462 liters at STP. It was placed in a different vessel with a volume of 1.601 liters, in which the pressure was measured as 722.0 torr. What was the temperature in the vessel at measurement time?

a. -32.6 °C
b. -10.6 °C
! c. 11.0 °C
d. 37.0 °C
e. 41.7 °C

28. A sample of a gas occupies a volume of 1.246 liters at STP. It was placed in a different vessel with a volume of 1.410 liters, in which the pressure was measured as 108.5 kPa. What was the temperature in the vessel at measurement time?

a. -47.7 °C
b. -14.7 °C
c. 15.5 °C
! d. 57.8 °C
e. 88.1 °C

29. A sample of a gas occupies a volume of 122.4 ml at STP. It was placed in a different vessel with a volume of 164.2 ml, in which the pressure was measured as 0.9915 atm. What was the temperature in the vessel at measurement time?

! a. 90.2 °C
b. 96.4 °C
c. 123.4 °C
d. 201.9 °C
e. 205.4 °C

30. A cylinder fitted with a movable piston and filled with a gas has a volume of 188.5 ml at 26.7 °C when the applied pressure is 755.2 torr. The temperature of the oil bath surrounding the cylinder was increased to 165.2 °C, and the load on the piston was changed. Careful measurement now gave a value of 210.5 ml for the volume. What was the final pressure in the system, in torr?

a. 462.6 torr
b. 576.9 torr
c. 715.5 torr
! d. 988.6 torr
e. 1233 torr

31. A cylinder fitted with a movable piston and filled with a gas has a volume of 16.44 liters at 22.4 °C when the applied pressure is 772.2 torr. The temperature of the oil bath surrounding it was increased to 184.4 °C, and the load on the piston was changed. Careful measurement now gave a value of 16.60 liters for the volume. What is the final pressure in the system, in torr?

a. 494.0 torr
b. 503.6 torr
! c. 1184 torr
d. 1207 torr
e. 6295 torr

32. A cylinder fitted with a movable piston and filled with a gas has a volume of $1.466 \times 10^{-2}$ cubic meters when temperature is 24.2 °C and the applied pressure is 97.03 kPa. The temperature of the oil bath surrounding it was increased to 146.4 °C. Careful measurement now gave a value of $1.479 \times 10^{-2}$ cubic meters for the volume. What is the final pressure in the system, in kilopascals?

a. 68.16 kPa
b. 69.38 kPa
! c. 135.7 kPa
d. 138.1 kPa
e. 581.8 kPa

33. A gas sample occupies volume of 18.86 liters when the temperature is 35.2 °C and the pressure is 735.5 torr. How many moles of gas are there in the sample?

! a. 0.7215 moles
b. 0.7702 moles
c. 1.298 moles
d. 5.411 moles
e. 6.319 moles

34. A gas sample containing 0.2820 moles of a compound is trapped in a vessel on a manifold at a temperature of 25.2 °C and a pressure of 642.0 torr. What is the volume of the vessel, in liters?

a. 0.01075 liters
b. 0.6903 liters
! c. 8.173 liters
d. 12.72 liters
e. 92.99 liters

35. A gas sample containing 0.3250 moles of a compound is confined in a glass vessel on a manifold at a temperature of 28.4 °C and a pressure of 112.8 kPa. What is the volume of the vessel, in cubic meters?

! a. $7.224 \times 10^{-3}$ m$^3$
b. $7.369 \times 10^{-1}$ m$^3$
c. $7.722 \times 10^{-2}$ m$^3$
d. $7.722 \times 10^{-2}$ m$^3$
e. $7.824 \times 10^{-3}$ m$^3$

36. A gas sample containing 0.2820 moles of a compound is trapped in a 2.461 liter vessel on a manifold at a temperature of 25.2 °C. What is the pressure in the vessel, in atmospheres, if it behaves as an ideal gas?

a. 0.2369 atm
b. 0.3740 atm
! c. 2.805 atm
d. 4.112 atm
e. 24.01 atm

37. A gas sample containing 0.3525 moles of a compound is trapped in a 2.641 liter vessel on a manifold at a temperature of 28.4 °C. What is the pressure in the vessel, in torr, if the gas behaves as an ideal gas?

a. 334.6 torr
b. 2007 torr
! c. 2510 torr
d. 2513 torr
e. 2694 torr

38. A gas sample containing 0.3250 moles of a compound is trapped in a chamber with a volume of 0.002460 cubic meters. If the temperature in the chamber is 26.8 °C, what is the pressure in the chamber, in kPa? Assume the gas exhibits ideal gas behavior.

a. 29.44 kPa
b. 307.0 kPa
! c. 329.5 kPa
d. 356.9 kPa
e. 3251 kPa

39. A gas sample occupies a volume of 1.662 liters when the temperature is 150.0 °C and the pressure is 842.0 torr. How many molecules are there in the sample?

a. $1.516 \times 10^{22}$
b. $2.602 \times 10^{22}$
! c. $3.194 \times 10^{22}$
d. $9.009 \times 10^{22}$
e. $9.419 \times 10^{21}$

40. A gas sample occupies a volume of 1.446 liters when the temperature is 185.0 °C and the pressure is 83.19 kPa. How many molecules are there in the sample?

! a. $1.902 \times 10^{22}$
b. $2.821 \times 10^{22}$
c. $4.710 \times 10^{22}$
d. $9.095 \times 10^{21}$
e. $9.095 \times 10^{22}$

41. A gas sample occupies a volume of $1.264 \times 10^{-3}$ cubic meters when the temperature is 168.0 °C and the pressure is 126.2 kPa. How many molecules are there in the sample?

a. $2.072 \times 10^{22}$
! b. $2.619 \times 10^{22}$
c. $2.654 \times 10^{23}$
d. $2.654 \times 10^{24}$
e. $6.877 \times 10^{22}$

42. A gas sample weighing 4.480 grams occupies a volume of 2.150 liters at STP. What is the apparent molecular weight of the sample?

a. 9.632 g/mole
b. 10.76 g/mole
! c. 46.70 g/mole
d. 112.8 g/mole
e. 215.9 g/mole

43. A student doing a laboratory study on determination of molecular weight of gases obtained the following data for one of his unknowns: 6.155 grams of the gaseous substance occupy a volume of 1,484 ml when the temperature was held at 27.3 °C and the pressure in the measurement system was recorded as 747.2 torr. Calculate the molecular weight of the substance, assuming ideal gas behavior.

a. 13.87 g/mole
b. 30.40 g/mole
! c. 104.0 g/mole
d. 106.7 g/mole
e. 110.0 g/mole

44. A chemical reaction, $A(s) \longrightarrow B(s) + C(g)$ occurs when substance A is strongly heated. The molecular weight of the gaseous substance was to be determined from the following experimental data. Mass of A: 4.962 g Mass of residue (B) after cooling and weighing when no more gas was evolved: 3.684 g. When collected and stored in a 658.5 ml glass vessel on a manifold at 30.4 °C, the gas exerted a pressure of 748.5 torr. From this data, determine the apparent molecular weight of C, assuming it behaves as an ideal gas.

a. 6.544 g/mole
b. 47.58 g/mole
! c. 49.08 g/mole
d. 71.68 g/mole
e. 141.5 g/mole

45. A specially designed gas containment vessel has a volume of 5.604 liters. It was charged with carbon monoxide gas at 28.2 °C. The pressure in the vessel now measured 868.5 torr. If the gas behaves as an ideal gas, how much should the gas sample weigh, in grams?

a. 0.7651 g
b. 1.289 g
c. 5.555 g
d. 6.286 g
! e. 7.254 g

46. A specially designed gas containment vessel has a volume of 6.504 liters. It was charged with propane gas, $C_3H_8$, at 28.3 °C. The pressure in the vessel now measured 486.3 torr. How much should the gas sample weigh, in grams, if it behaves as an ideal gas?

a. 4.674 g
! b. 7.419 g
c. 7.528 g
d. 18.12 g
e. 263.6 g

47. A mixture of 5.00 moles of neon and 3.00 moles of nitrogen occupy a volume of 36.0 liters in a vessel where the total pressure is 6.00 atm. What is the partial pressure of the neon in the container?

a. 0.375 atm
b. 0.625 atm
c. 0.833 atm
! d. 3.75 atm
e. 43.2 atm

48. What is the total pressure exerted by a gaseous mixture that consists of 8.00 g of methane and 12.00 g of ethane, $C_2H_6$, in a 3.50 liter container maintained at 35.20 °C?

a. 0.400 atm
b. 0.741 atm
c. 3.13 atm
! d. 6.49 atm
e. 16.5 atm

49. What is the total pressure exerted by a gaseous mixture that consists of 1.00 g of hydrogen and 8.00 g of neon in a 2.80 liter container maintained at 44.10 °C?

a. 1.15 atm
b. 3.77 atm
c. 3.95 atm
! d. 8.30 atm
e. 12.9 atm

50. What is the total pressure exerted by a gaseous mixture that consists of 8.00 g of methane and 8.00 g of ethane, $C_2H_6$, in a 3.50 liter container maintained at 35.20 °C?

a. 0.286 atm
b. 0.631 atm
c. 2.51 atm
! d. 5.53 atm
e. 30.7 atm

51. What is the mole fraction of methane in a gaseous mixture that consists of 8.00 g of methane and 12.00 g of ethane, $C_2H_6$, in a 3.50 liter container maintained at 35.20°C?

a. 0.400
b. 0.434
! c. 0.555
d. 0.800
e. 1.50

52. What is the mole fraction of hydrogen in a gaseous mixture that consists of 8.00 g of hydrogen and 12.00 g of neon in a 3.50 liter container maintained at 35.20°C?

a. 0.150
b. 0.400
c. 0.660
! d. 0.870
e. 0.930

53. A sample gas mixture occupies a volume of 1.662 liters when the temperature is 185.0 °C and the pressure is 624.4 torr. The mole fraction of oxygen in this sample is 0.3500. How many oxygen molecules are there in the sample?

a. $1.134 \times 10^{22}$
b. $1.896 \times 10^{22}$
c. $2.262 \times 10^{22}$
d. $5.803 \times 10^{24}$
! e. $7.656 \times 10^{21}$

54. A sample gas mixture has a volume of $1.288 \times 10^{-2}$ cubic meters when the temperature is 168.0 °C and the pressure is 109.7 kPa. The mole fraction of helium in this sample is 0.5224. How many helium molecules are there in the sample?

a. $1.212 \times 10^{20}$
! b. $1.212 \times 10^{22}$
c. $3.182 \times 10^{24}$
d. $4.441 \times 10^{23}$
e. $8.167 \times 10^{23}$

55. A sample of oxygen gas was collected by downward displacement of water in a large gas collection apparatus. The total pressure in the collection vessel was 744.2 torr, the temperature was 26.0 °C, and the vessel contained 522 ml of the collected gas. How many moles of oxygen were collected? At 26.0 °C, the vapor pressure of water is 25.2 torr.

! a. 0.0201 moles
b. 0.0215 moles
c. 0.0842 moles
d. 0.151 moles
e. 0.231 moles

56. A sample of hydrogen gas was collected by downward displacement of water in a very large gas buret. The total pressure in the buret was measured as 764.2 torr, the temperature was 23.0 °C, and the buret contained 511 ml of the collected gas. How many moles of hydrogen were collected? At 23.0 °C, the vapor pressure of water is 21.1 torr.

a. 0.0191 moles
! b. 0.0206 moles
c. 0.0211 moles
d. 0.0215 moles
e. 0.0217 moles

57. A sample of nitrogen gas was collected by downward displacement of water in a gas collection flask. The total pressure in the collection flask was measured as 754.2 torr, the temperature was 20.0 °C, and the measured volume of gas collected was 516 ml. How many grams should the nitrogen weigh? At 20.0 °C, the vapor pressure of water is 17.5 torr.

a. 0.291 g
b. 0.305 g
! c. 0.582 g
d. 0.596 g
e. 0.610 g

58. A sample of methane gas was collected by downward displacement of water in a gas collection flask. The total pressure in the collection flask was measured as 758.2 torr, the temperature was 25.0 °C, and the measured volume of gas in the flask was 526.0 ml. How many grams should the methane weigh. At 25.0 °C, the vapor pressure of water is 23.8 torr.

a. 0.01080 g
! b. 0.3333 g
c. 0.3441 g
d. 0.3549 g
e. 8.438 g

59. When $KClO_3$ is strongly heated, it decomposes to give KCl and oxygen gas. When 4.902 g of $KClO_3$ are heated until no more gas is evolved, how many liters of oxygen gas (measured at 25.0 °C and 771.0 torr) are produced, if the oxygen behaves as an ideal gas?

a. 0.6431 liters
b. 0.8314 liters
c. 0.9647 liters
! d. 1.447 liters
e. 1.925 liters

60. When $KClO_3$ is strongly heated, is decomposes to give KCl and oxygen gas. When 4.289 g of $KClO_3$ are heated until no more gas is evolved, how many liters of oxygen gas (measured at 22.4 °C and 754.0 torr) are produced, if the oxygen behaves as an ideal gas?

a. 0.8555 liters
b. 0.9625 liters
! c. 1.283 liters
d. 1.668 liters
e. 1.925 liters

61. When $Ag_2O$ is strongly heated, it decomposes to produce silver and oxygen gas. If a 6.894 g sample of $Ag_2O$ is heated until no more gas is evolved and all the oxygen is recovered, what volume of space, measured at 25.0 °C and 754.0 torr, should it occupy if it exhibits ideal gas behavior?

a. 0.2934 liters
! b. 0.3668 liters
c. 0.5502 liters
d. 0.7336 liters
e. 1.100 liters

62. When $Ag_2O$ is strongly heated, it decomposes to produce silver and oxygen gas. If a 8.964 g sample of $Ag_2O$ is heated until no more gas is evolved and all the oxygen is recovered, what volume of space, measured at 22.4 °C and 771.0 torr, should it occupy if it exhibits ideal gas behavior?

! a. 0.4624 liters
b. 0.6935 liters
c. 0.9247 liters
d. 1.156 liters
e. 1.387 liters

63. Silver oxide, $Ag_2O$, when heated strongly decomposes to yield silver metal and oxygen gas. A 4.262 g sample was heated to decompose it completely. What volume, in ml, would the gas collected occupy at 28.3 °C and 749.0 torr?

! a. 230.8
b. 346.2
c. 461.6
d. 615.5
e. 923.2

64. A gas mixture is 50.00% helium and 50.00% methane, by ***volume***. What is the mole fraction of methane in the mixture?

a. 0.1997
b. 0.2005
c. 0.2500
! d. 0.5000
e. 0.8003

65. How many liters of pure oxygen gas, measured at STP, are required for the complete combustion of 11.2 liters of methane gas, also measured at STP?

a. 11.2 liters
b. 16.8 liters
! c. 22.4 liters
d. 32.0 liters
e. 33.6 liters

66. How many liters of pure oxygen gas, measured at STP, are required for the complete combustion of 16.6 liters of ethylene gas ($C_2H_4$), also measured at STP?

a. 16.6 liters
b. 22.4 liters
c. 24.9 liters
d. 33.2 liters
! e. 49.8 liters

67. How many liters of pure oxygen gas, measured at 45.2 °C and 104.0 kPa, are required for the complete combustion of 5.60 liters of acetaldehyde gas ($C_2H_4O$), measured under the same conditions of temperature and pressure?

a. 4.93 liters
b. 12.3 liters
! c. 14.0 liters
d. 14.8 liters
e. 16.8 liters

68. According to the kinetic theory of gases, the average kinetic energy of the gas particles in a gas sample is directly proportional to the

a. pressure
b. volume
! c. temperature
d. molar mass
e. number of moles of gas

69. A gaseous substance diffuses twice as rapidly as sulfur dioxide gas. The gas could be

a. carbon monoxide
b. helium
c. hydrogen
! d. methane
e. oxygen

70. The average speed at which a nitrogen molecule effuses at 30.0 °C is 480 meters per second. The average speed at which a butene molecule ($C_4H_8$) effuses at this same temperature should then be

a. 120 m/s
b. 170 m/s
c. 240 m/s
! d. 339 m/s
e. 679 m/s

71. The average speed at which a methane molecule effuses at 28.5 °C is 631 meters per second. The average speed at which a krypton molecule effuses at this same temperature should therefore be

a. 121 m/s
! b. 276 m/s
c. 315 m/s
d. 417 m/s
e. 631 m/s

72. If container "A" is occupied by 1.00 mole of oxygen gas while container "B" is occupied by 20.0 grams of nitrogen gas and both containers are maintained at 0.00 °C and 650 torr then,

a. container "B" must be larger than container "A".
b. the average speed of the molecules in container "A" is greater than that of the molecules in container "B".
c. container "A" must have a volume of 22.4 L.
! d. the average kinetic energy of the molecules in "A" is equal to the average kinetic energy of the molecules in "B".
e. the number of atoms in container "B" is greater than the number of atoms in container "A".

73. The average speed at which a methane molecule effuses at 28.5 °C is 631 meters per second. The average speed at which an argon molecule effuses at this same temperature should therefore be

a. 253 m/s
b. 315 m/s
! c. 400 m/s
d. 502 m/s
e. 631 m/s

74. A mixture contains sand and $KClO_3$. The sand is not affected by heating, while the $KClO_3$ is decomposed to KCl and oxygen by heating. A 5.000 g sample, when heated until no further reaction occurred, gave off a gas which occupied a volume of 1.157 liters at 27.2 °C and 748.0 torr pressure. How many grams of sand are there in the mixture?

a. 0.6125 g
b. 0.817 g
! c. 1.225 g
d. 1.838 g
e. 3.775 g

75. A mixture contains sand and $KClO_3$. The sand is not affected by heating, while the $KClO_3$ is decomposed to KCl and oxygen by heating. A 5.000 g sample, when heated until no further reaction occurred, gave off a gas which occupied a volume of 0.9605 liters at 24.2 °C and 768.0 torr pressure. What is the per cent, by weight, of sand in the mixture?

a. 2.50 %
b. 12.73 %
c. 22.50 %
d. 25.46 %
! e. 35.00 %

76. A gas mixture is 50.00% helium and 50.00% methane, by ***volume***. What is the per cent, ***by weight***, of methane in the mixture?

a. 19.97 %
b. 20.05 %
c. 50.00 %
d. 75.00 %
! e. 80.03 %

## Fill in the Blanks

77. Air pressure at the top of a very high mountain was measured as 250 mm Hg. How many standard atmospheres is this? ______ (! 0.329 atm)

78. A mercury containing torricelli barometer is located in a controlled atmosphere chamber used for training astronauts. The chamber is maintained at an air pressure of $1.20 \times 10^4$ Pa. What height, in mm of mercury, should the mercury containing toricelli barometer in the chamber register? ______ ( 90.0 mm Hg)

79. A mixture of ethylene gas, $C_2H_4$, and methane is trapped in a container maintained at a pressure of 73.9 kPa and a temperature of 55.2 °C. If the mole fraction of methane in the mixture is 0.240, what is the partial pressure of methane in the mixture? ______ (! 17.7 kPa)

80. A gas sample has a volume of 12.42 liters at STP. How many moles of gas are there in the sample? ______ (! 0.5541 moles)

81. A gas sample has a volume of 1.424 cubic meters at STP. How many moles of gas are there in the sample? ______ (! 63.53 moles)

82. A gas sample occupies volume of 0.01500 cubic meters at a temperature is 30.0 °C and a pressure of 94.85 kPa. How many moles of gas are there in the sample? ______ (! 0.5645 moles)

83. A gas sample weighing 8.280 grams occupies a volume of 4.260 liters at STP. What is the apparent molecular weight of the sample, in grams per mole? ______ (! 43.57 g/mole)

84. A gas sample weighing 6.480 grams occupies a volume of 3.550 liters at STP. What is the apparent molecular weight of the sample, in grams per mole? ______ (! 40.91 g/mole)

85. In a measurement carried out to determine the molecular weight of a gaseous substance the following data was obtained: 6.268 g of the gaseous sample occupied a volume of 0.01148 cubic meters when the temperature was 29.5 °C and the applied pressure was 48.55 kPa. Calculate the molecular weight, in grams per mole, of the gaseous substance, assuming ideal gas behavior. ______ (! 28.30 g/mole)

86. A gas sample containing 0.3520 moles of a compound is trapped in a vessel on a manifold at a temperature of 28.4 °C and a pressure of 91.23 kPa. If the gas behaves as an ideal gas, what is the volume of the vessel, in liters? ______ (! 9.674 liters)

87. Calculate the density of oxygen gas, in g/liter at 1.000 atm and 25.00 °C. ______ (! 1.308 g/liter)

88. A gaseous element has a density of 1.098 g/liter when the temperature is 31.4 °C and the pressure is 744.5 torr. Which element best fits the description? ______ (! nitrogen, MW = 28.01)

89. A gaseous element has a density of 3.195 g/liter when the temperature is 35.8 °C and the pressure is 734.6 torr. Which element best fits the description? ______ (! krypton, MW = 83.80)

90. Results of measurements on a gas to determine its molecular weight gave a value of 1.614 g/liter for the density, at 27.2 °C and 749.4 torr. What is the molecular weight of the gas, in grams per mole? ______ (! 40.34 g/mole)

91. Results of measurements on a gas to determine its molecular weight gave a density of 2.416 $kg/m^3$, at 22.7 °C and 105.9 kPa. What is the molecular weight of the gas, in grams per mole? ______ (! 56.11 g/mole)

92. How many liters of pure oxygen gas, measured at STP, are required for the complete combustion of 6.01 grams of pentane ($C_5H_{12}$)? ______ ( ! 14.9 g)

93. How many cubic feet of pure oxygen gas, measured at 35.2 °C and 0.9474 atm, are required for the complete combustion of 11.2 cubic feet of butene gas ($C_4H_8$), under the same conditions of temperature and pressure? ______ ( ! 67.2 $ft^3$).

94. A sample gas mixture occupies a volume of 1.446 liters when the temperature is 155.0 °C and the pressure is 714.8 torr. The mole fraction of methane in this sample is 0.2250. How many methane molecules are there in the sample? ______ ( ! 5.245 x $10^{21}$)

## True and False

95. Air pressure in a deep mine shaft is the same (numerically equal) to the air pressure at the earth's surface next to the shaft opening. ( !F)

96. Gases usually possess negative volumes when the temperature falls below -273.15 K. ( !F)

97. The volume occupied by one mole of a gas which behaves as an ideal gas when the temperature is 25.0 °C and the pressure is 760.0 torr is 22.414 liters. ( !F)

98. Three 4.625 liter spheres in the same room are filled with gas at the same temperature and pressure. The gas is sphere A is pure hydrogen, the gas in sphere B is pure carbon monoxide, but the gas in sphere C is a *mixture* of hydrogen and carbon monoxide. According to the gas laws, all three spheres contain the same number of atoms. ( !T)

99. The density of a pure gaseous substance is inversely related to its temperature and directly related to its molecular weight. ( !T)

100. One pascal represents a greater pressure than one torr. ( !F)

101. If you have a gas and do some measurements, the number of grams you obtain by weighing divided by the number of moles you calculate from the temperature--pressure--volume measurements will give the molecular weight. This will indicate to you whether it is a pure substance or not. ( !F)

102. The value 8.31441 J/k-mole for the gas constant, R, can not be used in calculation involving the ideal gas law. ( !F)

103. The torricelli barometer is based on Charles' Law. ( !F)

104. Even though the average speed of a carbon monoxide molecule at 27.5°C is greater than that of a sulfur dioxide molecule at the same temperature, the average kinetic energy of the molecules is a sample of carbon monoxide at 27.5 °C is the same as the average kinetic energy of the molecules in a sample of sulfur dioxide gas at the same temperature. ( !T)

105. The ratio, pV/nT, for a gaseous substance has a fixed and reproducible value independent of the nature of the gas or the physical conditions provided it is in the gaseous state. ( !F)

## Critical Thinking -- Level 1

106. A gaseous mixture is 20.00 % argon, 25.00 % nitrogen, 30.00 % methane and 25.00 % ethylene, $C_2H_4$ ***by weight***. Calculate the percent, by volume, of methane in the mixture. ______ ( ! 45.01 %)

107. A mixture is composed of potassium chlorate, $KClO_3$, potassium chloride, and sand. The $KClO_3$, when heated strongly, decomposes to yield potassium chloride and oxygen gas, while both the potassium chloride and the sand are thermally stable and are not affected by heat. A 2.600 gram sample of the mixture was heated and the gas evolved was collected over water by downward displacement in a large gas collection tube. Heating was continued until no potassium chlorate remained, as evidenced by cessation of the evolution of bubbles of gas in the collection apparatus. The volume of the saturated gas collected was 183.5 ml and the pressure measured 757.5 torr. The temperature of the collection system was measured as 26.9 °C, at which temperature the saturation vapor pressure of water is 26.6 torr. From these data, determine the percent, by weight, of potassium chlorate in the sample. ______ ( ! 22.5%)

108. A gas mixture being used in a diving bell is 28.50% $O_2$, by weight and the rest is helium. Calculate the density of this mixture at 1.000 atm and 25.00 °C. ______ ( ! 0.2179 g/liter)

109. A gas mixture contains 31.50% CO, 22.50% argon and 46.00% oxygen, all by volume. What is the apparent molecular weight (g/mole) of the gas? ______ ( ! 32.53 g/mole)

110. A gas mixture contains $CH_4$, $C_2H_4$, and $C_3H_8$. Mole fractions of each component in the mixture are, $CH_4$: 0.3150; $C_2H_4$: 0.4750; $C_3H_8$: 0.2100. Calculate how many liters of $O_2$, measured at 25.0 °C and 742.5 torr, will be required for the complete combustion of 8.000 liters of the gas mixture, which is also at 25.0 °C and 742.5 torr. ______ ( ! 21.04 liters)

## Critical Thinking -- Level 2

111. Air can be purified by the reaction, $MgO(s) + CO_2(g) \longrightarrow MgCO_3(s)$. If the air to be recycled has a density of 1.185 g/liter at 25.0°C and 1.000 atm pressure, and contains 0.320% $CO_2$, by weight, how many cubic meters of air can be purified by 1.000 kg of $MgCO_3$ if the process is conservatively rated as 60.0% efficient? ______ ( ! 173 $m^3$)

112. An alloy contains zinc and magnesium as its only components. Both elements react with hydrochloric acid solution to produce hydrogen gas and the chloride salt of the metal cation. An alloy sample weighing 452 milligrams was treated with excess hydrochloric acid solution and the gas evolved was collected over water by downward displacement in a large gas collection tube. The volume of the saturated gas collected was 287.5 ml and the pressure measured 761.5 torr. The temperature of the collection system was measured as 26.9 °C, at which temperature the saturation vapor pressure of water is 26.6 torr. From these data, determine the per cent, by weight, of magnesium in the alloy. ______ ( ! 37.5%)

## Chemicals in Use 8:

113. Ozone in the atmosphere is a by-product of air pollution--air pollutants react with atmospheric oxygen to produce ozone. No naturally occurring chemical reaction will produce ozone in the absence of these pollutants. ( !F)

114. One reaction which produces $NO_2(g)$ in the upper atmosphere is,

$$N_2(g) + O_2(g) \xrightarrow{h\nu} N(g) + NO_2(g)$$ ( !F)

# Chapter 11 Intermolecular Attractions and the Properties of Liquids and Solids

## Multiple Choice

1. The strongest intermolecular forces between molecules of HCl are

a. covalent bonds
! b. dipole-dipole interactions
c. ionic bonds
d. hydrogen bonds
e. London forces

2. For a series of small molecules of comparable molecular weight, which of following lists the intermolecular forces in the correct increasing order?

a. hydrogen bonds < dipole-dipole forces < London forces
b. dipole-dipole forces < hydrogen bonds < London forces
c. London forces < hydrogen bonds < dipole-dipole forces
d. hydrogen bonds < London forces < dipole-dipole forces
! e. London forces < dipole-dipole forces < hydrogen bonds

3. What one of the following covalent compounds will not exhibit hydrogen bonding in the liquid state?

! a. $CH_3—CH_2—Br$
b. $CH_3—CH_2—NH_2$
c. $CH_3—CH_2—CH_2—OH$
d. $CH_3—NH—CH_3$
e. $NH_2—O—H$

4. Which one of the following covalent compounds will exhibit hydrogen bonding in the liquid state?

a. $CH_2F_2$
! b. $Cl_2NH$
c. $H_2PCl$
d. HBr
e. $NCl_3$

5. Which one of the following covalent compounds will exhibit hydrogen bonding in the liquid state?

a. $CCl_2F_2$
b. $H_2PCl$
c. HCl
! d. $NH_2OH$
e. $NF_3$

6. Which compound is expected to have the strongest intermolecular forces?

a. $CH_3—CH_2—H$
! b. $CH_3—CH_2—O—H$
c. $CH_3—CH_2—PH_2$
d. $CH_3—CH_2—S—H$
e. $CH_3—CH_2—Se—H$

7. Which compound is expected to have weakest intermolecular forces?

! a. $CH_3—CH_2—H$
b. $CH_3—CH_2—O—H$
c. $CH_3—CH_2—PH_2$
d. $CH_3—CH_2—S—H$
e. $CH_3—CH_2—Se—H$

8. Which compound is expected to exhibit hydrogen bonding forces?

a. $CH_3—CH_2—H$
! b. $CH_3—CH_2—O—H$
c. $CH_3—CH_2—PH_2$
d. $CH_3—CH_2—S—H$
e. $CH_3—CH_2—Se—H$

9. The property that measures or describes the magnitude of resistance to flow in a liquid is called

a. London forces
b. malleability
c. surface tension
d. vapor pressure
! e. viscosity

10. Which one of the following is not a property or phenomenon associated with the liquid state?

a. fluidity
b. meniscus
! c. sublimation
d. surface tension
e. viscosity

11. The ability of capillary tubing to draw liquids up against the action of gravity is as example of

a. fluidity
b. meiosis
! c. osmosis
d. scattering
e. viscosity

12. When a liquid undergoes a phase change to a gas, the process is called

a. condensation
b. deposition
c. fusion
d. sublimation
! e. vaporization

13. As we vary temperatures along isobars (lines of constant pressure) which lie below the triple point on a phase diagram for a substance, which of the processes below would **never** be observed to occur?

a. decrease in volume
b. deposition
c. expansion
! d. fusion
e. sublimation

14. As we vary temperatures along isobars (lines of equal pressure) which lie above the critical point pressure on a phase diagram for a substance, which of the processes below would **never** be observed to occur?

! a. condensation
b. deposition
c. expansion
d. scattering
e. sublimation

15. As we vary temperatures along isobars which lie above the critical point pressure on a phase diagram for a substance, which of the processes below would **never** be observed to occur?

a. deposition
b. expansion
! c. fusion
d. sublimation

16. Which one of the following listed compounds should have the lowest vapor pressure at a given temperature at which all these substances are in the liquid state?

a. $CH_3—CH_2—F$
b. $CH_3—CH_2—CH_3$
c. $CH_3—CH_2—CH_2—CH_3$
! d. $CH_3—CH_2—O—H$
e. $CH_3—O—CH_3$

17. Which one of the following listed compounds should have the highest vapor pressure at a given temperature at which all these substances are in the liquid state?

! a. $CH_3—CH_2—H$
b. $CH_3—O—CH_3$
c. $CH_3—CH_2—CH_2—CH_3$
d. $CH_3—CH_2—CH_3$
e. $CH_3—CH_2—S—H$

18. Which one of the following listed compounds should have the highest vapor pressure at a given temperature?

a. $CH_3$—Hg—$CH_3$
b. $CH_3$—O—$CH_3$
! c. $CH_3$—S—$CH_3$
d. $CH_3$—Se—$CH_3$
e. $CH_3$—Te—$CH_3$

19. Some temperatures (in °C) and the vapor pressure for water are: 20.0 °—17.4 torr, 25.0 °C —23.8 torr, 30.0 °C—31.6 torr, 35 °C—42.2 torr. If the temperature is 35.0 °C and the relative humidity is 50.0%, estimate the dew point temperature.

a. 20.0 °C
! b. 22.5 °C
c. 25.0 °C
d. 27.5 °C
e. 30.0 °C

20. Some temperatures (in °C) and the vapor pressure for water are: 20.0 °—17.4 torr, 25.0 °C —23.8 torr, 30.0 °C—31.6 torr, 35 °C—42.2 torr. If the temperature is 35.0 °C and the relative humidity is 50.0%, calculate the actual vapor pressure.

a. 17.5 torr
! b. 21.1 torr
c. 25.0 torr
d. 27.5 torr
e. 84.4 torr

21. Some temperatures (in °C) and the vapor pressure for water are: 20.0 °—17.4 torr, 25.0 °C —23.8 torr, 30.0 °C—31.6 torr, 35 °C—42.2 torr. If the temperature is 35.0 °C and the relative humidity is 78.0%, estimate the dew point temperature

! a. 20.0 °C
b. 22.5 °C
c. 25.0 °C
d. 27.5 °C
e. 30.5 °C

22. Which one of the following listed compounds should have the lowest boiling point temperature?

a. $CH_3$—Br
b. $CH_3$—Cl
c. $CH_3$—F
! d. $CH_3$—H
e. $CH_3$—I

23. Which one of the compounds listed below should have the highest boiling point temperature?

a. $CH_3$—Br
b. $CH_3$—Cl
c. $CH_3$—F
d. $CH_3$—H
! e. $CH_3$—I

24. Given the following substances and their normal boiling points, in Celsius degrees. C—43.8 D—93.7 M—56.7 T—83.5 R—63.6 Which set below correctly lists liquids in order of decreasing vapor pressure at 20 °C?

! a. C > R > D
b. D > T > R
c. R > T > D
d. C > D > M
e. D > R > M

25. Which one of the following factors has an effect on the value of the normal boiling point of a liquid?

a. atmospheric pressure
b. rate of condensation
c. rate of evaporation
! d. strength of the intermolecular forces
e. the external temperature

26. A liquid which is at room temperature is in equilibrium with its vapor, because there is a cover on the container. If the cover is removed, what is the immediate result?

a. the average kinetic energy increases
b. the evaporation rate decreases
c. the evaporation rate increases
! d. the re-condensation rate decreases
e. the re-condensation rate increases

27. Given the phase changes, condensation, deposition, freezing, fusion, sublimation, vaporization. Which (set) of these phase changes is(are) endothermic?

a. deposition, condensation, and freezing **only**
b. fusion **only**
c. fusion and vaporization **only**
d. fusion, deposition and sublimation **only**
! e. sublimation, fusion, and vaporization **only**

28. Substance, A, has a normal fusion point of -25.0 °C, an enthalpy of fusion = 1200 J/g; specific heats for the solid and the liquid are 3.00 and 6.20 joules per gram, respectively. To change 150 grams of A from -40 °C to +70 °C, still a liquid

a. will require $1.52 \times 10^5$ joules
b. will require $1.81 \times 10^5$ joules
c. will require $2.21 \times 10^3$ joules
d. will require $2.29 \times 10^5$ joules
! e. will require $2.75 \times 10^5$ joules

29. Substance, A, has a normal fusion point of +25.0 °C, an enthalpy of fusion = 1200 J/g; specific heats for the solid and the liquid are 3.00 and 6.20 joules per gram, respectively. To change 150 grams of A from -40 °C to +70 °C, still a liquid

a. will require $2.27 \times 10^5$ joules
b. will require $2.29 \times 10^5$ joules
! c. will require $2.51 \times 10^5$ joules
d. will require $2.75 \times 10^5$ joules
e. will require $2.83 \times 10^5$ joules

30. A substance, B, has a normal boiling point of +89.3 °C, an enthalpy of vaporization = 2600 J/g; specific heats for the liquid and the gas are 6.20 and 3.20 joules per gram, respectively. To change 150 grams of B from -10 °C to gas at +129 °C.

a. will require $4.83 \times 10^5$ joules
! b. will require $5.01 \times 10^5$ joules
c. will require $5.49 \times 10^5$ joules
d. will require $5.58 \times 10^5$ joules
e. will require $6.81 \times 10^5$ joules

31. A substance, B, has a normal boiling point of +89.3 °C, an enthalpy of vaporization = 2600 J/g; specific heats for the liquid and the gas are 6.20 and 3.20 joules per gram, respectively. To change 150 grams of B from +10 °C to gas at +129 °C.

! a. will require $4.83 \times 10^5$ joules
b. will require $5.01 \times 10^5$ joules
c. will require $5.49 \times 10^5$ joules
d. will require $5.58 \times 10^5$ joules
e. will require $6.81 \times 10^5$ joules

32. For substance A, the melting point is 16.61 °C and the specific heats of the solid and the liquid form are 2.662 and 5.188 joules/g-degree, respectively. It required 38,600 joules to move 120.0 g of A from 0.00 to 32.50 °C.

a. The molar enthalpy of fusion is 6.550 joules per gram.
b. The molar enthalpy of fusion is 151.3 joules per gram.
! c. The molar enthalpy of fusion is 195.0 joules per gram.
d. The molar enthalpy of fusion is 238.6 joules per gram.
e. The molar enthalpy of fusion is 321.7 joules per gram.

33. If 34.0 g of a solid with a molecular weight of 174.0 g/mole requires 21.3 kJ of heat to melt it, what is the molar enthalpy of fusion, in kJ per mole?

! a. (21.3)(174/34.0) kJ
b. (21.3)(174)(34.0) kJ
c. (21.3/174)(34.0) kJ
d. (21.3/174)/34 kJ
e. (174/21.3)/34 kJ

34. A liquid has a normal boiling point of 78.0 °C and its vapor pressure is 400 torr at 50.0 °C. To calculate the molar enthalpy of vaporization, one needs, in addition:

! a. nothing; enough information is given
b. the molecular weight of the substance
c. the specific heat of the substance
d. the vapor pressure at another temperature below the boiling point
e. the enthalpy of fusion

35. The normal boiling point of acetic acid, $HC_2H_3O_2$, is 117.9 °C, and its molar enthalpy of vaporization is 39,690 joules per mole. What is its vapor pressure, at 100.0 °C?

! a. 423.2 torr
b. 479.7 torr
c. 586.6 torr
d. 616.2 torr
e. 694.4 torr

36. The normal boiling point of 2,3,4-trimethypentane, $C_8H_{18}$, is 113.47 °C celsius, and its molar enthalpy of vaporization is 37,600 joules per mole. What is its vapor pressure, at 105.5 °C?

a. 479.7 torr
b. 586.6 torr
! c. 594.1 torr
d. 616.2 torr
e. 694.4 torr

37. The normal boiling point of 2,2,3-trimethypentane, $C_8H_{18}$, is 109.84 °C, and its molar enthalpy of vaporization is 37,070 joules per mole. What is its vapor pressure, at 102.5 °C?

a. 586.6 torr
b. 594.1 torr
! c. 605.4 torr
d. 616.2 torr
e. 694.4 torr

38. Find the normal boiling point temperature of an isomer of octane, $C_8H_{18}$, if its enthalpy of vaporization is 38,210 joules per mole and its vapor pressure at 110.0 °C is 638.43 torr and its vapor pressure at 125.0 °C is 1003.1 torr.

a. 111.52 °C
b. 113.22 °C
c. 115.00 °C
! d. 115.65 °C
e. 118.30 °C

39. The critical temperature of a substance is

! a. always higher than the triple point temperature
b. the temperature below which it cannot exist in the liquid state
c. the temperature which is always higher than the Kelvin point
d. the temperature below which it cannot be liquefied by increasing the pressure
e. the temperature below which it cannot be supercooled

40. Which one of the following noble gas elements will have the lowest critical point temperature?

a. Ar
! b. Ne
c. Kr
d. Rn
e. Xe

41. The triple point of a substance is the temperature and pressure at which

a. all three physical states cease to exist
! b. sublimation, fusion, and condensation are taking place simultaneously
c. the solid will always float on the liquid for all substances
d. the vapor pressure of the liquid is higher than the vapor pressure of the solid
e. the vapor pressure of the solid is higher than the vapor pressure of the liquid

42. Which one of the following cannot be liquefied by compression at temperature of 25.0 °C?

! a. $C_2H_4$, critical point data: 9.9 °C, 50.5 atm
b. $CH_3Cl$, critical point data: 144.0 °C, 66.0 atm
c. $C_2H_2$, critical point data: 35.5 °C, 61.6 atm
d. $C_2H_6$, critical point data: 32.2 °C, 48.2 atm
e. $SO_2$, critical point data: 158.0 °C, 78.0 atm

43. Which one of the following cannot be liquefied by compression at temperature of 25.0 °C?

a. $CH_3Cl$, critical point data: 144.0 °C, 66.0 atm
b. $C_2H_2$, critical point data: 35.5 °C, 61.6 atm
c. $C_2H_6$, critical point data: 32.2 °C, 48.2 atm
! d. $SiF_4$, critical point data: -14.1 °C, 36.7 atm
e. $SO_2$, critical point data: 158.0 °C, 78.0 atm

44. At 1.0 atm pressure, ice (solid $H_2O$) floats in water instead of sinking. The reason for this is

! a. When water freezes, it expands instead of contracting.
b. The fusion process is endothermic, therefore the solid will float.
c. The triple point has a lower temperature than the freezing point for water.
d. The critical temperature has a higher temperature than the normal boiling point.
e. The triple point corresponds to a pressure below 1 standard atmosphere.

45. How many lattice points does it require to fully define a body centered cubic unit cell?

a. 2
b. 4
c. 8
! d. 9
e. 14

46. How many lattice points does it require to fully define a face centered cubic unit cell?

a. 2
b. 4
c. 8
d. 9
! e. 14

47. How many tungsten atoms are there per unit cell in metallic tungsten if it forms a face centered cubic unit cell?

a. 2
! b. 4
c. 8
d. 9
e. 14

48. How many lanthanum atoms are there per unit cell in metallic lanthanum if it forms a body centered cubic unit cell?

! a. 2
b. 4
c. 8
! d. 9
e. 14

49. Gold crystallizes in a face centered cubic lattice. How many unit cells are there in a gold sample whose mass is 1.50 grams. The atomic weight of gold is 196.97 g/mole.

! a. $1.15 \times 10^{21}$ unit cells
b. $2.29 \times 10^{21}$ unit cells
c. $3.28 \times 10^{20}$ unit cells
d. $3.86 \times 10^{21}$ unit cells
e. $9.17 \times 10^{21}$ unit cells

50. How many grams would a sample of silver which has $5.00 \times 10^{22}$ unit cells weigh, if it crystallizes in a face centered cubic lattice?

a. 8.96 g
b. 14.2 g
! c. 35.8 g
d. 71.6 g
e. 128.8 g

51. Which set of properties below is associated with ionic crystalline substances?

a. broad range of melting points, soft to hard, conductors in liquid and solid states
! b. high melting point, brittle, non-conductor
c. high melting point, hard, not brittle, non-conductor
d. low melting point, soft, non-conductor
e. low melting point, soft, conductor in liquid and solid states

52. Which set of properties below is associated with molecular crystalline substances?

a. broad range of melting points, soft to hard, conductors in liquid and solid states
b. high melting point, brittle, non-conductor
c. high melting point, hard, not brittle, non-conductor
! d. low melting point, soft, non-conductor
e. low melting point, soft, conductor in liquid and solid states

53. Paradichlorobenzene is a white crystalline solid which is very soft, has a low melting point, and is a non-conductor of electricity. Which category of substance does it most likely fit under?

a. ionic
! b. molecular
c. metallic
d. covalent (network)
e. amorphous

54. Which one of the following types of substance shows poorly defined fusion point temperatures?

a. ionic
b. molecular
c. metallic
d. covalent (network)
! e. amorphous

55. Which one of the following substances is most likely composed of ions rather than distinct molecules when it is in the solid state?

! a. $Al_2O_3$, m.p. 2000°C, very hard substance
b. $CrO_3$, m.p. 196°C, very soft substance
c. $Mn_2O_7$, m.p. 5.9°C, soft in solid state
d. $OsO_4$, m.p. 40°C, soft substance
e. $SeO_3$, m.p. 118°C, relatively soft

56. Which one of the following substances is most likely composed of ions rather than distinct molecules when it is in the solid state?

! a. $CaCl_2$, m.p. 772°C
b. $SnCl_4$, m.p. -33°C
c. $SbCl_5$, m.p. 2.8°C
d. $TiCl_4$, m.p. -25°C
e. $VCl_4$, m.p. -248°C

57. Supercooling is defined as

a. the extremely rapid cooling of a vapor to form a liquid
b. the use of extremely cold refrigerants to achieve smaller crystal size when liquids are frozen
c. the extremely rapid cooling of a liquid to form a softer crystalline solid
! d. the cooling of a liquid to a temperture below its melting point with solidification
e. the cooling of a substance to absolute zero

## Fill in the Blanks

58. A solid chemical substance whose triple point pressure is 4.60 atmospheres will normally ______ when it is heated to a sufficiently high temperature in the open laboratory. ( ! sublime)

59. The basic building block of a crystalline ionic compound is called the ______. ( ! unit cell)

## True and False

60. The change in resistance to flow with temperature that many liquids exhibit is a chemical property which is based on intermolecular attracive forces. ( !F)

61. A cylinder contains a liquid which is in equilibrium with its vapor at 45.0°C. If the piston is moved down to constrict the volume of the cylinder without changing the temperature, the vapor pressure of the liquid will increase. ( !F)

62. When a substance decomposes before it vaporizes, it is an indication that the intermolecular bonds between the molecules are stronger than the covalent bonds within the molecule. ( !T)

## Special Topics

63. If we consider a substance being freeze dried as the system, the final stage in the freeze drying process

! a. is endothermic and involves enthalpy of sublimation
b. is endothermic and involves enthalpy of vaporization
c. is endothermic and involves enthalpy of fusion
d. is exothermic and involves enthalpy of vaporization
e. is exothermic and involves enthalpy of fusion

64. The working substance in a heat pump or an air condition unit must be a chemical substance whose critical temperature lies below the temperature at which the compressor operates. ( !F)

# Chapter 12 Solutions

## Multiple Choice

1. Concerning the process of separation of a solid substance into its component units (molecules or ions) where the interparticle distances are large compared to the particle sizes,

a. the process is exothermic and the potential energy increases
b. the process is exothermic and the potential energy decreases
! c. the process is endothermic and the potential energy increases
d. the process is endothermic and the potential energy decreases
e. the process is endothermic and occurs with no change in potential energy

2. Concerning the solvation step of the solution process,

a. the process is exothermic and the potential energy increases
! b. the process is exothermic and the potential energy decreases
c. the process is endothermic and the potential energy increases
d. the process is endothermic and the potential energy decreases
e. the process is endothermic and occurs with no change in potential energy

3. A solution in a beaker has some undissolved solute lying on the bottom of the beaker. If the rate of crystallization exceeds the rate of dissolution of the excess solute, the solution is described as

a. dilute
b. concentrated
c. unsaturated
d. saturated
! e. supersaturated

4. A solution in a beaker has some undissolved solute lying on the bottom of the beaker. If the rate of crystallization is equal to the rate of dissolution of the excess solute, the solution is described as

a. dilute
b. concentrated
c. unsaturated
! d. saturated
e. supersaturated

5. A solution in a beaker has some undissolved solute lying on the bottom of the beaker. If the rate of crystallization is exceeded by the rate of dissolution of the excess solute, the solution is described as

a. dilute
b. concentrated
! c. unsaturated
d. saturated
e. supersaturated

6. A solution is sitting undisturbed on a side shelf in the laboratory. A small crystal of the same solute of which the solution is made was gently dropped into the quiet solution. Suddenly, a mass of crystals formed and settled to the bottom of the container. The solution is, or must have been

a. dilute
b. concentrated
c. unsaturated
d. saturated
! e. supersaturated

7. Wax is a solid mixture of hydrocarbon compounds consisting of molecules with long chains of carbon atoms. Which solvent below would you expect to be most capable of dissolving wax by loosening the molecules in the solid and separating them?

a. 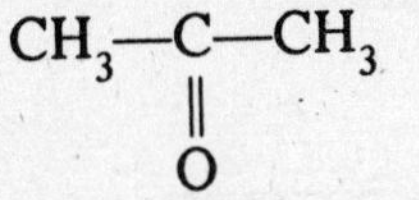
b. $CH_3$—O—H
c. 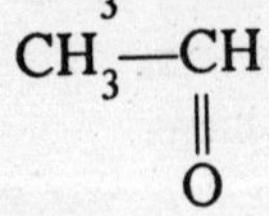
d. H—O—$CH_2$—$CH_2$—O—H
! e. $CH_3$—$CH_2$—$CH_2$—$CH_2$—$CH_2$—$CH_2$—$CH_2$—$CH_3$

8. Wax is a solid mixture of hydrocarbon compounds consisting of molecules with long chains of carbon atoms. Which solvent below would you expect to be most capable of dissolving wax by loosening the molecules in the solid and separating them?

a. HOH
b. $CH_3$—O—H
c. $CF_3$—O—H
d. H—O—$CH_2$—$CH_2$—O—H
! e. $CH_3$—$CH_2$—$CH_2$—$CH_2$—$CH_2$—$CH_2$—$CH_2$—$CH_3$

9. Which one of the following causes an increases in the solubility of a gas in a given solvent in the situation in which the gas does not react with the solvent to form a new substance?

a. increasing the temperature of the solvent and simultaneously decreasing the partial pressure of the gas in the space above the solvent
! b. decreasing the temperature of the solvent and simultaneously increasing the partial pressure of the gas in the space above the solvent
c. increasing the temperature of the solvent and simultaneously increasing the partial pressure of the gas in the space above the solvent
d. decreasing the temperature of the solvent and simultaneously decreasing the partial pressure of the gas in the space above the solvent
e. increasing the temperature of the solvent while maintaining the partial pressure of the gas in the space above the solvent at a set value

10. The solubility of $O_2$ in water is approximately 0.00380 g/liter of water when the temperature is .lh 0.210"
25.0 °C and the partial pressure of gaseous oxygen is 760 torr. What should the solubility of oxygen be if the oxygen pressure in readjusted to 1000 torr?

a. 0.00289 g/liter
! b. 0.00500 g/liter
c. 1.49 g/liter
d. $2.89 \times 10^3$ g/liter
e. $3.46 \times 10^3$ g/liter

11. The solubility of $O_2$ in water is approximately 0.00380 g/liter of water when the temperature is 25.0 °C and the partial pressure of gaseous oxygen is 760 torr. The oxygen gas above the water is replaced by air at the same temperature and pressure, in which the mole fraction of oxygen is 0.210. What should the solubility of oxygen in water under these new conditions?

a. $1.05 \times 10^{-6}$ g/liter
! b. $7.98 \times 10^{-4}$ g/liter
c. $1.33 \times 10^{-3}$ g/liter
d. 0.606 g/liter
e. 1.01 g/liter

12. A carbonated beverage is bottled and capped in a $CO_2$ atmosphere in which the pressure is maintained at 3.750 atm. At this pressure, the solubility of $CO_2$ in water is 0.65 g $CO_2$/100 g $H_2O$. If the bottle is opened, as the gas in the space above the liquid escapes, the partial pressure of the $CO_2$ falls to 0.30 torr, the value in the surrounding atmosphere of the room. What is the solubility of $CO_2$ in the beverage at this new pressure?

a. $5.2 \times 10^{-2}$ g/100 g beverage
b. $5.0 \times 10^{-4}$ g/100 g beverage
c. $6.7 \times 10^{-5}$ g/100 g beverage
! d. $6.8 \times 10^{-5}$ g/100 g beverage
e. $9.6 \times 10^{-4}$ g/100 g beverage

13. Which one of the following is a concentration unit whose value will be altered if the temperature of an aqueous solution is changed even if the solution remains unsaturated over the entire temperature range of the investigation.

a. mole fraction
! b. molarity
c. molality
d. mass fraction
e. percent by weight

14. A solution is made by mixing 138.2 grams of ethanol, $C_2H_6O$, (46.069 g/mole) 103.6 grams of water (18.015 g/mole) and 80.11 grams of methanol, $CH_4O$ (32.042 g/mole). What is the mole fraction of methanol in the mixture?

a. 0.02504
! b. 0.2222
c. 0.2493
d. 0.3333
d. 0.4490

15. Consider a 0.900 M $Al(NO_3)_3$ solution. This solution has a nitrate ion concentration of:

a. 0.300 M
b. 0.900 M
! c. 2.70 M
d. 3.60 M
e. 8.10 M

16. A solution of potassium nitrate is prepared by mixing 3.50 g of $KNO_3$ with 12.0 g of water. The percent, by mass, of $KNO_3$ is

! a. 22.6 %
b. 23.3 %
c. 28.0 %
d. 29.2 %
e. 41.8 %

17. A solution of sodium nitrite is prepared by mixing 3.25 g of $NaNO_2$ with 12.0 g of water. The percent, by mass, of $NaNO_2$ is

a. 28.0 %
b. 23.3 %
c. 27.0 %
! d. 21.3 %
e. 37.1 %

18. A solution of sodium hydroxide is prepared by mixing 2.00 g of LiOH with 10.0 g of water. The percent, by mass, of LiOH is

a. 10.7 %
b. 12.0 %
! c. 16.7 %
d. 20.0 %
e. 80.0 %

19. How many grams of $NaC_2H_3O_2$ (M = 82.034 g/mole) should be dissolved in 400.0 g of water to make a solution which is 1.550 molal?

a. 3.146 g
b. 7.558 g
c. 21.17 g
! d. 50.86 g
e. 127.15 g

20. A glucose solution is prepared by dissolving 5.10 g of glucose, $C_6H_{12}O_6$, in 110.5 g of water. What is the molality of the glucose solution?

a. 0.283 m
b. 0.000256 m
c. 0.245 m
! d. 0.256 m
e. 0.351 m

21. A glucose solution is prepared by dissolving 15.2 g of glucose, $C_6H_{12}O_6$, in 250.0 g of water. A student wants to make a sucrose solution ($C_{12}H_{22}O_{11}$) with the same molality. How many grams of sucrose should she use with 150.0 g of water, if that's how much solution she needs to prepare?

a. 4.80 g
b. 13.3 g
! c. 17.3 g
d. 18.0 g
e. 48.1 g

22. When 10.0 liters of a saturated aqueous solution of $CaSO_4$ was evaporated to dryness, 8.52 g of the dry, solid calcium sulfate dihydrate ($CaSO_4 \cdot 2H_2O$) was recovered. The density of the solution is 1.000 g/ml at this dilution. What is the calcium content of the water, in parts per million?

a. 19.8 ppm
b. 67.4 ppm
! c. 198 ppm
d. 674 ppm
e. 852 ppm

23. A liter of water is saturated with $PbCl_2$, which is very slightly soluble in water. When 10.0 liters of this solution was evaporated to dryness, 9.90 g of the dry, solid lead (II) chloride was recovered. The density of the solution is 1.000 g/ml at this dilution. What is the lead content of the water, in parts per million?

a. 0.990 ppm
b. 3.56 ppm
c. 4.78 ppm
! d. 738 ppm
e. 990 ppm

24. A water solution which contains magnesium chloride as the only solute measured 232 parts per million (ppm) chloride on an analytical instrument using a specific ion electrode as the sensor unit. When converted to moles per liter, what value would you get for the chloride ion concentration?

a. 0.00327 molar
! b. 0.00654 molar
c. 0.01309 molar
d. 0.0327 molar
e. 0.654 molar

25. A water solution is saturated with $MgC_2O_4$ (which is only very slightly soluble in water). When 10.0 liters of this solution was evaporated to dryness, 14.8 g of the dry, solid magnesium oxalate dihydrate ($MgC_2O_4 \cdot 2H_2O$ were recovered as the only residue. The magnesium content can be expressed in parts per million (ppm). What is the magnesium content of the solution? The density of the solution is 1.000 g/ml at this dilution.

a. 2.43 ppm
b. 24.3 ppm
c. 112 ppm
! d. 243 ppm
e. $1.12 \times 10^3$ ppm

26. A solution of ethylene glycol ($C_2H_6O_2$) in water is 3.981 molar and has a density of 1.0296 g/ml. Calculate the percent, by weight, of ethylene glycol in the solution.

a. 3.867 %
b. 4.099 %
c. 15.14 %
! d. 24.00 %
e. 25.45 %

27. A solution of sodium nitrate, $NaNO_3$, in water is 5.181 molar and its density is 1.2680 g/ml. Calculate the percent, by weight, of sodium nitrate in the solution.

a. 7.939 %
b. 17.21 %
c. 24.47 %
d. 29.56 %
! e. 34.73 %

28. An aqueous solution of phosphoric acid, $H_3PO_4$, has a measured density of 1.2089 g/ml and is 5.257 molal. How many moles of $H_3PO_4$ are there in exactly one liter of this solution?

a. 0.4261 moles
! b. 4.194 moles
c. 4.349 moles
d. 5.152 moles
e. 6.355 moles

29. An aqueous solution of ethanol, $C_2H_5OH$, is 19.00% ethanol by mass and has a density of 0.9700 g/ml. Calculate the molality of the ethanol solution.

a. 4.000 m
b. 4.124 m
c. 4.252 m
! d. 5.092 m
e. 14.48 m

30. An aqueous solution of ethanol, $C_2H_5OH$, is 19.00% ethanol by mass and has a density of 0.9700 g/ml. Calculate the molarity of the ethanol solution.

! a. 4.001 M
b. 4.124 M
c. 4.252 M
d. 5.092 M
e. 14.48 M

31. An aqueous solution of glycerol, $C_3H_8O_3$, is 48.0% glycerol by mass and has a density of 1.120 g/ml. Calculate the molarity of the glycerol solution.

a. 12.2 M
! b. 5.84 M
c. 0.584 M
d. 0.521 M
e. 0.465 M

32. An aqueous solution of glycerol, $C_3H_8O_3$, is 48.0% glycerol by mass and has a density of 1.120 g/ml. Calculate the molality of the glycerol solution.

a. 11.2 m
b. 5.84 m
c. 0.584 m
d. 0.521 m
! e. 10.0 m

33. A solution is prepared by mixing 0.3355 moles of $NaNO_3$ (84.995 g/mole) with 235.0 g of water (18.015 g/mole). Its density is 1.0733 g/ml. The solute is completely ionized. What is the percent by weight of $NaNO_3$ in the solution?

a. 10.16 %
! b. 10.82 %
c. 11.61 %
d. 14.19 %
e. 26.56 %

34. A solution is prepared by mixing 0.3355 moles of $NaNO_3$ (84.995 g/mole) with 235.0 g of water (18.015 g/mole). Its density is 1.0733 g/ml. The solute is completely ionized. What is the molality of the solution?

a. 0.6474 m
b. 0.7004 m
c. 1.320 m
! d. 1.428 m
e. 1.545 m

35. A solution is prepared by mixing 0.3355 moles of $NaNO_3$ (84.995 g/mole) with 235.0 g of water (18.015 g/mole). Its density is 1.0733 g/ml. The solute is completely ionized. What is the molarity of the solution?

a. 1.186 M
b. 1.273 M
c. 1.350 M
! d. 1.366 M
e. 1.428 M

36. An aqueous solution is prepared by mixing 0.2750 moles of NaOH (40.00 g/mole) with 189.0 g of water. Its density is 1.065 g/ml. The solute is completely ionized. What is the percent by weight of NaCl in the solution?

a. 0.001375 %
! b. 5.500 %
c. 5.858 %
d. 13.75%
d. 20.00 %

37. An aqueous solution of nitric acid has a density of 1.084 g/ml and a measured concentration of 2.580 molar. What is the percent by weight of nitric acid in the solution?

a. 2.380 %
b. 13.82 %
! c. 15.00 %
d. 22.29 %
e. 44.38 %

38. The vapor pressure of a solution containing a nonvolatile solute is directly proportional to the

! a. mole fraction of the solvent
b. mole fraction of the solute
c. molality of the solvent
d. molarity of the solvent
e. osmotic pressure of the solute

39. When a solute such as ammonium sulfate is dissolved in a solvent like water, one of the observed effects is

! a. a decrease in the vapor pressure of the solvent.
b. an increase in the vapor pressure of the solute.
c. an increase in the melting point of the liquid.
d. a decrease in the boiling point of the liquid.
e. scattering of light beams by the solute particles in the solution

40. At 23.0 °C, the vapor pressure of acetonitrile, $CH_3CN$, is 81.0 torr while that of acetone, $C_3H_6O$, is 184.5 torr. What is the vapor pressure of a solution which contains 0.550 moles of acetonitrile and 0.350 moles of acetone? (Assume the mixture behaves as an ideal solution.)

a. 109 torr
! b. 121 torr
c. 130 torr
d. 144 torr
e. 239 torr

41. At 28.0 °C, the vapor pressure of *n*-propyl mercaptan, $C_3H_7SH$, is 175 torr, while that of acetonitrile, $CH_3CN$, is 102 torr. What is the vapor pressure, at 28.0 °C, of a solution made by mixing 100.0 g of $C_3H_7SH$ and 100.0 g $CH_3CN$ if Raoult's Law is obeyed?

a. 35.7 torr
! b. 128 torr
c. 139 torr
d. 149 torr
e. 277 torr

42. At 28.0 °C, the vapor pressure of *n*-propyl mercaptan, $C_3H_7SH$, is 175 torr, while that of acetonitrile, $CH_3CN$, is 102 torr. What is the vapor pressure, at 28.0 °C, of a solution made by mixing 120.0 g of $C_3H_7SH$ and 80.0 g $CH_3CN$ if Raoult's Law is obeyed?

a. 131 torr
! b. 135 torr
c. 142 torr
d. 146 torr
e. 277 torr

43. At 28.0 °C, the vapor pressure of acetonitrile, $CH_3CN$, is 102.0 torr while that of acetone, $C_3H_6O$, is 228.9 torr and for $CS_2$ the value is 378.7 torr. A three component solution is made by adding 0.300 moles of $CH_3CN$ and 0.400 moles of $C_3H_6O$ to 0.350 moles of $CS_2$. The mixture behaves as an ideal mixture. What is the vapor pressure of the solution.

a. 203 torr
! b. 243 torr
c. 262 torr
d. 275 torr
e. 610 torr

44. At 28.0 °C, the vapor pressure of *n*-propyl mercaptan, $C_3H_7SH$, is 175 torr, while that of acetonitrile, $CH_3CN$, is 102 torr. What is the vapor pressure, at 28.0 °C, of a solution made by mixing 80.0 g of $C_3H_7SH$ and 120.0 g $CH_3CN$ if Raoult's Law is obeyed?

! a. 121 torr
b. 131 torr
c. 139 torr
d. 146 torr
e. 156 torr

45. At 28.0 °C, the vapor pressure of pure carbon disulfide, $CS_2$, is 378.7 torr, while that of acetone, $C_3H_6O$, is 228.9 torr. What is the vapor pressure, at 28.0 °C, of a solution made by mixing 0.250 moles of carbon disulfide and 0.450 moles of acetone if Raoult's Law is obeyed?

a. 198 torr
b. 228 torr
! c. 282 torr
d. 325 torr
e. 425 torr

46. Which one of the following listed properties of a solution is not a colligative property?

! a. solubility, in grams of solute per 100 grams of solvent
b. freezing-point depression
c. boiling point elevation
d. osmotic pressure
e. vapor pressure lowering

47. What is the freezing point of an aqueous 1.00 molal CsCl solution?

a. 0.00 °C
b. -0.93 °C
c. +1.86 °C
d. -1.86 °C
! e. -3.72 °C

48. What is the freezing point of a solution that contains 25.0 g of fructose, $C_6H_{12}O_6$, in 250.0 g of $H_2O$? $K_f$ = 1.86 °C/m.

a. -0.10 °C
b. +0.10 °C
c. -0.186 °C
d. +0.186 °C
! e. -1.03 °C

49. A non-ionic solute with a molecular weight of 50.0 grams/mole is dissolved in 500 g of water and the resulting solution has a boiling point of 101.53 °C. How many grams of solute were in the solution?

a. 30.0 grams
! b. 75.0 grams
c. 100 grams
d. 125 grams
e. 150 grams

50. What is the freezing point of an aqueous solution of a nonvolatile solute that has a boiling point of 102.45° C? $K_f$ = 1.86 °C/m and $K_b$ = 0.51 °C/m.

a. -0.67 °C
b. -0.99 °C
c. -2.45 °C
d. -2.99 °C
! e. -8.94 °C

51. Pure benzene, $C_6H_6$, has a molecular weight of 78.114 g/mole, a density of 0.8765 g/ml, a freezing point of 5.45 °C, and a boiling point of 80.2 °C. Its freezing point depression and boiling point elevation constants are: $K_f$ = 5.07 °C/m; $K_b$ = 2.53 °C/m. A solution was made by taking 24.20 g of an unknown non-electrolyte and dissolving it in 125.0 g of benzene. The measured freezing point of the solution was -1.65 °C. Calculate the molecular weight of the unknown substance.

! a. 138 g/mole
b. 145 g/mole
c. 258 g/mole
d. 272 g/mole
e. 595 g/mole

52. Pure cyclohexane, $C_6H_{12}$, has a molecular weight of 84.161 g/mole, a density of 0.7785 g/ml, a freezing point of 6.53 °C, and a boiling point of 80.72 °C. Its freezing point depression and boiling point elevation constants are: $K_f$ = 20.0 °C/m; $K_b$ = 2.69 °C/m. A solution was made by taking 11.40 g of an unknown non-electrolyte and dissolving it in 150.0 g of cyclohexane. The measured freezing point of the solution was -0.78 °C. Calculate the molecular weight of the unknown substance.

a. 27.8 g/mole
b. 46.8 g/mole
! c. 208 g/mole
d. 264 g/mole
e. 1949 g/mole

53. Pure cyclohexane, $C_6H_{12}$, has a molecular weight of 84.161 g/mole and a density of 0.7785 g/ml, a freezing point of 6.53 °C, and a boiling point of 80.72 °C. Its freezing point depression and boiling point elevation constants are: $K_f$ = 20.0 °C/m; $K_b$ = 2.69 °C/m. A solution was made by taking 18.55 g of an unknown non-electrolyte and dissolving it in 150.0 g of cyclohexane. The measured freezing point of the solution was -4.28 °C. Calculate the molecular weight of the unknown substance.

a. 61.8 g/mole
b. 66.8 g/mole
! c. 229 g/mole
d. 578 g/mole
e. 1099 g/mole

54. Pure glacial acetic acid, $HC_2H_3O_2$, has a molecular weight of 60.052 g/mole and a density of 1.0492 g/ml, a freezing point of 16.62 °C, and a boiling point of 118.3 °C. Its freezing point depression and boiling point elevation constants are: $K_f$ = 3.57 °C/m; $K_b$ = 3.07 °C/m. A solution was made by taking 9.755 g of an unknown non-electrolyte and dissolving it in 90.50 g of glacial acetic acid. The measured freezing point of the solution was 8.64 °C. Calculate the molecular weight of the unknown substance.

a. 24.4 g/mole
b. 30.5 g/mole
c. 45.3 g/mole
! d. 48.2 g/mole
e. 174 g/mole

55. Pure benzene, $C_6H_6$, has a molecular weight of 78.114 g/mole, a density of 0.8765 g/ml, a freezing point of 5.45 °C, and a boiling point of 80.2 °C. Its freezing point depression and boiling point elevation constants are: $K_f$ = 5.07 °C/m; $K_b$ = 2.53 °C/m. A solution was made by taking 10.00 g of an unknown non-electrolyte and dissolving it in 105.0 g of benzene. The measured freezing point of the solution was -1.05 °C. Calculate the molecular weight of the unknown substance

! a. 74.3 g/mole
b. 82.7 g/mole
c. 110 g/mole
d. 122 g/mole
e. 460 g/mole

56. Pure chloroform, $CHCl_3$, has a molecular weight of 119.38 g/mole, a density of 1.4980 g/ml, a freezing point of -63.59 °C and a boiling point of 61.23 °C. Its boiling point elevation constant, $K_b$, is 3.63 °C/m. A solution was made by taking 11.25 g of an unknown non-electrolyte and dissolving it in 115.5 g of chloroform. The measured boiling point of the solution was 65.46 °C. Calculate the molecular weight of the unknown substance.

! a. 83.6 g/mole
b. 113 g/mole
c. 114 g/mole
d. 120 g/mole
e. 158 g/mole

57. A solution contains 221 g of glycerol ($C_3H_8O_3$) in 600 grams of water. The $K_f$ is 1.86 °C/m and $K_b$ = 0.51 °C/m. What should the boiling point of the solution be?

a. 100.02 °C
b. 100.73 °C
c. 101.65 °C
! d. 102.04 °C
e. 103.62 °C

58. An aqueous ethylene glycol solution being considered for use as a radiator coolant is 16.0 % $C_2H_6O_2$ by weight. What would be the expected boiling point of this solution? For water, $K_f$ is 1.86 °C/m and $K_b$ = 0.51 °C/m.

a. 101.3 °C
! b. 101.6 °C
c. 105.1 °C
d. 106.0 °C
e. 156.5 °C

59. A solution, which was made by dissolving 62.07 g of a non-electrolyte in 500 g of water exhibits a freezing point of -1.86 °C. What is the molecular weight of this nonelectrolyte compound? For water, $K_f$ is 1.86 °C/m and $K_b$ = 0.51 °C/m.

a. 57.7 g/mole
b. 62.07 g/mole
c. 115 g/mole
! d. 124 g/mole
e. 231 g/mole

60. How many moles of the nonelectrolyte, propylene glycol ($C_3H_8O_2$) should be dissolved in 800 g of water ro prepare a solution whose freezing point is -3.72 °C? For water, $K_f$ is 1.86 °C/m and $K_b$ = 0.51 °C/m.

! a. 1.60 moles
b. 2.00 moles
c. 2.50 moles
d. 2.98 moles
e. 4.65 moles

61. How many grams of glycerol ($C_3H_8O_3$, a nonelectrolyte) should be dissolved in 600 g of water to prepare a solution whose freezing point is -4.65 °C? For water, $K_f$ is 1.86 °C/m and $K_b$ = 0.51 °C/m.

a. 22.1 grams
b. 93.6 grams
! c. 138 grams
d. 384 grams
e. 478 grams

62. During osmosis:

! a. Pure solvent diffuses through a membrane but solutes do not.
b. Pure solutes diffuse through a membrane but solvent does not.
c. Pure solvent moves in one direction through the membrane while the solution moves through the membrane in the other direction.
d. Pure solvent moves in one direction through the membrane while the solute moves through the membrane in the other direction.
e. Pure solute moves in one direction through the membrane while the solution moves through the membrane in the other direction.

63. Which of the equations below can be applied to calculation of osmotic pressure in solutions?

a. de Broglie equation
b. Clapeyron equation
c. Peters equation
d. van der Waals equation
! e. van't Hoff equation

64. A very dilute solution contains 116 mg of fructose (M = 180.16 g/mole) in 1.000 liter of solution. It is placed in a semipermeable membrane bladder which is then suspended by a support in some pure water. What osmotic pressure would develop across the membrane if the temperature is 26.0 °C?

a. 3.36 torr
! b. 12.0 torr
c. 151 torr
d. 475 torr
e. 1217 torr

65. An aqueous solution made by dissolving 168 mg of an unknown compound ( a nonelectrolyte) in enough water to make 500.0 ml of solution registered an osmotic pressure of 5.22 torr when the temperature was 23.5 °C. What is the molecular weight of this unknown compound?

a. 94.3 g/mole
b. 124 g/mole
c. 943 g/mole
! d. 1.19 x $10^3$ g/mole
e. 1.57 x $10^3$ g/mole

66. A solution is prepared by mixing 0.3355 moles of $NaNO_3$ (84.995 g/mole) with 235.0 g of water (18.015 g/mole). Its density is 1.0733 g/ml. The solute is completely ionized. If $K_f$ = 1.86 °C/m, what is the expected freezing point of the solution?

a. -2.65 °C
b. -2.87 °C
! c. -5.31 °C
d. -5.75 °C
e. -7.97 °C

67. Which one of the following aqueous solutions will have the lowest freezing point temperature?

a. 0.100 molal NaBr(*aq*)
b. 0.100 molal $MgSO_4$(*aq*)
c. 0.150 molal $KClO_3$(*aq*)
! d. 0.150 molal $MgCl_2$(*aq*)
e. 0.250 molal sucrose(*aq*)

68. Which one of the following aqueous solutions will have the highest freezing point temperature?

a. 0.100 molal $MgBr_2$(*aq*)
! b. 0.100 molal $MgSO_4$(*aq*)
c. 0.150 molal $KClO_3$(*aq*)
d. 0.150 molal $MgCl_2$(*aq*)
e. 0.250 molal sucrose(*aq*)

69. Which one of the following aqueous solutions will have the highest boiling point temperature?

a. 0.100 molal $NiBr_2$(*aq*)
b. 0.100 molal $MgSO_4$(*aq*)
c. 0.150 molal $NH_4NO_3$(*aq*)
! d. 0.150 molal $Na_2SO_4$(*aq*)
e. 0.250 molal methanol, $CH_3OH$(*aq*)

70. Which one of the following aqueous solutions will have the highest boiling point temperature?

a. 0.100 molal $NiBr_2$(*aq*)
b. 0.200 molal $MgSO_4$(*aq*)
c. 0.150 molal $NH_4NO_3$(*aq*)
d. 0.150 molal $Na_2SO_4$(*aq*)
! e. 0.250 molal methanol, $CH_3OH$(*aq*)

71. Which one of the following aqueous solutions has the highest boiling point? $K_f$ = 0.51 °C/m.

a. 0.200 m KCl
b. 0.200 m $Na_2SO_4$
c. 0.200 m $Ca(NO_3)_2$
d. 0.200 m glycerol ($C_3H_8O_3$)
! e. 0.200 m $Na_3PO_4$

72. Which one of the following solutions has the highest osmotic pressure at 25 °C?

a. 0.200 M $(NH_4)_2SO_4$
b. 0.200 M $KClO_3$
c. 0.200 M ethanol ($C_2H_6O$)
d. 0.200 M $NiF_2$
! e. 0.200 M $Fe(NO_3)_3$

73. Arrange the following aqueous solutions in order of increasing boiling points: 0.100 m $Mg(NO_3)_2$; 0.200 m ethylene glycol, $C_2H_6O_2$; 0.175 m LiCl.

! a. ethylene glycol < $Mg(NO_3)_2$ < LiCl
b. $Mg(NO_3)_2$ < LiCl < ethylene glycol
c. ethylene glycol < LiCl < $Mg(NO_3)_2$
d. LiCl < ethylene glycol < $Mg(NO_3)_2$
e. $Mg(NO_3)_2$ < ethylene glycol < LiCl

74. A solution was made by mixing 24.40 g of NaCl, 19.77 g of $MgCl_2$, 38.22 g of $Na_2SO_4$ and 1.665 kg of water. It had a measured density of 1.042 g/ml. What is the colligative molality (moles of particles per kilogram of solvent) of the solution?

a. 0.5371 molal
! b. 1.360 molal
c. 1.418 molal
d. 1.452 molal
e. 4.297 molal

75. A solution was made by mixing 24.40 g of NaCl, 19.77 g of $MgCl_2$, 38.22 g of $Na_2SO_4$ and 1.665 kg of water. It had a measured density of 1.042 g/ml. What is the expected freezing point of the solution?

a. -1.00 °C
! b. -2.53 °C
c. -2.64 °C
d. -2.70 °C
e. -7.99 °C

76. A solution is made by dissolving 0.840 moles of sodium hydroxide in 300.0 g of water. If the i factor for this particular concentration is 1.70, what is the expected freezing point of this solution? $K_f$ = 1.86 °C/m.

a. -0.80 °C
b. -2.66 °C
c. -3.06 °C
! d. -8.85 °C
e. -9.97 °C

77. 9.92 grams of a compound with a molecular weight of 124.0 grams/mole when dissolved in 150.0 grams of water gave a solution which had a freezing point of -2.69 °C. Calculate the experimental value of the i factor

a. 1.21
b. 1.45
c. 1.54
d. 2.69
! e. 2.71

78. A solution is made by dissolving 1.25 moles of magnesium nitrate in 300.0 g of water. If the i factor for this particular concentration is 2.25, what is the expected freezing point of this solution? $K_f$ = 1.86 °C/m.

a. -0.45 °C
b. -1.57 °C
c. -3.44 °C
d. -5.23 °C
! e. -17.4 °C

79. An aqueous solution which is 12.00% sodium hydroxide by weight has a freezing point of -10.40 °C. What is the observed value of the i factor in this solution? $K_f$ = 1.86 °C/m.

a. 1.59
b. 1.86
! c. 1.64
d. 1.75
e. 1.70

80. Which one of the following aqueous solutions has the highest osmotic pressure?

a. 0.100 molar $Al(NO_3)_3$
! b. 0.150 molar $Ba(NO_3)_2$
c. 0.100 molar $CaCl_2$
d. 0.150 molar NaCl
e. 0.200 molar $NH_3$

81. A dilute aqueous solution of $CaCl_2$ contains 0.159 grams of solute per liter of solution. It is fully dissociated. What is the osmotic pressure, in torr, at 20.0 °C?

a. 1.79 torr
b. 3.49 torr
c. 26.2 torr
! d. 78.6 torr
e. $2.65 \times 10^3$ torr

82. An aqueous solution of calcium chloride contains 32.5 parts per million of calcium and is fully dissociated. What is its osmotic pressure, in torr, at 20.0 °C?

a. 14.8 torr
b. 26.2 torr
c. 29.7 torr
! d. 44.5 torr
e. 97.5 torr

83. A dilute aqueous solution of potassium sulfate contains 8.84 parts per million of sulfur. If the solution is 100% dissociated and the temperature is 25.0 °C, what is the osmotic pressure of the solution, in torr?

a. 5.1 torr
b. 10.3 torr
c. 11.2 torr
! d. 15.4 torr
e. 18.8 torr

84. When a liquid is dispersed in another liquid, the resulting colloid is called a(n)

a. aerosol
! b. emulsion
c. foam
d. gel
e. sol

## True and False

85. One driving force toward formation of homogeneous gas mixtures is the spontaneity of random mixing through random motion of small molecules. ( !T)

86. Liquids which are mutually miscible possess intermolecular forces of similar type and magnitude. ( !T)

87. The process, $McSO_4(s) \longrightarrow McSO_4(aq)$, is an endothermic process. The solubility of maclurium sulfate, $McSO_4$, in water should increase as the solvent temperature is increased ( !T)

88. Colligative properties are similar in that they all depend on the concentration of solute particles in the solution. ( !T)

## Fill in the Blanks

89. An aqueous solution of calcium chloride contains 420 parts per million of calcium. Calculate the molarity of this solution. (Assume the density of so dilute a solute is 1.000 g/ml) ________ ( ! $1.05 \times 10^{-2}$ M)

90. An aqueous solution of calcium hydroxide is $2.00 \times 10^{-2}$ molar. Calculate how many part per million of calcium are present in this solution. (Assume the density of so dilute a solution is 1.000 g/ml) ________ ( ! 802 ppm)

91. A solution contains 25.50 grams of $NaNO_3$ (M = 84.99 g/mole) in 250.0 grams of water. Its density is 1.0620 g/ml. Calculate the molality of the solution. ________ ( ! 1.200 m)

92. What is the mole fraction of ethylene glycol, $C_2H_6O_2$, in an aqueous solution which is 50.0% ethylene glycol by mass? ( ! 0.225)

93. A solution made by dissolving 16.25 grams of p-toluic acid ($C_7H_7CO_2H$), 136.15 g/mole) in 125.5 ml of benzene (f.p. = 5.45 °C, $K_f$ = 5.07 °C/m) gave a freezing point of 3.00 °C. What is the calculated molecular weight of $C_7H_7CO_2H$ from this data? ________ ( ! 268 g/mole)

94. A solution made by dissolving 11.65 grams of butyric acid ($C_3H_7CO_2H$), 88.106 g/mole) in 110.5 ml of benzene (f.p. = 5.45 °C, $K_f$ = 5.07 °C/m) exhibited a freezing point of 2.40 °C. What is the calculated molecular weight of $C_3H_7CO_2H$ from this data? ________ (! 175 g/mole)

## Critical Thinking

95. A very dilute solution contains sodium chloride and magnesium chloride as the only solutes. There are 8.05 parts per million of sodium and 6.08 parts per million of magnesium in the solution. If the solutes are fully dissociated, calculate the osmotic pressure of this solution, in torr, at 25.0 °C. ________ (! 27.0 torr)

96. An aqueous solution of fructose, $C_6H_{12}O_6$, has an osmotic pressure of 10.50 atm at 25 °C? What is the boiling point this solution? $K_f$ = 1.86 °C/m and $K_b$ = 0.51 °C/m. The density of the solution is 1.077 g/ml.

! a. 100.22 °C
b. 100.41 °C
c. 100.43 °C
d. 100.52 °C
e. 100.75 °C

97. A solution was made by mixing 24.40 g of NaCl, 19.77 g of $MgCl_2$, 38.22 g of $Na_2SO_4$ and 1.665 kg of water. It had a measured density of 1.042 g/ml. What is the chloride ion concentration, in moles per liter?

a. 0.2476 molar
b. 0.3728 molar
c. 0.3755 molar
! d. 0.4966 molar
e. 0.5002 molar

98. A solution was made by mixing 24.40 g of NaCl, 19.77 g of $MgCl_2$, 38.22 g of $Na_2SO_4$ and 1.665 kg of water. It had a measured density of 1.042 g/ml. What is the colligative molarity (moles of particles per liter of solution) of the solution?

a. 0.5332 molar
b. 0.5371 molar
! c. 1.351 molar
d. 1.360 molar
e. 4.266 molar

99. An aqueous solution of glycerol, $C_3H_8O_3$, in which the mole fraction of $C_3H_8O_3$ is 0.07070 was found to have a density of 1.0350 g/ml by direct measurement. What is the molarity of this solution?

! a. 3.147 molar
b. 4.371 molar
c. 4.223 molar
d. 2.938 molar
e. 3.651 molar

100. A solution is made by dissolving 48.07 g of $MgSO_4 \cdot 7H_2O$ in 250.0 grams of water. What is the expected freezing point of this solution if the i factor is 1.90? $K_f$ = 1.86 °C/m.

_______ ( -2.51 °C)

101. A solution is made by mixing acetone and methyl alcohol, whose structures are shown as I and II below. We took 0.250 moles of acetone and 0.750 moles of methyl alcohol. Which statement below describes the most probable state of affairs?

I. 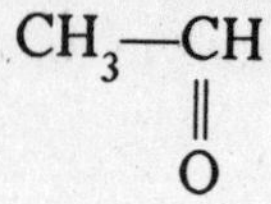

II. $CH_3$—O—H

a. Since the intermolecular forces for acetone—acetone, methanol—methanol, and acetone—methanol interactions are comparable, the mixture behaves as an ideal solution.

b. Since the intermolecular forces for acetone—methanol interactions are weaker than either the acetone—acetone or methanol—methanol interactions we should observe a positive deviation from ideal solution behavior.

! c. Since the intermolecular forces for acetone—methanol and methanol—methanol interactions are stronger than the acetone—acetone interactions due to hydrogen bonding, the observed vapor pressure would be smaller than the theoretical vapor pressure and a negative deviation would be observed.

d. Since the intermolecular forces for methanol—methanol interactions are weaker than either the acetone—acetone or acetone—methanol interactions due to hydrogen bonding, the vapor pressure would be greater than expected and we should observe a positive deviation from ideal solution behavior.

e. Since the intermolecular forces for methanol—methanol interactions are weaker than either the acetone—acetone or acetone—methanol interactions due to hydrogen bonding, the vapor pressure would be less than expected and we should observe a negative deviation from ideal solution behavior.

## Chemicals in Use 10: Concerning Pure Water

102. The purest rain water from rain falling through a clean atmosphere to the surface in remote areas of the Southwest Pacific is

a. neutral, in the acid base sense
b. slightly basic due to the effect of dissolved hydrocarbons
c. slightly acidic due to the effect of dissolved carbon monoxide
! d. slightly acidic due to the effect of dissolved carbon dioxide
e. slidhtly acidic due to the effect of dissolved hydrocarbons

103. Gases produced by the numerous industrial plants and internal combustions of aircraft and wheeled vehicles of many kinds and sizes in industrialized countries are thought to be largely responsible for acid rain pollution. The major gases are

a. carbon monoxide, carbon dioxide and sulfur dioxide
b. carbon dioxide, sulfur dioxide and phosphorus oxides
c. sulfur dioxide, phosphorus oxides and chlorine oxides
! d. nitrogen oxides and sulfur dioxide
e. chlorine oxides, carbon dioxide and sulfur dioxide

104. The best and most cost effective way to control acid rain is to

a. locate offending industrial machines in places with minimal annual rainfall
b. remove the elements which give rise to the offending gases from the fuels prior to their combustion
! c. remove the offending gases produced by the fuels from the exhaust prior to their release into the atmosphere
d. recirculate the exhaust gases through the engines so they have additional opportunity to consume these substances
e. use reducing agents in the exhaust stream to change the oxidation state of the elements in the offending gases and greatly reduce their acidity

105. Normally, water from even the purest of natural sources is slightly acidic due to

a. calcium carbonate which is dissolved in the water
b. interaction of water with limestone in deep soil and wells
c. interaction of pure ground water with aluminum containing silicate compounds in the deep water table
! d. carbon dioxide dissolved in rain water as it falls through the air
e. interaction of the water with marble chips in the deep water table

106. Which statement is true about sulfur dioxide which eventually affects water in the environment?

a. A principal source is the $SO_2$ gases trapped in oil wells which is released when oil and natural gas are pumped out.
! b. Sulfur dioxide dissolves in water by forming hydrate, $SO_2 \cdot nH_2O$.
c. Oxygen and ozone are both cpable of converting $SO_3$ to $SO_2$.
d. $SO_2$ does not contribute to the acidity of rain, because it is a reducing agent.
e. Less than a few million tons of $SO_2$ are produced yearly from fossil fuel combustion.

107. The threat posed by a certain mercury species, $CH_3Hg^+$, is due to its ability to selectively concentrate in the food chain to a far greater extent than its general concentration in the aquatic environment. ( !T)

108. All of the following gases are responsible for acid rain except

a. $CO_2(g)$
b. $SO_2(g)$
c. $NO_2(g)$
d. $SO_3(g)$
! e. $NF_3(g)$

109. A general term which encompasses all methods by which acidic materials enter the surface environment of the earth is

a. acid rain
! b. acid deposition
c. acid hydration
d. acid volatization
e. acid concentration

110. Mercury in the aquatic environment enters it from

a. limestone and marble containing some $Hg^{2+}$ ions in place of the $Ca^{2+}$ ions
! b. natural sources and combustion products of coal which contained mercury compounds
c. decomposition of algae and mold in sluicewys of industrial plants
d. aquatic bacteria
e. brain and nerve tissue in aquatic plants and animals

111. Mercury persists for a very long time in the aquatic environment because

a. mercurous chloride (formed with the chloride ion in water) is a very inert substance
b. mercury is one of the least reactive of the metals
c. mercury compounds are unpalatable and hence living organisms tend to ignore them as possible food
! d. once there, mercury compounds continue to be recycled into the aquatic environment by natural processes
e. mercury compounds act as a preservative and this in turn means they will persist within the preserved tissue

112. What form of mercury binds to proteins in cells rendering them incapable of carrying our their functions and thus posing a threat to humans?

! a. $CH_3Hg^+$
b. $HgCO_3$
c. $Hg_2CO_3$
d. $Hg(NO_3)_2$
e. $Hg(l)$, the elemental state

## Chapter 13 Thermodynamics

### Multiple Choice

1. The standard enthalpy of reaction, $\delta H^{o}_{rxn}$, for $C_2H_2(g)$ + $2\ H_2(g) \longrightarrow C_2H_6(g)$ is -311.5 kJ/mole. Determine the value of $\delta E^{o}_{rxn}$ for this reaction.

! a. -306.5 kJ/mole
b. -309.0 kJ/mole
c. -314.0 kJ/mole
d. -316.46 kJ/mole
e. 4646 kJ/mole

2. The standard enthalpy of reaction, $\delta H^{o}_{rxn}$, for $NH_3(g)$ + $HCl(g) \longrightarrow NH_4Cl(s)$ is -175.9 kJ/mole. Determine the value of $\delta E^{o}_{rxn}$ for this reaction.

a. -164.8 kJ/mole
! b. -170.9 kJ/mole
c. -173.4 kJ/mole
d. -180.9 kJ/mole
e. 5134 kJ/mole

3. The standard enthalpy of reaction, $\delta H^{o}_{rxn}$, for the reaction, $C_4H_{10}(g)$ + $\frac{13}{2}\ O_2(g) \longrightarrow$ $4\ CO_2(g)$ + $5\ H_2(l)$, is -2877 kJ/mole. Determine the value of $\delta E^{o}_{rxn}$ for this reaction.

! a. -2871 kJ/mole
b. -2880 kJ/mole
c. -2883 kJ/mole
d. -3320 kJ/mole
e. 3320 kJ/mole

4. The standard enthalpy of reaction, $\delta H^{o}_{rxn}$, for the reaction, $CaO(s)$ + $SO_3(g) \longrightarrow$ $CaSO_4(s)$, is -401.5 kJ/mole. Determine the value of $\delta E^{o}_{rxn}$ for this reaction.

a. -362.2 kJ/mole
b. -399.0 kJ/mole
! c. -404.0 kJ/mole
d. -2880 kJ/mole
e. 2077 kJ/mole

5. Which one of the processes below is the one which is accompanied by an increase in entropy?

a. setting up a stack of dominos
b. setting up decorations on a Christmas tree
c. filing correspondence in file folders and placing them in hanging file folders
! d. dropping a glass pane on the front walk of your residence
e. restocking a canned goods shelf display in a supermarket

6. Which one of the following reactions is accompanied by an *increase* in entropy?

a. $ZnS(s) + \frac{3}{2} O_2(g) \longrightarrow ZnO(s) + SO_2(g)$
! b. $CH_4(g) + H_2O(g) \longrightarrow CO(g) + 3H_2(g)$
c. $BaO(s) + CO_2(g) \longrightarrow BaCO_3(s)$
d. $Na_2CO_3(s) + CO_2(g) + H_2O(g) \longrightarrow 2\,NaHCO_3(s)$
e. $N_2(g) + 3\,H_2(g) \longrightarrow 2\,NH_3(g)$

7. Which one of the following reactions is accompanied by an *increase* in entropy?

a. $2\,H(g) \longrightarrow H_2(g)$
b. $NiCl_2(s) + 6\,NH_3(g) \longrightarrow NiCl_2{\cdot}6NH_3(s)$
! c. $I_2(g) \longrightarrow 2\,I(g)$
d. $ZnO(s) + CO_2(g) \longrightarrow ZnCO_3(s)$
e. $C_2H_4(g) + Cl_2(g) \longrightarrow C_2H_4Cl_2(l)$

8. Which one of the following reactions is accompanied by an *increase* in entropy?

a. $C_{12}H_{20}(l) + 17\,O_2(g) \longrightarrow 12\,CO_2(g) + 10\,H_2O(l)$
! b. $NH_4Cl(s) \longrightarrow NH_3(g) + HCl(g)$
c. $2\,C_2H_2(g) + 5\,O_2(g) \longrightarrow 4\,CO_2(g) + 2\,H_2O(s)$
d. $Ba(OH)_2(s) + 2\,HCl(g) \longrightarrow BaCl_2{\cdot}2H_2O(s)$
e. $(CH_3)_2CO(l) + 4\,O_2(g) \longrightarrow 3\,CO_2(g) + 3\,H_2O(l)$

9. Which one of the following reactions is accompanied by an *increase* in entropy?

a. $C_8H_{16}(l) + 12\,O_2(g) \longrightarrow 8\,CO_2(g) + 8\,H_2O(l)$
b. $N_2(g) + 3\,H_2(g) \longrightarrow 2\,NH_3(g)$
c. $2\,C_2H_2(g) + 5\,O_2(g) \longrightarrow 4\,CO_2(g) + 2\,H_2O(s)$
d. $Ba(OH)_2(s) + CO_2(g) \longrightarrow BaCO_3(s) + H_2O(l)$
! e. $NH_4NO_2(s) \longrightarrow N_2(g) + 2\,H_2O(l)$

10. Which one of the following statements is true?

a. Spontaneous changes are ***always*** accompanied by an increase in the entropy of the system.
b. Spontaneous changes are ***always*** accompanied by a decrease in the entropy of the system.
c. Spontaneous changes are ***always*** accompanied by an increase in the enthalpy of the system.
d. Spontaneous changes are ***always*** accompanied by a decrease in the enthalpy of the system.
! e. Most highly exothermic chemical reactions are also spontaneous chemical reactions.

11. Which one of the following processes is accompanied by a decrease in the entropy of the system?

a. the mixing of one liter of water with one liter of ethylene glycol to produce one liter of an antifreeze solution.
b. the breaking of a large rock into very many smaller pieces of crushed gravel.
c. the thawing of the frozen orange juice concentrate which was left in the car.
d. the spontaneous chemical reaction of TNT (a solid chemical compound) wherein it decomposes into several simple compounds, some of which are gaseous.
! e. the absorption of odorous gaseous compounds by the charcoal filter in your home central air cleaning unit.

12. Which statement below is always true for a spontaneous chemical reaction

a. $\delta S_{sys} + \delta S_{surr} = 0$
b. $\delta S_{sys} + \delta S_{surr} < 0$
! c. $\delta S_{sys} + \delta S_{surr} > 0$
d. $\delta S_{sys} - \delta S_{surr} = 0$
e. $\delta S_{sys} - \delta S_{surr} < 0$

13. For a chemical reaction, $\delta H$ is $< 0$ and $\delta S$ is $< 0$ also. This means that

a. we conclude the reaction must be spontaneous regardless of temperature and becomes even more so at higher temperatures.
b. we conclude the reaction must be spontaneous regardless of temperature and becomes even more so at lower temperatures.
! c. we conclude the reaction may or may not be spontaneous, but spontaneity is favored, or at least enhanced, by low temperatures rather than by high temperatures.
d. we conclude the reaction may or may not be spontaneous, but spontaneity is favored, or at least enhanced, by high temperatures rather than by low temperatures.
e. we cannot make any conclusion about spontaneity or even tendencies from the limited information presented.

14. For a chemical reaction, $\delta H$ is $> 0$ while $\delta S$ is $< 0$. This means that

! a. we conclude the reaction must be non-spontaneous regardless of temperature and becomes even more so at higher temperatures.
b. we conclude the reaction must be spontaneous regardless of temperature and becomes even more so at lower temperatures.
c. we conclude the reaction may or may not be spontaneous, but spontaneity is favored, or at least enhanced, by low temperatures rather than by high temperatures.
d. we conclude the reaction may or may not be spontaneous, but spontaneity is favored, or at least enhanced, by high temperatures rather than by low temperatures.
e. we cannot make any conclusion about spontaneity or even tendencies from the limited information presented.

15. Of the species listed below, which one has the highest standard entropy ($S^o$)?

a. $(CH_3)_2CO(l)$
! b. $C_4H_{10}(g)$
c. $K_2SO_4(s)$
d. $H_2O(l)$
e. $Ar(g)$

16. Of the species listed below, which one has the lowest standard entropy (S°)?

a. $CH_4(g)$
b. $COCl_2(g)$
c. $NH_3(g)$
d. $H_2O(g)$
! e. $Ar(g)$

17. Of the species listed below, which one has the highest standard entropy (S°)?

a. $CH_3—CH_2—O—H(l)$
b. $SnCl_4(l)$
c. $TiCl_4(l)$
d. $H_2O(l)$
! e. $CH_3—CH_2—CH_2—CH_2—CH_2—Cl(l)$

18. Which one of the sets below has the species listed in order of increasing standard entropy, S°?

a. $Au(s) < C_6H_5NO_2(l) < CaCO_3(s) < H_2O(l)$
! b. $Au(s) < CaCO_3(s) < H_2O(l) < C_6H_5NO_2(l)$
c. $CaCO_3(s) < H_2O(l) < Au(s) < C_6H_5NO_2(l)$
d. $Au(s) < H_2O(l) < CaCO_3(s) < C_6H_5NO_2(l)$
e. $C_6H_5NO_2(l) < CaCO_3(s) < Au(s) < H_2O(l)$

19. Which one of the sets below has the species listed in order of increasing standard entropy, S°?

a. $NaHCO_3(s) < C_2H_5OH(l) < Cr(s) < N_2(g)$
b. $Cr(s) < N_2(g) < NaHCO_3(s) < C_2H_5OH(l)$
c. $Cr(s) < C_2H_5OH(l) < NaHCO_3(s) < N_2(g)$
! d. $Cr(s) < NaHCO_3(s) < C_2H_5OH(l) < N_2(g)$
e. $N_2(g) < NaHCO_3(s) < Cr(s) < C_2H_5OH(l)$

20. Which one of the sets below has the species listed in order of increasing standard entropy, S°?

! a. $CaSO_4(s) < CH_3CH_2—O—H(l) < Ar(g) < CH_3—CH_2—Cl(g)$
b. $CH_3CH_2—O—H(l) < Ar(g) < CaSO_4(s) < CH_3—CH_2—Cl(g)$
c. $CaSO_4(s) < CH_3—CH_2—Cl(g) < CH_3CH_2—O—H(l) < Ar(g)$
d. $CaSO_4(s) < Ar(g) < CH_3CH_2—O—H(l) < CH_3—CH_2—Cl(g)$
e. $CH_3—CH_2—Cl(g) < CH_3CH_2—O—H(l) < CaSO_4(s) < Ar(g)$

21. Using the standard entropy values: $H_2(g)$, S° = 130.6 J/mol K; $I_2(s)$, S° = 116.12 J/mol K; HI(g), S° = 206.0 J/mol K; calculate the standard entropy change, δS°, for the reaction:
$H_2(g) + I_2(s) \longrightarrow 2\,HI(g)$.

a. -40.8 kJ
b. +40.8 kJ
c. -165.3 kJ
! d. +165.3 kJ
e. 206.0 kJ

22. Using the standard entropy values: NO(*g*), S° = 210.6 J/mol K; $O_2$(*g*), S° = 205.0 J/mol K; $NO_2$(*g*), S° = 240.5 J/mol K; calculate the standard entropy change, δS°, for the reaction: NO(*g*) + $\frac{1}{2}$ $O_2$(*g*) ⟶ $NO_2$(*g*).

! a. -72.6 kJ
b. -175.1 kJ
c. -246.1 kJ
d. +246.1 kJ
e. -656.1 kJ

23. Using the standard entropy values: $C_2H_4$(*g*), S° = 219.8 J/mol K; $H_2$(*g*), S° = 130.6 J/mol K; $C_2H_6$(*g*), S° = 229.5 J/mol K; calculate the standard entropy change, δS°, for the reaction: $C_2H_4$(*g*) + $H_2$(*g*) ⟶ $C_2H_6$(*g*).

a. +89.2 kJ
b. -98.9 kJ
! c. -120.9 kJ
d. +140.3 kJ
e. +579.9 kJ

24. Using the standard entropy values: $SO_2$(*g*), S° = 248.0 J/mol K; $SO_3$(*g*), S° = 256.0 J/mol K; NO(*g*), S° = 210.6 J/mol K; $NO_2$(*g*), S° = 240.5 J/mol K; calculate the standard entropy change, δS°, for the reaction: $NO_2$(*g*) + $SO_2$(*g*) ⟶ NO(*g*) + $SO_3$(*g*).

a. -6.2 kJ
b. -18.8 kJ
! c. -21.9 kJ
d. -37.9 kJ
e. 52.9 kJ

25. The criterion for a spontaneous chemical reaction is

a. δG = 0
b. δH > 0
c. δS = 0
d. δE > 0
! e. δG < 0

26. As electrical energy is withdrawn from an automobile storage battery, which one of the following occurs?

a. δG decreases
! b. δG increases
c. δG does not change
d. δE increases
e. δH increases

27. Using the standard free energies of formation: $NO_2(g)$, $\delta G^o_f$ = +51.84 kJ/mol; $NO(g)$ $\delta G^o_f$ = +86.69 kJ/mol; $SO_2(g)$, $\delta G^o_f$ = -300.0 kJ/mol; $SO_3(g)$, $\delta G^o_f$ = -370.0 kJ/mol; calculate the standard free energy change, $\delta G^o$, for the reaction: $NO_2(g) + SO_2(g) \longrightarrow NO(g) + SO_3(g)$.

! a. -35.15 kJ
b. -104.9 kJ
c. -429.2 kJ
d. -619.6 kJ
e. -808.5 kJ

28. Using the standard free energies of formation: $BaCO_3(s)$, $\delta G^o_f$ = -1139.0 kJ/mol; $BaSO_4(s)$, $\delta G^o_f$ = -1353.0 kJ/mol; $CO_2(g)$, $\delta G^o_f$ = -394.4 kJ/mol; $SO_3(g)$, $\delta G^o_f$ = -370.0 kJ/mol; calculate the standard free energy change, $\delta G^o$, for the reaction: $BaCO_3(s) + SO_3(g) \longrightarrow BaSO_4(s) + CO_2(g)$.

a. +189.6 kJ
! b. -238.4 kJ
c. +472.4 kJ
d. +1727.6 kJ
e. -2516.4 kJ

29. Given the data:

$H_2(g)$, $\delta H^o_f$ = 0.00 kJ/mol, $S^o$ = 130.6 J/mol K;
$I_2(s)$, $\delta H^o_f$ = 0.00 kJ/mol, $S^o$ = 116.12 J/mol K;
$HI(g)$, $\delta H^o_f$ = +26 kJ/mol, $S^o$ = 206 J/mol K;

calculate the standard free energy change, $\delta G^o$, for the reaction: $H_2(g) + I_2(s) \longrightarrow 2\,HI(g)$.

! a. 2.7 kJ
b. -46.5 kJ
c. 64.1 kJ
d. 128.2 kJ
e. -165.3 kJ

30. Given the data:

$Ag_2O(s)$, $\delta H^o_f$ = -31.1 kJ/mol, $S^o$ = 121.3 J/mol K;
$O_2(g)$, $\delta H^o_f$ = 0.00 kJ/mol, $S^o$ = 205 J/mol K;
$Ag(s)$, $\delta H^o_f$ = 0.00 kJ/mol, $S^o$ = 42.55 J/mol K;

calculate the standard free energy change $\delta G^o$, for the reaction: $Ag_2O(s) \longrightarrow 2\,Ag(s) + \frac{1}{2}O_2(g)$.

a. 6.5 kJ
b. -7.4 kJ
c. -24.0 kJ
! d. -38.2 kJ
e. -68.7 kJ

31. Which one of observables associated with the following symbols is the one which is dependent on how a chemical reaction is carried out, not on just the initial and final states?

a. $\delta S$
b. $\delta H$
c. $\delta E$
! d. w
e. $\delta G$

32. Given the data:

$N_2(g)$, $\delta H^\circ_f$ = 0.00 kJ/mol, S° = 191.5 J/mol K;
$H_2(g)$, $\delta H^\circ_f$ = 0.00 kJ/mol, S° = 130.6 J/mol K;
$NH_3(g)$, $\delta H^\circ_f$ = -46.0 kJ/mol, S° = 192.5 J/mol K;

calculate the standard free energy change, $\delta G^\circ$ for the reaction: $N_2(g) + 3\ H_2(g) \longrightarrow 2\ NH_3(g)$

a. -7.4 kJ
! b. -32.9 kJ
c. -84.6 kJ
d. +112.3 kJ
e. -151.1 kJ

33. Given the data:

$NH_3(g)$, $\delta H^\circ_f$ = -46.0 kJ/mol, S° = 192.5 J/mol K;
$NO(g)$, $\delta H^\circ_f$ = +90.4 kJ/mol, S° = 210.6 J/mol K;
$H_2O(l)$, $\delta H^\circ_f$ = -286 kJ/mol, S° = 69.96 J/mol K;
$O_2(g)$, $\delta H^\circ_f$ = 0.00 kJ/mol, S° = 205 J/mol K;

calculate the standard free energy change, $\delta G^\circ$, for the reaction:

$2\ NH_3(g) + \frac{5}{2}O_2(g) \longrightarrow 2\ NO(g) + 3\ H_2O(g)$.

a. -100.8 kJ
b. -206.7 kJ
c. -276.5 kJ
! d. -505.8 kJ
e. -664.3 kJ

34. Assuming that the values of $\delta H$ and $\delta S$ do not change as we shift temperature, and using,

$CdO(s)$, $\delta H^\circ_f$ = -258.2 kJ/mol, S° = 54.8 J/mol K;
$CdSO_4(s)$, $\delta H^\circ_f$ = -933.5 kJ/mol, S° = 123 J/mol K;
$SO_3(g)$, $\delta H^\circ_f$ = -396 kJ/mol, S° = 256 J/mol K;

calculate a value for the free energy change, $\delta G^{1\ atm}$ for the reaction,

$CdSO_4(s) \longrightarrow CdO(s) + SO_3(g)$ at 750 K.

a. +223.3 kJ
! b. +262.6 kJ
c. +296.0 kJ
d. +313.3 kJ
e. +335.3 kJ

35. Assuming that the values of $\delta H$ and $\delta S$ do not change as we shift temperature, and using,

$CaO(s)$, $\delta H^\circ_f$ = -635.5 KJ/mol, S° = 40 J/mol K;
$CaCO_3(s)$, $\delta H^\circ_f$ = -1207 kJ/mol, S° = 92.9 J/mol K;
$CO_2(g)$, $\delta H^\circ_f$ = -394 kJ/mol, S° = 213.6 J/mol K;

calculate a value for the free energy change, $\delta G^{1\ atm}$ for the reaction,

$CaCO_3(s) \longrightarrow CaO(s) + CO_2(g)$ at 815 °C.

! a. +2.6 kJ
b. +465.0 kJ
c. -1182.7 kJ
d. +1182.7 kJ
e. -1255.5 kJ

36. Given the data:

$PbO(s)$, $\delta H_f^o$ = -217.3 kJ/mol, S° = 68.7 J/mol K;
$PbO_2(s)$, $\delta H_f^o$ = -277.0 kJ/mol, S° = 68.6 J/mol K;
$O_2(g)$, $\delta H_f^o$ = 0.00 kJ/mol, S° = 205 J/mol K;

calculate the standard free energy change, $\delta G°$, for the reaction: $PbO(s) + O_2(g) \longrightarrow PbO_2(s)$.

a. +1.6 kJ
! b. -28.9 kJ
c. -68.3 kJ
d. -90.5 kJ
e. -121.0 kJ

37 Assuming that the values of $\delta H$ and $\delta S$ do not change as we shift temperature, and using,

$N_2(g)$, $\delta H_f^o$ = 0.00 kJ/mol, S° = 191.5 J/mol K;
$H_2(g)$, $\delta H_f^o$ = 0.00 kJ/mol, S° = 130.6 J/mol K;
$NH_3(g)$, $\delta H_f^o$ = -46.0 kJ/mol, S° = 192.5 J/mol K;

calculate a value for the free energy change, $\delta G^{1\text{ atm}}$ for the reaction,

$N_2(g) + 3\,H_2(g) \longrightarrow 3\,NH_3(g)$ at 500 °C.

a. +7.2 kJ
b. +18.8 kJ
c. +54.2 kJ
! d. +61.3 kJ
e. +83.3 kJ

38. Assuming that the values of $\delta H$ and $\delta S$ do not change as we shift temperature, and using,

$NH_3(g)$, $\delta H_f^o$ = -46.0 kJ/mol, S° = 192.5 J/mol K;
$NO(g)$, $\delta H_f^o$ = +90.4 kJ/mol, S° = 210/6 J/mol K;
$H_2O(l)$, $\delta H_f^o$ = -286 kJ/mol, S° = 69.96 J/mol K;
$O_2(g)$, $\delta H_f^o$ = 0.00 kJ/mol, S° = 205 J/mol K;

calculate a value for the free energy change, $\delta G^{1\text{ atm}}$ for the reaction,

$2\,NH_3(g) + \frac{5}{2}\,O_2(g) \longrightarrow 2\,NO(g) + 3\,H_2O(g)$ at 350 °C.

a. -76.73 kJ
b. -108.7 kJ
c. -492.0 kJ
! d. -419.2 kJ
e. -678.8 kJ

39. Using the data:

$C_2H_4(g)$, $\delta H_f^o$ = +51.9 kJ/mol, S° = 219.8 J/mol K;
$CO_2(g)$, $\delta H_f^o$ = -394.0 kJ/mol, S° = 213.6 J/mol K;
$H_2O(l)$, $\delta H_f^o$ = -286.0 kJ/mol, S° = 69.96 J/mol K;
$O_2(g)$, $\delta H_f^o$ = 0.00 kJ/mol, S° = 205 J/mol K;

calculate the maximum amount of useful work that can be obtained, at 25.0 °C, from the process:

$C_2H_4(g) + 3\,O_2(g) \longrightarrow 2\,CO_2(g) + 2\,H_2O(l)$.

! a. 1332 kJ
b. 1380 kJ
c. 1451 kJ
d. 1492 kJ
e. 2422 kJ

40. Given the data:

$Ni(CO)_4(g)$, $\delta H_f^o$ = -220.0 kJ/mol, S° = +399 J/mol K;
$Ni(s)$, $\delta H_f^o$ = -0.00 kJ/mol, S° = 30 J/mol K;
$CO(g)$, $\delta H_f^o$ = -110.0 kJ/mol, S° = 197.9 J/mol K;

calculate the standard free energy change, $\delta G^o$, for the reaction, $Ni(CO)_4(g) \longrightarrow Ni(s) + 4\ CO(g)$.

a. -94 kJ
b. -169 kJ
c. -271 kJ
! d. -346 kJ
e. -412 kJ

41. Assuming that the values of $\delta H$ and $\delta S$ do not change as we shift temperature, and using,

$Ni(CO)_4(g)$, $\delta H_f^o$ = -220.0 kJ/mol, S° = +399 J/mol K;
$Ni(s)$, $\delta H_f^o$ = -0.00 kJ/mol, S° = 30 J/mol K;
$CO(g)$, $\delta H_f^o$ = -110.0 kJ/mol, S° = 197.9 J/mol K;

calculate a value for the free energy change, $\delta G^{1\ atm}$ for the reaction,

$CaCO_3(s) \longrightarrow CaO(s) + CO_2(g)$ at 815 °C.

a. -36 kJ
b. -404 kJ
! c. -673 kJ
d. +234 kJ
e. +404 kJ

42. Using the data:

$C_2H_6(g)$, $\delta H_f^o$ = -84.5 kJ/mol, S° = 229.5 J/mol K;
$CO_2(g)$, $\delta H_f^o$ = -394.0 kJ/mol, S° = 213.6 J/mol K;
$H_2O(l)$, $\delta H_f^o$ = -286.0 kJ/mol, S° = 69.96 J/mol K;
$O_2(g)$, $\delta H_f^o$ = 0.00 kJ/mol, S° = 205 J/mol K;

calculate the maximum amount of useful work that can be obtained, at 25.0 °C, from the process:

$2\ C_2H_6(g) + 7\ O_2(g) \longrightarrow 4\ CO_2(g) + 6\ H_2O(l)$, per mole of $C_2H_6$

a. -1426 kJ
! b. -1469 kJ
c. -1654 kJ
d. -2166 kJ
e. -3029 kJ

43. Using the data:

$Ag_2O(s)$, $\delta H_f^o$ = -31.1 kJ/mol, S° = 121.3 J/mol K
$Ag(s)$, $\delta H_f^o$ = 0.00 kJ/mol, S° = 42.55 J/mol K;
$O_2(g)$, $\delta H_f^o$ = 0.00 kJ/mol, S° = 205.0 J/mol K;

if we assume that $\delta H^{1\ atm}$ and $\delta S^{1\ atm}$ are independent of temperature between -50.0 °C and 950.0 °C, calculate the value of $\delta G^{1\ atm}$ for the reaction, $Ag_2O(s) \longrightarrow 2\ Ag(s) + \frac{1}{2}\ O_2(g)$ at 650.0 °C.

a. -12.0 kJ
! b. -30.1 kJ
c. -41.5 kJ
d. -51.0 kJ
e. -85.5 kJ

44. The reaction, $M_2O_3(s) + C(s) \longrightarrow M(s) + CO_2(g)$ is spontaneous at high temperatures but non-spontaneous at high temperatures. If we assume that the values of $\delta H^{1\ atm}$ and $\delta S^{1\ atm}$ are constant over a wide temperature range, including 25.0°C, we can deduce that,

a. $\delta H < 0$ and $\delta S > 0$
! b. $\delta H < 0$ and $\delta S < 0$
c. $\delta H > 0$ and $\delta S < 0$
d. $\delta H > 0$ and $\delta S > 0$
e. the information is insufficient to make any judgment as to the signs of $\delta H$ and $\delta S$.

45. The reaction, $2\ MO(s) + CO_2(g) \longrightarrow M_2O_3(s) + CO(s)$ is spontaneous at high temperatures but non-spontaneous at low temperatures. If we assume that the values of $\delta H^{1\ atm}$ and $\delta S^{1\ atm}$ are constant over a wide temperature range, including 25.0°C, we can deduce that,

a. $\delta H < 0$ and $\delta S > 0$
b. $\delta H < 0$ and $\delta S < 0$
c. $\delta H > 0$ and $\delta S < 0$
! d. $\delta H > 0$ and $\delta S > 0$
e. the information is insufficient to make any judgment as to the signs of $\delta H$ and $\delta S$.

46. Using the data:

$Ag_2O(s)$, $\delta H_f^o$ = -31.1 kJ/mol, $S^o$ = 121.3 J/mol K
$Ag(s)$, $\delta H_f^o$ = 0.00 kJ/mol, $S^o$ = 42.55 J/mol K;
$O_2(g)$, $\delta H_f^o$ = 0.00 kJ/mol, $S^o$ = 205.0 J/mol K;

if we assume that $\delta H^{1\ atm}$ and $\delta S^{1\ atm}$ are independent of temperature between -50.0 °C and 950.0 °C, calculate the temperature at which $\delta G^{1\ atm} = 0$ for the reaction, $Ag_2O(s) \longrightarrow 2\ Ag(s) + \frac{1}{2}O_2(g)$.

! a. 196 °C
b. 246 °C
c. 423 °C
d. 610 °C
e. 818 °C

## Fill in the Blanks

47. Using the data,

$I_2(g)$, $\delta H_f^o$ = +62.4 kJ/mol, $S^o$ = 260.7 J/mol K;
$I_2(s)$, $\delta H_f^o$ = 0.00 kJ/mol, $S^o$ = 116.12 J/mol K,

calculate the temperature at which solid iodine should have a vapor pressure of 1.00 atm, based on the reaction, $I_2(g)$ < = = > $I_2(s)$. This is the temperature at which $\delta G^{1\ atm} = 0$. ______ (! 158-159 °C)

48. For a chemical reaction, $\delta G^o$ = +21.16 kJ/mol. The value of $\delta S^o$ for the same reaction was 74.50 J/mol K. If the value of $\delta H$ and $\delta S$ do not change with temperature, what should be the value of $\delta G^{1atm}$ at 550.0 °C? ______ (! 40.17 kJ/mol)

49. For a chemical reaction, $\delta G^o$ = +31.16 kJ/mol. The value of $\delta S^o$ for the same reaction was 74.50 J/mol K. If the value of $\delta H$ and $\delta S$ do not change with temperature, at what temperature should $\delta G^{1atm}$ = 0.0, in °C? ______ (! 145.1 °C)

## True and False

50. The standard entropy, S° for an element in its standard state is always zero ( !F)

51. A negative value for the standard molar entropy, S°, of a chemical substance indicates that the substance is ***unstable*** and may decompose or detonate if mishandled. ( !F)

52. The entropy of a solid amorphous substance at absolute zero is 0 according to the third law of thermodynamics. ( !F)

53. Three factors which, acting together determine whether a reaction is spontaneous or not are entropy, internal energy, and enthalpy. ( !F) (temperature is also needed)

54. The equation, $\delta H^o_{reaction} = \Sigma\, \delta H^o_{f\,(products)} - \Sigma\, \delta H^o_{f\,(reactants)}$ is a statement of the second law of thermodynamics. ( !F)

## Critical Thinking

55. The reaction, $Al_2O_3(s) + \frac{3}{2} C(s) \longrightarrow \frac{3}{2} CO_2(g) + 2\, Al(s)$ is endothermic. At 25.0°C, $\delta G^o$ = +990.4 kJ/mole, while at 1000 °C, the value of the corresponding parameter is $\delta G^{1\ atm}$ = 680.8 kJ/mole. If we make the assumption that the values of $\delta H^{1\ atm} = \delta H^o$ and $\delta S^{1\ atm} = \delta S^o$ throughout this temperature range and beyond, calculate values for $\delta H^o$ and $\delta S^o$ for the reaction.

______ ( ! $\delta H$ = 1085 kJ/mol, $\delta S$ = 318 J/mol K)

56. The reaction, $Fe_2O_3(s) + \frac{3}{2} C(s) \longrightarrow \frac{3}{2} CO_2(g) + 2\, Fe(s)$ is endothermic. At 25.0°C, $\delta G^o$ = +148.9 kJ/mole, while at 750 °C, the value of the corresponding parameter is $\delta G^{1\ atm}$ = -51.2 kJ/mole. If we make the assumption that the values of $\delta H^{1\ atm} = \delta H^o$ and $\delta S^{1\ atm} = \delta S^o$ throughout this temperature range and beyond, calculate values for $\delta H^o$ and $\delta S^o$ for the reaction.

______ ( ! $\delta H$ = 231.2 kJ/mol, $\delta S$ = 276 J/mol K)

57. The species in the reaction, $2\, KNO_3(s) + 2\, C(s) \longrightarrow K_2CO_3(s) + N_2(g) + \frac{3}{2} CO_2(g)$ have the values for standard enthalpies of formation,

$KNO_3(s)$, $\delta H^o_f$ = -492.7 kJ/mol     $CO_2(g)$, $\delta H^o_f$ = -394.0 kJ/mol

$K_2CO_3(s)$, $\delta H^o_f$ = -1146 kJ/mol

Make the assumption that the values of $\delta H^{1\ atm} = \delta H^o$ and $\delta S^{1\ atm} = \delta S^o$ throughout a broad temperature range, and use this information to determine which of the following conditions may apply.

! a. The reaction is spontaneous at all temperatures.
b. The reaction is non-spontaneous at all temperatures.
c. The reaction is spontaneous at low temperatures but non-spontaneous at high temperatures.
d. The reaction is non-spontaneous at low temperatures but spontaneous at high temperatures.
e. It is impossible to make any judgment about spontaneity because insufficient information can be gleaned from the data and equations presented.

58. The species in the reaction, $KClO_3(s) \longrightarrow KCl(s) + \frac{3}{2}O_2(g)$ have the values for standard enthalpies of formation,

$KClO_3(s)$, $\delta H^o_f$ = -391.2 kJ/mol  $KCl(s)$, $\delta H^o_f$ = -436.8 kJ/mol

Make the assumption that the values of $\delta H^{1\,atm} = \delta H^o$ and $\delta S^{1\,atm} = \delta S^o$ throughout a broad temperature range, and use this information to determine which of the following conditions may apply.

! a. The reaction is spontaneous at all temperatures.
b. The reaction is non-spontaneous at all temperatures.
c. The reaction is spontaneous at low temperatures but non-spontaneous at high temperatures.
d. The reaction is non-spontaneous at low temperatures but spontaneous at high temperatures.
e. It is impossible to make any judgment about spontaneity because insufficient information can be gleaned from the data and equations presented.

## Chemicals in Use 11: Synthetic Diamonds and Diamond Coatings

59. Diamonds can be made from graphite. The equation for the process is,

$$C_{(graphite)} \longrightarrow C_{(diamond)}$$

Analysis of the data gave $\delta H^o$ = 1.88 kJ/mol and $\delta S^o$ = -3.29 J/mol k. Further, it is noted that diamond is more than 1.6 times as dense as graphite. If we assume $\delta H^{1\,atm} = \delta H^o$ and $\delta S^{1\,atm} = \delta S^o$ at temperatures to 3000K and beyond, what set of conditions would most greatly favor the process?

a. very high temperatures and very low pressures
! b. very high temperatures and very high pressures
c. very low temperatures and very low pressures
d. very low temperatures and very high pressures

60. Diamond coatings for devices used in high tech industries are made by crushing the diamonds to a very fine powder at high pressures to prevent them from reverting back to graphite, the using special adhesives to make to coatings secure under rough usage. ( !F)

61. For the process, $C(s)_{graphite} \longrightarrow C(s)_{diamond}$, to occur, which condition below but be met?

a. $\delta G/T$ must vary with temperature for a given pressure.
b. The pressure must be one atmosphere.
! c. $\delta G$ must be negative.
d. T must be negative.
e. $\delta G/T$ must be positive.

62. When a diamond film is formed, which one of the following statements described the situation?

a. Carbon and hydrogen atoms form a film of $CH_4$ with a tetrahedral structure.
b. Carbon atoms form a film of carbon with $sp^2$ hydridization
c. At high temperatures and pressures, methane and graphite unite to produce and deposit a smooth carbom film at a surface.
d. Low pressures and high temperatures produce the best condition for preparation of diamond coatings on a surface.
e. Carbon atoms form a film in which the carbon atoms are in a tetrahedral structure.

63. At the present time, the preferred method for manufacturing reasonably sized diamonds in quantity is

a. expose graphite to high intensity x-rays at high temperatures
b. expose graphite to high intensity x-rays in a vacuum chamber at temperatures near absolute zero
c. irradiate graphite in a high pressure chamber with neutrons at temperatures near absolute zero
! d. subject graphite to very high pressures at very high temperatures
e. irradiate graphite with radiation from a pulsed x-ray laser in a high pressure chamber

64. All of the following are applications of diamond coated surfaces except,

a. heat sinks
b. high frequency audio speaker (tweeter) cones
c. scratch resistant coatings
! d. dental caps
e. transistor contacts in semiconductor devices

65. In a diamond, the structure is that of a network covalent solid. The hydridization of the carbon in this diamond structure is

a. sp
b. $sp^2$
! c. $sp^3$
d. $sp^4$
e. $sp^3d$

66. A property, normally associated with metals, in which diamond excells over the best metals is

a. malleability
b. ductility
c. thermal conductivity
! d. electrical conductivity
e. elasticity

# Chapter 14 Kinetics: The Study of Rates of Reaction

## Multiple Choice

1. A catalyst alters the rate of a chemical reaction by

! a. providing an alternate pathway which has a different activation energy
b. changing the products formed in the reaction
c. changing the frequency of collisions between molecules
d. always providing a surface on which molecules react
e. changing the enthalpy of reaction for the reaction

2. Which one of the following statements concerning the rate of a chemical reaction is *false*?

! a. It will be very rapid if the activation energy is large.
b. It will be slow if one or more of the steps is slow.
c. It may be inhibited sometimes by certain catalytic agents.
d. It is dependent on temperature.
e. It often increases when the concentrations of one of the reactants is increased.

3. Which statement below is true concerning a negative catalyst?

a. It lowers the energy of activation of the rate determining step.
b. It increases the enthalpy of reaction.
c. It never undergoes a chemical change at any time during a chemical reaction.
d. It blocks the path with the highest energy of activation for the rate determining step.
! e. It blocks the path with the lowest energy of activation for the rate determining step.

4. If a reaction involving a single reactant is first order with a rate constant of $4.50 \times 10^{-2}$ $s^{-1}$, how much time is required for 75.0% initial concentration of reactant to be used up?

a. 16.7 seconds
! b. 30.8 seconds
c. 23.1 seconds
d. 25.3 seconds
e. 11.6 seconds

5. A reaction has the rate law, rate = $k[A][B]^2$. What is the overall order of the reaction?

a. 2
b. 4
c. 1
! d. 3
e. 0

6. A reaction has the rate law, rate = $k[A][B]^2$. Which one of the following will cause the greatest increase in the reaction rate?

a. decreasing the temperature without changing the concentrations
b. doubling the concentration of B
c. quadrupling the concentration of A
! d. tripling the concentration of B
e. doubling the concentration of A

7. Given these data in a study on how the rate of a reaction was affected by the concentration of the reactants

| run # | [A] | [B] | [C] | initial rate, mol $L^{-1}$ $hr^{-1}$ |
|---|---|---|---|---|
| 1 | 0.200 | 0.100 | 0.600 | 5.0 |
| 2 | 0.200 | 0.400 | 0.400 | 80.0 |
| 3 | 0.600 | 0.100 | 0.200 | 15.0 |
| 4 | 0.200 | 0.100 | 0.200 | 5.0 |
| 5 | 0.200 | 0.200 | 0.400 | 20.0 |

! a. The rate constant is 2500 $L^2$ $mol^{-2}$ $hr^{-1}$
b. The rate constant is 208 $L^2$ $mol^{-2}$ $hr^{-1}$
c. The rate constant is 139 $L^2$ $mol^{-2}$ $hr^{-1}$
d. The rate constant is 2083 $L^2$ $mol^{-2}$ $hr^{-1}$
e. The rate constant is 6667 $L^2$ $mol^{-2}$ $hr^{-1}$

8. For the reaction, A + 2B ⟶ C + 2 D, some intitial rate measurements were carried out

| run # | [A] | [B] | rate, mol $L^{-1}$ $s^{-1}$ |
|---|---|---|---|
| 1 | 0.100 | 0.200 | 0.000360 |
| 2 | 0.200 | 0.200 | 0.000720 |
| 3 | 0.100 | 0.400 | 0.000720 |

a. The rate law is therefore, rate = k[A][B]$^2$
b. The rate law is therefore: rate = k[B]
c. The rate law is therefore: rate = k[A]
! d. The rate law is therefore: rate = k[A][B]
e. The rate law is therefore: rate = k[A]$^2$[B]

9. For the reaction, 2 M + 2 N ⟶ 2 P + Q, some studies on how the intitial rate of the reaction varied with concentration were carried out. Some data are given below.

| run # | [M] | [N] | rate, mol $L^{-1}$ $s^{-1}$ |
|---|---|---|---|
| 1 | 0.100 | 0.100 | 0.000230 |
| 2 | 0.100 | 0.200 | 0.000920 |
| 3 | 0.200 | 0.200 | 0.000920 |

! a. The rate law is therefore: rate = k[N]$^2$
b. The rate law is therefore: rate = k[M][N]$^2$
c. The rate law is therefore: rate = k[M][N]
d. The rate law is therefore: rate = k[M]$^2$
e. The rate law is therefore: rate = k[M]$^2$[N]$^2$

10. Given the reaction, aA + bB $\xrightarrow{C}$ dD + eE, where C is a catalyst. If we try, rate = $k[A]^q[B]^r[C]^s$ for a generic rate law statement, which one of the statements below is **false?**

a. The exponents q, r, and s are often integers.
b. The exponent s must be determined experimentally.
! c. The exponents q and r are equal to the coefficients a and b, respectively.
d. The overall order of the reaction is q + r + s.
e. The symbol k represents the rate constant.

11. For the reaction, 2 XO + $O_2$ $\longrightarrow$ 2 $XO_2$, some data obtained from measurement of the initial rate of reaction at varying concentrations are given below.

| run # | [XO] | [$O_2$] | rate, mmol $L^{-1}$ $s^{-1}$ |
|---|---|---|---|
| 1 | 0.010 | 0.010 | 2.5 |
| 2 | 0.010 | 0.020 | 5.0 |
| 3 | 0.030 | 0.020 | 45.0 |

The rate law is therefore

a. rate = $k[XO]^2[O_2]$
! b. rate = $k[XO][O_2]^2$
c. rate = $k[XO][O_2]$
d. rate = $k[XO]^2[O_2]^2$
e. rate = $k[XO]^2/[O_2]^2$

12. For the reaction, A + 2 B $\longrightarrow$ C + 2 D, some measurements of the initial rate of reaction at varying concentration gave the following data.

| run # | [A] | [B] | rate, mol $L^{-1}$ $s^{-1}$ |
|---|---|---|---|
| 1 | 0.100 | 0.200 | 0.000360 |
| 2 | 0.150 | 0.200 | 0.000540 |
| 3 | 0.150 | 0.250 | 0.001055 |

a. The rate law is therefore, rate = $k[A]^2[B]$.
b. The rate law is therefore: rate = $k[A][B]^2$.
c. The rate law is therefore: rate = $k[A]^2[B]^2$.
d. The rate law is therefore: rate = $k[A][B]$.
! e. The rate law is therefore: rate = $k[A][B]^3$.

13. For the reaction, 2 $A_2X_4(g)$ $\longrightarrow$ 2 $A_2X_3(g)$ + $X_2(g)$, the reaction was found to be first order. The rate law, therefore, should be,

a. rate = $k([A_2X_3]^2[X_2])/[A_2X_4]^2$
b. rate = $k[A_2X_4]^2$
c. rate = $k([A_2X_3][X_2])/[A_2X_4]$
d. rate = $k[A_2X_4]$
e. rate = $k[A_2X_4]^2/([A_2X_3]^2[X_2])$

14. In a reaction, reagents A and B are mixed and the time interval until a certain *quantity* of product, C, is produced is measured. The data are

| run # | [A] | [B] | time, secs |
|---|---|---|---|
| 1 | 0.100 | 0.140 | 25 |
| 2 | 0.050 | 0.140 | 50 |
| 3 | 0.100 | 0.070 | 100 |

! a. The reaction is first order with respect to substance A.
b. The reaction is zero order with respect to substance A.
c. The reaction is one-half order with respect to substance A.
d. The reaction is second order with respect to substance A.
e. The reaction is third order with respect to substance B.

15. The data below were obtained in a study on how the rate of a reaction was affected by the concentration of the reactants.

| run # | [A] | [B] | [C] | rate, mol $L^{-1}$ $hr^{-1}$ |
|---|---|---|---|---|
| 1 | 0.200 | 0.100 | 0.600 | 5.0 |
| 2 | 0.200 | 0.400 | 0.400 | 80.0 |
| 3 | 0.600 | 0.100 | 0.200 | 15.0 |
| 4 | 0.200 | 0.100 | 0.200 | 5.0 |
| 5 | 0.200 | 0.200 | 0.400 | 20.0 |

a. The order of the reaction with respect to C cannot be determined this data alone.
b. The reaction is second order with respect to C.
! c. The reaction is zero order with respect to C.
d. The reaction is first order with respect to C.
e. The reaction order for C is minus one (Rate prop to 1/[C]).

16. Given these data in a study on how the rate of a reaction was affected by the concentration of the reactants,

| run # | [A] | [B] | [C] | rate, mol $L^{-1}$ $hr^{-1}$ |
|---|---|---|---|---|
| 1 | 0.200 | 0.100 | 0.600 | 5.0 |
| 2 | 0.200 | 0.400 | 0.400 | 80.0 |
| 3 | 0.600 | 0.100 | 0.200 | 15.0 |
| 4 | 0.200 | 0.100 | 0.200 | 5.0 |
| 5 | 0.200 | 0.200 | 0.400 | 20.0 |

a. The reaction is zero order with respect to B.
b. The reaction is first order with respect to B.
c. The reaction order for B cannot be determined from just this data alone .
! d. The reaction is second order with respect to B.
e. The reaction order for B is minus one (Rate prop to 1/[B]).

17. Given these data in a study on how the rate of a reaction was affected by the concentration of the reactants,

| run # | [A] | [B] | [C] | rate, mol $L^{-1}$ $hr^{-1}$ |
|---|---|---|---|---|
| 1 | 0.200 | 0.100 | 0.600 | 5.0 |
| 2 | 0.200 | 0.400 | 0.400 | 80.0 |
| 3 | 0.600 | 0.100 | 0.200 | 15.0 |
| 4 | 0.200 | 0.100 | 0.200 | 5.0 |
| 5 | 0.200 | 0.200 | 0.400 | 20.0 |

! a. The reaction is first order with respect to A.
b. The reaction is second order with respect to A.
c. The reaction is zero order with respect to A.
d. The reaction order for A is minus one (Rate proportional to 1/[A]).
e. The reaction order for A cannot be determined from just this data alone.

18. Nitric oxide reacts with bromine gas at elevated temperatures according to the equation, $2\ NO(g) + Br_2(g) \longrightarrow 2\ NOBr(g)$. The experimental rate law is rate = $k[NO][Br_2]$. In a certain reaction mixture the rate of formation of NOBr(*g*) was found to be $4.50 \times 10^{-4}$ mol $L^{-1}$ $s^{-1}$. What is the rate of consumption of $Br_2(g)$, also in mol $L^{-1}$ $s^{-1}$?

a. $-4.50 \times 10^{-4}$ mol $L^{-1}$ $s^{-1}$
! b. $-2.25 \times 10^{-4}$ mol $L^{-1}$ $s^{-1}$
c. $-9.00 \times 10^{-4}$ mol $L^{-1}$ $s^{-1}$
d. $-2.12 \times 10^{-4}$ mol $L^{-1}$ $s^{-1}$
e. $-2.03 \times 10^{-3}$ mol $L^{-1}$ $s^{-1}$

19. Nitric oxide reacts with bromine gas at elevated temperatures according to the equation, $2\ NO(g) + Br_2(g) \longrightarrow 2\ NOBr(g)$. The experimental rate law is rate = $k[NO][Br_2]$. In a certain reaction mixture the rate of formation of NOBr(*g*) was found to be $4.50 \times 10^{-4}$ mol $L^{-1}$ $s^{-1}$. Which unit below is the correct unit for the rate law in this case?

a. mol $L^{-1}$ $s^{-1}$
b. $s^{-1}$
c. $mol^2$ $L^{-2}$ $s^{-1}$
! d. $mol^{-1}$ L $s^{-1}$
e. $mol^{-2}$ $L^2$ $s^{-1}$

20. For a first order reaction with a single reactant, after 230.0 seconds, 10.0% of the reactant remains. The rate constant for the reaction is therefore

a. 0.000640 $s^{-1}$
! b. 0.0100 $s^{-1}$
c. 100 $s^{-1}$
d. 0.0510 $s^{-1}$
e. 0.0915 $s^{-1}$

21. For the reaction, 3 B + C ⟶ E + 2 F, some intitial rate measurements were carried out and data for three runs are shown below

| run # | [B] | [C] | rate, mol $L^{-1}$ $s^{-1}$ |
|---|---|---|---|
| 1 | 0.100 | 0.250 | 0.000250 |
| 2 | 0.200 | 0.250 | 0.000500 |
| 3 | 0.100 | 0.500 | 0.00100 |

The rate law, therefore, is

a. rate = $k[B]^3[C]$
b. rate = $k[B][C]$
c. rate = $k[B]^2[C]^2$
d. rate = $k[B]^2[C]$
! e. rate = $k[B][C]^2$

22. In a first order reaction with only one reagent, the reaction was started with with a concentration of reactant equal to 0.0800 molar. After exactly two hours, the concentration had fallen to 0.0400 molar. What is the molarity after exactly three hours?

a. 0.0300 molar
b. 0.0267 molar
c. 0.0340 molar
! d. 0.0283 molar
e. 0.0200 molar

23. The half life of a chemical reaction was found to be independent of the quantity of material which the researcher employed. The reaction is therefore

a. possibly first order
! b. definitely first order
c. zero order
d. possibly second order
e. definitely second order

24. The decomposition of an aldehyde solution in carbon tetrachloride is a first order reaction with a rate constant of $1.20 \times 10^{-3}$ $min^{-1}$. If we start with [aldehyde] = 0.0500 molar, what will the concentration be 150 minutes later?

a. 0.00900 molar
! b. 0.0418 molar
c. 0.00926 molar
d. 0.00499 molar
e. 0.000333 molar

25. The rate constant for a first order decomposition reaction is 0.0111 $min^{-1}$. What is the half life of the reaction?

a. 111 min
! b. 62.4 min
c. 5000 sec
d. 31.25 min
e. 27.1 min

26. Given a reaction, 2 A + B $\longrightarrow$ P, for which the observed rate law is, rate = k[A]. Which one of the following is true?

a. [A] = 1/kt

b. *ln*[A] = k/t

c. 1/[A] = kt

! d. the half life is 0.693/k

e. $e^{[A]}$ = -kt

27. In a first order reaction, what fraction of the material will remain after 4 half lives?

! a. 1/16
b. 1/8
c. 1/9
d. 1/4
e. 1/3

28. The initial concentration of a reactant in a first order reaction is 0.620 molar. What will be its concentration after 3 half lives?

a. 0.0865 molar
b. 0.310 molar
c. 0.0775 molar
d. 0.103 molar
e. 0.207 molar

29. For the reaction, A $\longrightarrow$ B + C, the rate law is k[A]. If it takes 80.0 seconds for 70.0% of a 10.0 gram sample of A to be transformed into products, what is the value of the rate constant

a. 0.00450 $s^{-1}$

b. 0.0290 $s^{-1}$

c. 0.00530 $s^{-1}$

d. 0.0150 $s^{-1}$

e. 5.40 $s^{-1}$

30. The reaction of substance A with substance C was carefully studied under conditions where the [C] remained essentially constant. The graphs of [A] *vs* time and that of *ln*[A] *vs* time both gave curves, but the graph of 1/[A] *vs* time gave a straight line.

a. The reaction is therefore zero order with respect to A.
b. The reaction is therefore one-half order with respect to A.
c. The reaction is therefore first order with respect to A.
! d. The reaction is therefore second order with respect to A.
e. The reaction is therefore third order with respect to A.

31. The reaction of substance A with substance C was carefully studied under conditions where the [C] remained essentially constant. The graph of [A] *vs* time gave a straight line while the graph of *ln*[A] *vs* time and that of 1/[A] *vs* time both gave curves.

! a. The reaction is therefore zero order with respect to A.
b. The reaction is therefore one-half order with respect to A.
c. The reaction is therefore first order with respect to A.
d. The reaction is therefore second order with respect to A.
e. The reaction is therefore third order with respect to A.

32. A reaction is first order overall. For a given sample, if its initial rate is 0.0200 M/sec and 25.0 days later its rate dropped to $6.25 \times 10^{-4}$ mol $L^{-1}$ $s^{-1}$, what is its half life?

a. 25.0 days
b. 50.0 days
c. 12.5 days
d. 5.0 days
e. 37.5 days

33. A first order reaction has a rate constant of 0.00318 $min^{-1}$. The half life of this reaction is therefore

a. 94.7 minutes
! b. 218 minutes
c. 31.4 minutes
d. 5.24 seconds
e. 68.6 minutes

34. When a catalyst is used in a reaction which of the statements below describes the situation which occurs?

a. The forward reaction rate is increased while the reverse reaction rate is retarded.
b. The enthalpy change for the reaction becomes more exothermic.
! c. It does not affect the final (equilibrium) amounts of reactants and products.
d. The activation energy for the reverse reaction is increased.
e. The activation energy for the forward reaction is not altered.

35. For a one step reaction, the activation energy for the forward reaction is 40.0 kJ per mole, and the enthalpy of reaction is -20.0 kJ per mole. Which statement below is true?

a. The activation energy of the forward reaction would be affected more than the activation energy of the reverse reaction by addition of a catalyst.
b. The value for the enthalpy of reaction would be decreased by addition of a catalyst
c. The reaction is endothermic.
! d. The reverse reaction is slower than the forward reaction (smaller rate constant)

36. For a one step reaction, the activation energy for the forward reaction is 40.0 kJ per mole, and the enthalpy of reaction is -20.0 kJ per mole. Calculate the activation energy for the reverse reaction.

! a. 60.0 kJ per mole
b. -20.0 kJ per mole
c. -1200 kJ per mole
d. 20.0 kJ per mole
e. 1200 kJ per mole

37. For a chemical reaction, the rate constant at 250.0 °C is 0.00383 $s^{-1}$, and the activation energy is 22.40 kilojoules. Calculate the value of the rate constant at 335.0 °C.

a. 0.00513 $s^{-1}$
b. 0.00946 $s^{-1}$
! c. 0.00787 $s^{-1}$
d. 0.0224 $s^{-1}$
e. 0.000640 $s^{-1}$

38. For a chemical reaction, the rate constant at 42.0 °C is 0.00395 $s^{-1}$, while the rate constant at 67.4 °C is 0.0133 $s^{-1}$. Calculate the value of the energy of activation, in kilojoules.

! a. 42.65
b. 1.13
c. 0.421
d. 18.5
e. 0.617

39. The rate constant for a certain chemical reaction is 0.00250 L $mol^{-1}$ $s^{-1}$ at 25.0 °C and 0.0125 L $mol^{-1}$ $s^{-1}$ at 50.0 °C. What is the activation energy for the reaction, expressed in kJ?

a. 25.1 kJ
! b. 51.6 kJ
c. 37.6 kJ
d. 45.3 kJ
e. 60.3 kJ

40. For a particular chemical reaction, the rate constant at 30.0 °C is $1.38 \times 10^{-4}$ L $mol^{-1}$ $s^{-1}$, while the value at 49.0 °C is $1.21 \times 10^{-3}$ L $mol^{-1}$ $s^{-1}$. What is the activation energy for this reaction?

! a. 92.8 kJ
b. 200 kJ
c. 40.4 kJ
d. 343 kJ
e. 56.4 kJ

41. The activation energy for a reaction can be found by finding the slope of a plot of *ln*(k) *vs* $T^{-1}$ and

a. adding this slope to -R
b. multiplying this slope by 2.303
c. dividing this slope by -R
d. multiplying this slope by 2.303R
! e. multiplying this slope by -R

42. A chemical reaction has been the subject of intense study, and a rate law, rate = $k[A][B]^2$, was developed which summarized the mechanistic findings. Which one of the following changes to the system will not cause an increase in the rate constant under any circumstances?

a. raising the temperature by 25.0 degrees
b. adding a positive catalyst
c. adding a negative catalyst
! d. doubling the concentration of B
e. lowering the temperature by 0.500 degrees

43. The equation, $A + 2B \longrightarrow C + D$ describes an elementary reaction, which takes place in a single step. The rate law must therefore be

! a. rate = $k[A][B]^2$
b. rate = $k([C][D])/([A][B]^2)$
c. rate = $k[A]/[C]$
d. rate = $k[A]^2$
e. rate = $k[A][B]$

44. The reaction: $A + 3B \longrightarrow D + F$ was studied and the following mechanism was finally determined

$A + B \longrightarrow C$ (fast)
$C + B \longrightarrow D + E$ (slow)
$E + B \longrightarrow F$ (very fast)

The step with smallest activation energy is

a. the first step
! b. the second step
c. the third step
d. None of the steps has an activation energy.
e. All of the steps have the same activation energy.

45. Suppose the reaction: $A + B \longrightarrow D$ followed the mechanism

$A + B \Longleftrightarrow C$ (fast)
$C \longrightarrow D$ (slow)

The rate law for the reaction would be

a. rate = $k[A]$
b. rate = $k[A]^2$
! c. rate = $k[A][B]$
d. rate = $k[A][B]/[D]$
e. rate = $k[A][B][C]$

46. The reaction: $A + 3B \longrightarrow D + F$ was studied carefully and the following mechanism was finally determined.

$A + B \longrightarrow C$ (fast)
$C + B \longrightarrow D + E$ (slow)
$E + B \longrightarrow F$ (very fast)

The rate law for the reaction would therefore be

a. rate = $k[A]^2[B]$
! b. rate = $k[A][B]^2$
c. rate = $k[C][B]$
d. rate = $k[A][B]^3$
e. rate = $k[A][B]$

47. Which statement about the slow step in the mechanism for a reaction is true?

a. It has a rate that is independent of the activation energy for the reaction.
b. It limits the effectiveness of a catalyst.
! c. It controls the rate at which the products are produced.
d. It almost always involves the breaking of hydrogen bonds.
e. It determines the value of the standard enthalpy of reaction for the reaction.

48. The decomposition of hydrogen peroxide in solution was studied in laboratory, and the following mechanism proposed based on the experimental data.

$$H_2O_2 + I^- \longrightarrow H_2O + IO^- \quad \text{slow}$$
$$H_2O_2 + IO^- \longrightarrow H_2O + O_2 + I^- \quad \text{fast}$$

Which one of the following statements is false?

a. The reaction is first order with respect to $H_2O_2$.
b. The reaction is first order with respect to $I^-$.
c. $I^-$ is a catalyst.
d. $IO^-$ is an intermediate.
e. The reaction is first order with respect to $IO^-$.

49. The reaction, $A + 3B \longrightarrow D + F$ was followed and the following mechanism was finally determined.

$$A + B \longrightarrow C \quad \text{(fast)}$$
$$C + B \longrightarrow D + E \quad \text{(slow)}$$
$$E + B \longrightarrow F \quad \text{(very fast)}$$

The species, E, is properly described as

! a. an intermediate
b. a co-factor
c. a catalyst
d. an inhibitor
e. an enzyme

50. The reaction: $A + 3B \longrightarrow D + F$ was studied and the following mechanism was finally determined.

$$A + B \longrightarrow C \quad \text{(fast)}$$
$$C + B \longrightarrow D + E \quad \text{(slow)}$$
$$E + B \longrightarrow F \quad \text{(very fast)}$$

The species, C, is properly described as

a. a co-factor
b. an inhibitor
c. a catalyst
d. an enzyme
! e. an intermediate

51. A variable which has **no** effect on the rate of a chemical reaction under any circumstances is,

a. energy of activation
b. catalyst
c. concentration of the reactants
d. temperature
! e. standard enthalpy of reaction for the system

52. The reaction mechanism that has been proposed for the decomposition of $H_2O_2$ is

$$H_2O_2 + I^- \longrightarrow H_2O + IO^- \quad \text{slow}$$
$$H_2O_2 + IO^- \longrightarrow H_2O + O_2 + I^- \quad \text{fast}$$

which one of the following is true?

a. The reaction is second order with respect to $I^-$.
b. $I^-$ is an intermediate.
! c. The reaction is first order with respect to $I^-$.
d. $IO^-$ is a catalyst.
e. The reaction is zero order with respect to $I^-$.

53. If the reaction, $H_2(g) + Cl_2(g) \longrightarrow 2HCl(g)$ occurred in just two steps what would the overall order of the reaction be?

a. 1
b. 2
c. 3
d. 4
! e. It impossible to determine the order based on this limited information.

54. The reaction, $2\,NO(g) + O_2(g) \longrightarrow 2\,NO_2(g)$, was found to be first order in each of the two reactants and second order overall. The rate law should therefore be written as

a. rate = $k[NO]^2$
! b. rate = $k([NO][O2]$
c. rate = $k[NO_2]^2[NO]^{-2}[O_2]^{-1/2}$
d. rate = $k[NO]^2[O_2]^2$
e. rate = $k([NO][O_2])^2$

55. The units in which the rate of a chemical reaction in solution is measured is (could be)

a. $L^2\ mol^{-1}\ s^{-1}$
! b. $mol\ L^{-1}\ s^{-1}$
c. $s^{-2}$
d. $mol\ s\ L^{-1}$
e. $sec\ L^{-1}\ mol^{-1}$

## Fill in the Blanks

56. What effect does a negative catalyst have on the value of the rate constant for a particular reaction? ____________ (! It causes the value to decrease)

57. What effect does the magnitude of the standard enthalpy of reaction have on the rate constant for a reaction, if the temperature is kept constant? ____________ ( ! no effect)

## True and False

58. In order to carry out chemical kinetics studies, all reactants must be in a catalyzed form or else they will not form products at a regular rate. ( !F)

59. The half-life of a second order reaction depends on the initial concentration of the reactants. ( !T)

60. Nitric oxide reacts with bromine gas at elevated temperatures according to the equation, $2\ NO(g) + Br_2(g) \longrightarrow 2\ NOBr(g)$. The experimental rate law is rate = $k[NO][Br_2]$. The half life of either of the reactant species is independent of concentration and is equal to 0.693/k. ( !F)

61. The activation energy is equal to the difference between the standard enthalpy of the reactants and the standard enthalpy of the products. ( !F)

62. The final step in a multistep reaction mechanism is called the rate determining step. ( !F)

63. If tripling the concentration of a particular reactant in a reaction mixture fails to show any measurable change in the rate of a chemical reaction, the order with respect to the particular reagent is zero. ( !T)

64. The activation energy is the energy required to initiate a reaction, and the reaction will continue without further outside assistance. ( !F)

65. A positive catalyst that lowers the activation of the forward reaction in a system to one-half its uncatalyzed value will also lower the activation energy of the reverse reaction in the system to one-half its uncatalyzed value. ( !F)

## Critical Thinking

66. Suppose the reaction, $2\ A + 2\ B \longrightarrow 2\ D + F$ followed the mechanism

$$A + B \rightleftharpoons C + D \quad \text{(fast)}$$
$$A + C \longrightarrow E \quad \text{(slow)}$$
$$E + B \longrightarrow D + F \quad \text{(very fast)}$$

The rate law for the reaction would be

! a. rate = $k[A]^2[B]$
b. rate = $k[A][B][C]$
c. rate = $k[A][B]$
d. rate = $k[A]^2[B]^2$
e. rate = $k[A][B]^2$

67. The half-life of a first order chemical reaction is 24.0 minutes at 55.0 °C and 16.4 minutes at 63.5 °C. What will be the half-life, in minutes, at 70.0 °C?
______ ( ! 12.4 minutes)

# Chapter 15 Chemical Equilibrium - General Concepts

## Multiple Choice

1. A chemical system is considered to have reached equilibrium when,

   a. the frequency of collisions between the reactant molecules is equal to the frequency of collisions between the product molecules
   b. the sum of the concentrations of the reactant species is equal to the sum of the concentrations of the product species
   c. the activation energy of the forward reaction is equal to the activation energy of the reverse reaction
   ! d. the rate of production of the product species is equal to the rate of production of the reactant species by the reverse reaction
   e. the rate of production of the product species is equal to the rate of consumption of the reactant species

2. In a study of the system, $H_2(g) + I_2(g) \Longleftrightarrow 2\ HI(g)$, at 699 K, several different reaction mixes, described below, were thrown together, placed in a 5.00 liter container, and allowed to attain equilibrium at 699 K. Despite having different starting compositions, as shown, **four** of the five mixtures had the **identical** composition at equilibrium. Which one of the systems attains a different equilibrium composition than the others?

   a. System 1: 0.100 moles of $H_2$, 0.000 moles of $I_2$ and 0.600 moles of HI
   b. System 2: 0.175 moles of $H_2$, 0.075 moles of $I_2$ and 0.450 moles of HI
   c. System 3: 0.400 moles of $H_2$, 0.300 moles of $I_2$ and 0.000 moles of HI
   ! d. System 4: 0.150 moles of $H_2$, 0.100 moles of $I_2$ and 0.500 moles of HI
   e. System 5: 0.300 moles of $H_2$, 0.200 moles of $I_2$ and 0.200 moles of HI

3. In a study of the system, $Cl_2(g) + Br_2(g) \Longleftrightarrow 2\ BrCl(g)$, at 350 K, several different reaction mixes, described below, were thrown together, placed in a 5.00 liter container, and allowed to attain equilibrium at 350 K. Despite having different starting compositions, as shown, **four** of the five mixtures had the **identical** composition at equilibrium. Which one of the systems attains a different equilibrium composition than the others?

   a. System 1: 0.200 moles of $Cl_2$, 0.000 moles of $Br_2$ and 0.600 moles of BrCl
   b. System 2: 0.250 moles of $Cl_2$, 0.150 moles of $Br_2$ and 0.300 moles of BrCl
   c. System 3: 0.400 moles of $Cl_2$, 0.300 moles of $Br_2$ and 0.000 moles of BrCl
   ! d. System 4: 0.300 moles of $Cl_2$, 0.250 moles of $Br_2$ and 0.250 moles of BrCl
   e. System 5: 0.350 moles of $Cl_2$, 0.250 moles of $Br_2$ and 0.100 moles of BrCl

4. In a study of the system, $N_2O_4(g) + \Longleftrightarrow 2\ NO_2(g)$, at 585 K, several different reaction mixes, described below, were thrown together, placed in a 5.00 liter container, and allowed to attain equilibrium at 480 K. Despite having different starting compositions, as shown, **four** of the five mixtures had the **identical** composition at equilibrium. Which one of the systems attains a different equilibrium composition than the others?

   a. System 1: 0.400 moles of $N_2O_4$ and 0.800 moles of $NO_2$
   b. System 2: 0.300 moles of $N_2O_4$ and 1.000 moles of $NO_2$
   c. System 3: 0.800 moles of $N_2O_4$ and 0.000 moles of $NO_2$
   d. System 4: 0.600 moles of $N_2O_4$ and 0.400 moles of $NO_2$
   ! e. System 5: 0.400 moles of $N_2O_4$ and 0.700 moles of $NO_2$

5. In a study of the system, $Cl_2(g)$ + 2 $NO(g)$ <===> 2 $NOCl(g)$, at 425 K, several different reaction mixes, described below, were thrown together, placed in a 5.00 liter container, and allowed to attain equilibrium at 425 K. Despite having different starting compositions, as shown, **four** of the five mixtures had the **identical** composition at equilibrium. Which one of the systems attains a different equilibrium composition than the others?

a. System 1: 0.200 moles of $Cl_2$, 0.000 moles of NO and 0.600 moles of NOCl
b. System 2: 0.250 moles of $Cl_2$, 0.150 moles of NO and 0.300 moles of NOCl
c. System 3: 0.400 moles of $Cl_2$, 0.300 moles of NO and 0.000 moles of NOCl
! d. System 4: 0.300 moles of $Cl_2$, 0.250 moles of NO and 0.250 moles of NOCl
e. System 5: 0.350 moles of $Cl_2$, 0.250 moles of NO and 0.100 moles of NOCl

6. For the reaction, $XBr_2(g)$ <===> $X(g)$ + $Br_2(g)$, if a container is charged to an initial concentration, $[XBr_2]$ = 2.00 mol/L and the system allowed to react and come to equilibrium, if $K_c$ = $4.0 \times 10^{-8}$, what is the concentration of $X(g)$ at equilbrium?

a. 0.000056 M
b. 0.00020 M
! c. 0.00028 M
d. 0.000000040 M
e. 0.000000080 M

7. Given the reaction, 2 $NO(g)$ + $O_2(g)$ <===> 2 $NO_2(g)$, for which the enthalpy of reaction is -118.9 kJ. Which one of the following will cause an increase in the equilibrium concentration of NO in a closed reaction chamber?

a. adding some more $O_2(g)$ through an injection nozzle
! b. increasing the temperature of the system
b. removing the $NO_2$ from the system by absorbing it in a species specific zeolite
c. increasing the pressure of the system while temperature is kept constant
d. adding a catalyst

8. For a specific reaction, which of the following statements can be made about the equilibrum constant?

! a. it can change with temperature
b. it may be changed by addition of a catalyst
c. it increases if the concentration of one of the products is increased
d. it increases if the concentration of one of the products is decreased
e. it always remains the same

9. The equilbrium constant for the reaction, $R_2$ <===> $D_2$ is $6.8 \times 10^{-10}$. Which one of the following statements is true?

! a. The equilbrium concentration of $D_2$ is always greater than that of $R_2$.
b. The equilibrium concentration of $R_2$ is always greater than that of $D_2$.
c. Adding more $R_2$ will increase the value of the equilibrium constant.
d. Adding a catalyst will increase the equilibrium concentration of $D_2$.
e. Adding a catalyst will increase the value of the equilibrium constant.

10. All of the following statements about chemical equilibrium are true except

a. At equilibrium, the reactant and the product concentrations show no further change with time.
! b. A true chemical equilibrium can only be attained starting from the reactant side of the equation.
c. At equilibrium, the forward reaction rate equals the reverse reaction rate.
d. The same equilibrium state is attained starting either from the reactant or product side of the equation.
e. At equilibrium, the reactant and product concentrations are constant.

11. The concentration of a pure solid or a pure liquid is left out of the expression for the equilibrium constant because

a. solids and liquids drive the reaction in an undetermined fashion.
b. solids and liquids do not react.
c. their concentrations cannot be determined.
d. solids and liquids react too slowly.
! e. their activity is constant and independent of the amount of solid or liquid present.

12. The equilbrium constant, $K_c$, for the system, $CaO(s) + CO_2(g) <\!\!\Longrightarrow CaCO_3(s)$, is

a. $K_c = [CO_2]$
b. $K_c = [CaCO_3]/([CaO] \times [CO_2])$
! c. $K_c = 1/[CO_2]$
d. $K_c = [CaCO_3]/[CaO]$
e. $K_c = ([CaO] \times [CO_2])/[CaCO_3]$

13. For the reaction, $2\ SO_2(g) + O_2(g) <\!\!\Longrightarrow 2\ SO_3(g)$, at equilibrium, the concentrations are $SO_3$: 0.120M, $SO_2$: 0.860M, $O_2$: 0.330M Calculate the value of $K_c$ for this reaction.

a. 1.31
b. 2.51
c. 0.423
d. 0.872
! e. 0.0590

14. For the reaction, $2\ SO_2(g) + O_2(g) <\!\!\Longrightarrow 2\ SO_3(g)$, at 900.0 K the equilibrium constant, $K_c$ has a value of 13.0. Calculate the value of $K_p$ at the same temperature

a. $97.3 \times 10^3\ atm^{-1}$
! b. $0.176\ atm^{-1}$
c. $960\ atm^{-1}$
d. $0.00174\ atm^{-1}$
e. $0.077\ atm^{-1}$

15. For the reaction, $2\ SO_2(g) + O_2(g) <\!\!==\!\!> 2\ SO_3(g)$, at 450.0K the equilibrium constant, $K_c$ has a value of 4.62. A system, at equilibrium has the following concentrations: $[SO_3] = 0.254$ M $[O_2] = 0.00855$ M. What is the equilibrium concentration of $SO_2$?

! a. 1.28 M
b. 0.216 M
c. 0.465 M
d. 6.43 M
e. 41.3 M

16. For the reaction, $2\ SO_2(g) + O_2(g) <\!\!==\!\!> 2\ SO_3(g)$, at 450.0 K the equilibrium constant, $K_c$ has a value of 4.62. A system was charged to give these initial concentrations, $[SO_3] = 0.254$ M $[O_2] = 0.00855$ M, $[SO_2] = 0.500$ M. In which direction will it go?

a. to the right or the left depending on the pressure
b. to the left
c. it will remain at the same concentrations
! d to the right

17. For the reaction, $2\ SO_2(g) + O_2(g) <\!\!==\!\!> 2\ SO_3(g)$, at 450.0K the equilibrium constant, $K_c$ has a value of 4.62 M. A system was charged to give these initial concentrations: $[SO_3] = 0.500$ M, $[O_2] = 0.00855$ M, $[SO_2] = 0.254$ M. In which direction will it go?

a. to the right
b. it will remain at the same concentrations
! c. to the left
d. to the right or the left depending on the pressure

18. For the reaction, $2\ SO_2(g) + O_2(g) <\!\!==\!\!> 2\ SO_3(g)$, at 450.0K the equilibrium constant, $K_c$ has a value of 4.62. A system was charged to give these initial concentrations: $[SO_3] = 0.0254$ M $[O_2] = 0.0855$ M $[SO_2] = 0.500$ M In which direction will it go

! a. to the right
b. to the left
c. to the right or the left depending on the pressure
d. it will remain at the same concentrations

19. For the reaction, $H_2(g) + I_2(g) <\!\!==\!\!> 2\ HI(g)$, $K_c = 54.9$ at 699.0 K. A system, at equilibrium contains 2.50 moles of HI and 2.12 moles of $I_2$ in a 5.00 liter vessel. How many moles of $H_2$ should there be in the container?

! a. 0.0537 moles
b. 0.380 moles
c. 0.0215 moles
d. 0.0107 moles
e. 2.12 moles

20. For the reaction, $H_2(g)$ + $I_2(g)$ <===> 2 HI($g$), $K_c$ = 54.9 at 699.0 K. A system was charged with 2.50 moles of HI initially in a 5.00 liter vessel. How many moles of $H_2$ should there be in the container when equilibrium is attained?

a. 0.337 moles
b. 1.25 moles
c. 0.297 moles
d. 0.500 moles
! e. 0.266 moles

21. For the reaction, $H_2(g)$ + $I_2(g)$ <===> 2 HI($g$), $K_c$ = 54.9 at 699.0K. A system was charged with 2.50 moles of $H_2$ and 2.50 moles $I_2$ initially in a 5.00 liter vessel. How many moles of $H_2$ should there be in the container when equilibrium is attained?

a. 2.50 moles
b. 0.674 moles
c. 0.594 moles
d. 1.00 moles
! e. 0.266 moles

22. For the reaction, $H_2(g)$ + $I_2(g)$ <===> 2 HI($g$), $K_c$ = 54.9 at 699.0K. A system was charged with 2.50 moles of $H_2$, 2.50 moles $I_2$ and 5.00 moles of HI initially in a 5.00 liter vessel. How many moles of $H_2$ should there be present at equilibrium?

a. 1.35 moles
b. 2.00 moles
c. 5.00 moles
d. 1.19 moles
! e. 1.06 moles

23. Given the following system, 4 $NH_3(g)$ + 7 $O_2(g)$ <===> 2 $N_2O_4(g)$ + 6 $H_2O(g)$. Initially $[NH_3]$ = $[O_2]$ = 3.60 M (the others are absent then) and at equilibrium, $[N_2O_4]$ is 0.60 M. Calculate the equilbrium concentration of $NH_3$.

a. 3.00 M
b. 2.10 M
c. 3.30 M
d. 1.80 M
! e. 2.40 M

24. Given the following system, 4 $NH_3(g)$ + 7 $O_2(g)$ <===> 2 $N_2O_4(g)$ + 6 $H_2O(g)$. Initially $[NH_3]$ = $[O_2]$ = 3.60 M (the others are absent then) and at equilibrium, $[N_2O_4]$ is 0.60 M. Calculate the equilbrium concentration of $O_2$.

a, 3.00 M
b. 2.40 M
! c. 1.50 M
d. 2.10 M
e. 3.30 M

25. Given the following system, $4\ NH_3(g)\ +\ 7\ O_2(g)$ <==> $2\ N_2O_4(g)\ +\ 6\ H_2O(g)$. Initially $[NH_3] = [O_2] = 3.60$ M (the others are absent then) and at equilibrium, $[N_2O_4]$ is 0.60 M. Calculate the equilbrium constant, $K_c$, for the reaction.

a. 8.10
b. 0.00000930
c. 0.300
! d. 0.0216
e. 3.33

26. Given the following system, $4\ NH_3(g)\ +\ 7\ O_2(g)$ <==> $2\ N_2O_4(g)\ +\ 6\ H_2O(g)$. Initially $[N_2O_4] = [H_2O] = 3.60$ M (the others are absent then) and at equilibrium, $[H_2O]$ is 0.60 M. Calculate the equilbrium concentration of $NH_3$.

a. 1.50
b. 0.90 M
c. 3.00 M
d. 2.40 M
! e. 2.00 M

27. Given the following system, $4\ NH_3(g)\ +\ 7\ O_2(g)$ <==> $2\ N_2O_4(g)\ +\ 6\ H_{20}(g)$. Initially $[N_2O_4] = [H_2O] = 3.60$ M (the others are absent then) and at equilibrium, $[H_2O]$ is 0.60 M. Calculate the equilbrium concentration of $O_2$.

a. 0.70 M
b. 3.00 M
c. 1.00 M
d. 2.40 M
! e. 3.50 M

28. The equilibrium constant for the reaction, $H_2(g)\ +\ I_2(g)$ <==> $2\ HI(g)$ is 54.9 at 699.0 K. What is the equilibrium constant for $4\ HI(g)$ <==> $2\ H_2(g)\ +\ 2\ I_2(g)$ under the same conditions?

a. 109.8
b. 0.00911
! c. 0.000332
d. -109.8
e. 0.0182

29. The reaction, $2\ SO_3(g)$ <==> $2\ SO_2(g)\ +\ O_2(g)$ is exothermic. Predict what will happen if the temperature is increased.

a. $K_c$ remains the same
b. $K_c$ decreases
c. the pressure decreases
d. more $SO_3(g)$ is produced
! e. $K_c$ increases

30. Consider the following system, which is at equilbrium,

$$CO(g) + 3\,H_2(g) \;<\!\!==\!\!>\; CH_4(g) + H_2O(g).$$

The result of removing some $CH_4(g)$ and $H_2O(g)$ from the system will be:

! a. more $CH_4(g)$ and $H_2O(g)$ are produced to replace that which is removed
b. $K_c$ decreases
c. more $CO(g)$ is produced
d. more $H_2O(g)$ is consumed to restore the equilibrium
e. more $CH_4$ is consumed to restore the equilibrium

31. Consider the following system, which is at equilbrium,

$$3C(s) + 3\,H_2(g) \;<\!\!==\!\!>\; CH_4(g) + C_2H_2(g)$$

The result of removing some $CH_4(g)$ and $C_2H_2(g)$ from the system will be,

a. no further change occurs
b. $K_c$ increases
c. more $C(s)$ is produced
d. more $C_2H_2(g)$ is consumed to restore the equilibrium
! e. more $CH_4(g)$ and $C_2H_2(g)$ are produced to replace that which is removed

32. Consider the following system, which is at equilbrium,

$$3\,C(s) + 3\,H_2(g) \;<\!\!==\!\!>\; CH_4(g) + C_2H_2(g).$$

The result of removing some $C(s)$ from the system will be:

a. $K_c$ increases
b. more $C(s)$ is produced
! c. no further change occurs
d. more $CH_4(g)$ and $C_2H_2(g)$ are produced to restore the equilibrium
e. more $C_2H_2(g)$ is consumed to restore the equilibrium

33. Given, $2\,H_2O(g) + 2\,Cl_2(g) \;<\!\!==\!\!>\; 4\,HCl(g) + O_2(g)$, for which $K_p = 8.00$ atm
Initially, the partial pressures of $H_2O$ and $Cl_2$ are set at 0.100 atm, while those of HCl and $O_2$ are set at 0.250 atm. Which statement below is true?

a. $Q_p > K_p$ and the reaction proceeds to the right to reach equilibrium
b. $Q_p < K_p$ and the reaction proceeds to the left to reach equilibrium
c. The reaction system is already at equilibrium
! d. $Q_p > K_p$ and the reaction proceeds to the left to reach equilibrium
e. $Q_p < K_p$ and the reaction proceeds to the right to reach equilibrium

34. $NH_2CO_2NH_4(s)$ when heated to 450 K undergoes the following reaction to produce a system at equilibrium:

$$NH_2CO_2NH_4(s) \; <\!\!==\!\!> \; 2\,NH_3(g) + CO_2(g).$$

The total pressure in the closed container under these condition is found to be 0.843 atm. Calculate a value for the equilibrium constant, $K_p$.

a. 0.00701 $atm^3$
! b. 0.0888 $atm^3$
c. 0.843 $atm^3$
d. 0.599 $atm^3$
c. 0.222 $atm^3$

35. For the chemical reaction, $H_2(g) + I_2(g) \; <\!\!==\!\!> \; 2\,HI(g)$, $K_c = 62.9$ at 750 K. 2.80 moles of HI were placed in a 10.0 liter container and it was brought up to 750 K and allowed to come to equilibrium. Which situation below is true, at equilibrium?

a. $[HI] = 2 \times [H_2]$
b. $[HI] = [H_2]$
c. $[HI] < [H_2]$
! d. $[HI] > [H_2]$, $[HI] \neq 2 \times [H_2]$
e. $[H_2] > [I_2]$

36. For the chemical reaction, $N_2(g) + 3\,H_2(g) \; <\!\!==\!\!> \; 2\,NH_3(g)$, $K_c = 4.2 \times 10^2$ at some temperature which we call $T_1$. 4.00 moles of $NH_3$ were placed in a 50.0 liter container and it was brought up to $T_1$ and allowed to come to equilibrium. Which situation below is true, at equilibrium?

a. $[NH_3] = 3 \times [H_2]$
! b. $[NH_3] > [H_2]$
c. $[H_2] > [NH_3]$
d. $[NH_3] = [H_2]$
e. $[N_2] = [NH_3]$

37. For the reaction system, $H_2(g) + 2\,X_2(g) \; <\!\!==\!\!> \; 2\,HX(g)$, $K_c = 24.4$ at 300 K. A system which is at equilibrium contains 0.150 moles of $H_2$ and 0.600 moles of HX in a 3.00 liter container. Calculate the number of mole of $X_2(g)$ present at equilibrium.

! a. 0.098 mol
b. 0.164 mol
c. 0.197 mol
d. 0.324 mol
e. 0.393 mol

38. For the reaction system, $H_2(g)$ + $2\ X_2(g)$ <===> $2\ HX(g)$, $K_c = 24.4$ at 300 K. A system which is at equilibrium contains 0.200 moles of $X_2$ and 0.600 moles of HX in a 4.00 liter container. Calculate the number of mole of $H_2(g)$ present at equilibrium.

a. 0.059 mol
! b. 0.074 mol
c. 0.123 mol
d. 0.148 mol
e. 0.295 mol

39. For the reaction system, $H_2(g)$ + $2\ X_2(g)$ <===> $2\ HX(g)$, $K_c = 24.4$ at 300 K. A system was charged with 0.150 moles of $H_2$, 0.150 moles of $X_2$, and 0.600 moles of HX in a 3.00 liter container. The catalyst was introduced using a remote unit, and the system was allowed to come to equilibrium. Which statement below describes the situation?

! a. The reaction goes to the right, $Q < K$.
b. The reaction goes to the left, $Q < K$.
c. The reaction goes to the right, $Q > K$.
d. The reaction goes to the left, $Q > K$.
e. It is not possible to predict in which direction the system will travel to arrive at equilibrium.

40. For the reaction system, $H_2(g)$ + $2\ X_2(g)$ <===> $2\ HX(g)$, $K_c = 24.4$ at 300 K. A system was charged with 2.00 of HX in a 3.00 liter container. The catalyst was introduced using a remote unit, and the system was allowed to come to equilibrium. How many moles of $H_2$ will be present when the system reaches equilibrium?

a. 0.268 moles
! b. 0.337 moles
c. 0.388 moles
d. 0.404 moles
e. 0.424 moles

41. For the reaction system, $H_2(g)$ + $2\ X_2(g)$ <===> $2\ HX(g)$, $K_c = 24.4$ at 300 K. A system was charged with 1.00 moles of $H_2$ and 1.00 moles of $I_2$ in a 3.00 liter container. The catalyst was introduced using a remote unit, and the system was allowed to come to equilibrium. How many moles of $H_2$ will be present when the system reaches equilibrium?

a. 0.268 moles
! b. 0.337 moles
c. 0.388 moles
d. 0.404 moles
e. 0.442 moles

42. Given the pair of reactions shown with the equilibrium constants,

$PCl_3(g) + Cl_2(g) \longrightarrow PCl_5(g) \quad K_1$

$2\ NO(g) + Cl_2(g) \longrightarrow 2\ NOCl(g) \quad K_2$

what is the equilibrium constant for the reaction,

$PCl_5(g + 2\ NO(g) <\!\!==\!\!> PCl_3(g) + 2\ NO(g)$

a. $K_1K_2$
! b. $K_2/K_1$
c. $K_1/K_2$
d. $(K_1K_2)^{-1}$
e. $K_2 - K_1$

43. Given the pair of reactions shown with the equilibrium constants,

$PCl_3(g) + \frac{1}{2}O_2(g) \longrightarrow POCl_3(g)\ K_1$

$NO(g) + \frac{1}{2}O_2(g) \longrightarrow NO_2(g) \quad K_2$

what is the equilibrium constant for the reaction,

$PCl_3(g + NO_2(g) <\!\!==\!\!> POCl_3(g) + NO(g)$

a. $K_1K_2$
b. $K_2/K_1$
! c. $K_1/K_2$
d. $(K_1K_2)^{-1}$
e. $K_2 - K_1$

## Fill in the Blanks

44. The correct expression for the equilibrium constant for the reaction below is
$CH_4(g) + 2\ NO_2(g) <\!\!==\!\!> CO_2(g) + 2\ H_2O(g) + N_2(g)$

$$! \ K_p = \frac{(p_{CH_3OH})}{(p_{CO})(p_{H_2})^2}$$

45. Write the expression for the equilibrium constant, $K_p$, for the reaction,
$N_2(g) + O_2(g) <\!\!==\!\!> 2\ NO(g)$

$$! \ K_p = \frac{(p_{NO})^2}{(p_{N_2})(p_{O_2})}$$

46. Write the expression for the equilibrium constant, $K_p$, for the reaction,
$CO(g) + 2\ H_2(g) <\!\!==\!\!> CH_3OH(g)$

$$! \ K_p = \frac{(p_{CH_3OH})}{(p_{CO})(p_{H_2})^2}$$

47. Write the expression for the equilibrium constant, $K_p$, for the reaction,
$4\,HCl(g) + O_2(g) <\!\!==\!\!> 2\,H_2O(g + 2\,Cl_2(g)$

$$! \; K_p = \frac{(p_{H_2O})^2(p_{Cl_2})^2}{(p_{HCl})^4(p_{O_2})}$$

48. Write the expression for the equilibrium constant, $K_c$, for the reaction,
$2\,AsF_3(g) + 3\,CCl_4(g) <\!\!==\!\!> 2\,AsCl_3(g) + 3\,CCl_2F_2(g)$

$$! \; K_c = \frac{[AsCl_3]^2[CCl_4]^3}{[AsF_3]^2[CCl_2F_2]^3}$$

49. Write the expression for the equilibrium constant, $K_c$, for the reaction,
$N_2H_4(g) + 6\,H_2O_2(g) <\!\!==\!\!> 2\,NO_2(g) + 8\,H_2O(g)$

$$! \; K_c = \frac{[NO_2]^2[H_2O]^8}{[N_2H_4][H_2O]^6}$$

50. Write the expression for the equilibrium constant, $K_c$, for the reaction,
$SO_2Cl_2(g) + 2\,NO(g) <\!\!==\!\!> SO_2(g) + 2\,NOCl(g)$

$$! \; K_c = \frac{[SO_2][NOCl]^2}{[SO_2Cl_2][NO]^2}$$

51. Write the expression for the equilibrium constant, $K_p$, for the reaction,
$NH_4Cl(s) <\!\!==\!\!> NH_3(g) + HCl(g)$

$$( ! \; K_p = (p_{NH_3})(p_{HCl})^2$$

52. Write the expression for the equilibrium constant, $K_p$, for the reaction,
$SnCl_2(s) + Cl_2(g) <\!\!==\!\!> SnCl_4(g)$

$$! \; K_p = \frac{(p_{SnCl_4})}{(p_{Cl_2})}$$

53. Write the expression for the equilibrium constant, $K_c$, for the reaction,
$4\,Cr(s) + 3\,CCl_4(g) <\!\!==\!\!> 4\,CrCl_3(g) + 4\,C(s)$

$$! \; K_c = \frac{[CrCl_3]^4}{[CCl_4]^3}$$

54. Write the expression for the equilibrium constant, $K_c$, for the reaction,
$C(s)$ + $H_2O(g)$ <===> $HCHO(g)$

$$! \; K_c = \frac{[HCHO]}{[H_2O]}$$

55. For the system, 2 $NO_2(g$ <===> $N_2O_4(g)$, $\delta G^o_{298}$ = -5.40 kJ. Calculate the value of the equilibrium constant, $K_p$, for this system. ( ! 8.83 $atm^{-1}$)

## True and False

56. In a chemical reaction system which has reached equilibrium, the concentrations of the various species are constantly changing even though the total number of molecules is unchanging. ( !F)

57. In a chemical reaction system which has reached equilibrium the net concentration of each species is unchanging even though particular molecules continue to react throughout the system. ( !T)

## Critical Thinking

58. Given the following values for components of a system, at 530 K. For $PCl_5$, $\delta H_f$ = -370.0 kJ $mol^{-1}$, S = 375.0 J $K^{-1}$ $mol^{-1}$. For $PCl_3$, $\delta H_f$ = -293.0 kJ $mol^{-1}$, S = 320.5 J $K^{-1}$ $mol^{-1}$. For $Cl_2$, $\delta H_f$ = 000.0 kJ $mol^{-1}$, S = 230.0 J $K^{-1}$ $mol^{-1}$. Calculate a value of $K_p$ at 540 K for
$PCl_5(g)$ <===> $PCl_3(g)$ + $Cl_2(g)$ ______ ( ! $K_p$ = 1.7)

59. Given the following values for components of a system, at 410 K. For $PCl_5$, $\delta H_f$ = -370.0 kJ $mol^{-1}$, S = 375.0 J $K^{-1}$ $mol^{-1}$. For $PCl_3$, $\delta H_f$ = -293.0 kJ $mol^{-1}$, S = 320.5 J $K^{-1}$ $mol^{-1}$. For $Cl_2$, $H_f$ = 000.0 kJ $mol^{-1}$, S = 230.0 J $K^{-1}$ $mol^{-1}$. Calculate a value of $K_p$ at 410 K for
$PCl_5(g)$ <===> $PCl_3(g)$ + $Cl_2(g)$ ______ ( ! $K_p$ = 4.9 x $10^{-3}$)

60. Given the following values for components of a system, at 410 K. For $PCl_5$, $\delta H_f$ = -370.0 kJ $mol^{-1}$, S = 375.0 J $K^{-1}$ $mol^{-1}$. For $PCl_3$, $\delta H_f$ = -293.0 kJ $mol^{-1}$, S = 320.5 J $K^{-1}$ $mol^{-1}$. For $Cl_2$, $\delta H_f$ = 000.0 kJ $mol^{-1}$, S = 230.0 J $K^{-1}$ $mol^{-1}$. Calculate a value of $K_c$ at 410 K for
$PCl_5(g)$ <===> $PCl_3(g)$ + $Cl_2(g)$ ______ ( ! $K_p$ = 1.5 x $10^{-4}$)

61. For the system, 2 $NO_2(g$ <===> $N_2O_4(g)$, $\delta G^o_{298}$ = -5.40 kJ. If the system was initialized using enough $NO_2$ to create a partial pressure of 0.50 atm and enough $N_2O_4$ to create a partial pressure of 0.35 atm, calculate the value of $\delta G_{298}$ for the other than standard conditions prevailing at the start.
______ ( ! -6.23 kJ)

62. For the system, $PCl_5(g)$ + 2 $NO(g)$ <===> $PCl_3(g)$ + 2 $NOCl(g)$, $\delta G^o$ = -9.720 kJ. If a system is charged with 0.200 atm of $PCl_5$, 0.800 atm of $PCl_3$, 0.800 atm of NOCl, and 0.100 atm of NO, what is the initial value of $\delta G_{298}$ for the other than standard conditions given above?
______ ( ! -23.47 kJ)

## Chemicals In Use 12: Ammonia

63. Which one of the following assertions about ammonia is true?

a. Low pressures favor the conversion of nitrogen and hydrogen into ammonia in higher yield than high pressures.
b. At very low temperatures (< -150 °C), ammonia decomposes into its elements rather quickly.
c. To make ammonia from hitrogen and hydrogen in favorable yield at reasonable temperatures to keep costs down, metal or metal oxide catalysts are needed.
d. Ammonia cannot be made directly from its constituent elements on a practical scale.
e. The formation of ammonia is an endothermic process, and therefore is an inherently low yield process.

64. The thermochemical equation representing one process used for the synthesis of ammonia involves the equilibrium shown,

$$N_2(g) + 3\,H_2(g) \Longleftrightarrow 2\,NH_3(g) \qquad \delta H^o = -92.2\ \text{kJ},\ \delta G^o = -33.4\ \text{kJ}$$

and leads one to conclude that

a. The coefficients give us the moe ratios but not the volume ratios of the reacting species.
! b. An increase in pressure favors an increase in the yield of ammonia.
c. Carrying out the reaction at a higher temperature shifts the position of equilibrium to the right, thus favoring an increase in the yield of ammonia.
d. Cooling the mixture to remove ammonia from the reaction mixture unfortunately also decreases the overall yield of ammonia.
e. The mixture does not need heating, in fact, cooling the mixture assists the rapid establishment of equilibrium.

65. The raw materials that enter a Haber-Bosch plan to make ammonia are

a. pure nitrogen and hudrogen
b. pure hydrogen chloride and nitrogen
c. water and nitrogen
! d. water, air, and natural gas
e. air, water and carbon dioxide

66. All of the following are needed for the Haber-Bosch reaction except

! a. carbon monoxide
b. methane
c. doped iron
d. water
e. air

# Chapter 16 Acid Base Equilibrium

## Multiple Choice

1. A solution has a $H^+$ concentration of $4.0 \times 10^{-2}$ molar. What is the $OH^-$ concentration in the same solution, in moles per liter?

a. $4.0 \times 10^{-2}$
b. $4.0 \times 10^{-9}$
c. $4.0 \times 10^{-12}$
! d. $2.5 \times 10^{-13}$
e. 25.0

2. If the $H^+$ ion concentration in an aqueous solution has a value of 0.100 molar, then the $OH^-$ concentration in moles per liter is

a. 0.100
b. $1.00 \times 10^{-7}$
c. $100 \times 10^{-12}$
! d. $1.00 \times 10^{-13}$
e. $4.00 \times 10^{12}$

3. If the $OH^-$ ion concentration in an aqueous solution is $6.6 \times 10^{-4}$ M, then the $H^+$ concentration in moles per liter in the same solution is,

a. $1.5 \times 10^{-1}$
b. $1.5 \times 10^{-4}$
c. $6.6 \times 10^{-10}$
! d. $1.5 \times 10^{11}$
e. $6.6 \times 10^{-11}$

4. If the $H^+$ ion concentration in a solution is measured as $6.60 \times 10^{-4}$ M, then the pH is

a. 3.0.
! b. 3.18
c. 6.60
d. 9.55
e. 10.82

5. Which one of the following would not be acidic?

a. grapefruit juice
b. household ammonia
c. carbonated water
d. vinegar
! e. lemon juice

6. Calculate the pH of a 0.0200 M solution of $Ca(OH)_2$:

a. 1.40
b. 0.040
c. 1.69
! d. 12.6
e. 12

7. If the dissociation constant, $K_a$, for macnic acid is $5.0 \times 10^{-5}$, what is the $pK_a$ of this acid?

a. $2 \times 10^4$
! b. 4.3
c. 5.7
d. $1.75 \times 10^{-1}$
e. 10.7

8. A 0.100 M solution of an acid, HA, has a pH = 2.00. What is the value of the dissociation constant, $K_a$ for this acid?

a. $1.1 \times 10^{-2}$
! b. $1.1 \times 10^{-3}$
c. $1.1 \times 10^{-4}$
d. $1.0 \times 10^{-3}$
e. $1.0 \times 10^{-4}$

9. A 0.200 M solution of an acid, H*Brun*, has a pH = 1.00. What is the value of the dissociation constant, $K_a$, for this acid?

a. 0.200
b. 0.0400
c. $1.00 \times 10^{-3}$
d. 1.00
! e. $1.00 \times 10^{-1}$

10. A 0.200 M solution of a weak base in water has a pH = 10.4. Calculate the value of $K_b$ for this base?

a. $1.0 \times 10^{-5}$
! b. $2.5 \times 10^{-4}$
c. $2.2 \times 10^{-5}$
d. $4.0 \times 10^{-11}$
e. $5.0 \times 10^{-5}$

11. What is the pH of a $1.00 \times 10^{-9}$ molar solution of HCl?

a. 5.00
b. 6.00
c. 6.80
! d. 7.00
e. 9.00

12. Given the following: 0.200M $NH_3$(*aq*) with $pK_b$ = 4.76 0.200M $HC_2H_3O_2$(*aq*) with $pK_a$ = 4.76 0.200M NaOH(*aq*) and 0.200M HCl(*aq*). A mixture is made using 50.0 ml of the HCl and 50.0 ml of water. What is its pH?

a. 0.70
! b. 1.00
c. 1.30
d. 1.61
e. 2.00

13. What is the pH of a 1.00 molar solution of NaCN(*aq*)? The $K_a$ for HCN = $6.2 \times 10^{-10}$.

a. 4.60
b. 9.21
c. 9.40
! d. 11.60
e. 13.00

14. Given the following: 0.200M $NH_3$(*aq*) with $pK_b$ = 4.76 0.200M $HC_2H_3O_2$(*aq*) with $pK_a$ = 4.76 0.200M NaOH(*aq*) and 0.200M HCl(*aq*). A mixture is made using 50.0 ml of the HCl and 50.0 ml of the $NH_3$. What is its pH?
,lh 0.175"

a. 1.00
b. 5.13
c. 5.29
! d. 8.87
e. 9.02

15. Given the following: 0.200M $NH_3$(*aq*) with $pK_b$ = 4.76 0.200M $HC_2H_3O_2$(*aq*) with $pK_a$ = 4.76 0.200M NaOH(*aq*) and 0.200M HCl(*aq*). A mixture is made using 50.0 ml of the $HC_2H_3O_2$ and 50.0 ml of the NaOH. What is its pH?

a. 1.00
b. 5.13
c. 8.71
! d. 8.87
e. 13.00

16. What is the pH of a 1.00 molar solution of $NaNO_2$(*aq*)? The $K_a$ for nitrous acid is $7.0 \times 10^{-4}$

a. 1.57
b. 3.15
c. 5.43
! d. 8.57
e. 10.85

17. What is the pH of a 1.00 molar solution of $NaNO_3$(*aq*)?

a. 6.00
b. 6.90
! c. 7.00
d. 7.10
e. 13.00

18. What is the pH of a 1.00 molar solution of KCl?

a. 0.00
b. 6.90
! c. 7.00
d. 7.10
e. 13.00

19. What is the pH of a 1.00 molar solution of $NH_4Cl(aq)$? The $K_b$ for $NH_3$ is $1.8 \times 10^{-5}$.

a. 2.37
! b. 4.63
c. 9.26
d. 9.37
e. 11.63

20. What is the pH of a 1.00 molar solution of $NaC_2H_3O_2(aq)$? The $K_a$ for acetic acid is $1.8 \times 10^{-5}$

a. 2.37
b. 4.63
c. 4.75
! d. 9.37
e. 11.63

21. The $pK_a$ of acetic acid, $HC_2H_3O_2$, is 4.76. A buffer solution contains $HC_2H_3O_2$ and $NaC_2H_3O_2$, has a pH of 4.76, and a ***total*** acetate concentration (***all forms***) of 0.400 molar. What is the acetate ion concentration $[C_2H_3O_2^-]$ in this buffer, in moles per liter?

a. 0.100
b. 0.150
! c. 0.200
d. 0.250
e. 0.300

22. The $pK_a$ of acetic acid, $HC_2H_3O_2$, is 4.76. A buffer solution contains $HC_2H_3O_2$ and $NaC_2H_3O_2$, has a pH of 4.60, and a ***total*** acetate concentration (***all forms***) of 0.400 molar. What is the acetate ion concentration $[C_2H_3O_2^-]$ in this buffer (molarity)?

a. 0.124
! b. 0.164
c. 0.200
d. 0.250
e. 0.324

23. The $pK_a$ of acetic acid, $HC_2H_3O_2$, is 4.76. A buffer solution was made using an unspecified amount of acetic acid and 0.300 moles of $NaC_2H_3O_2$ in enough water to make 2.000 liters of solution. Its pH was measured as 4.40 on a meter. How many moles of $HC_2H_3O_2$ were used?

a. 0.131
b. 0.180
c. 0.600
! d. 0.687
e. 0.830

24. The $pK_a$ of acetic acid, $HC_2H_3O_2$, is 4.76. A buffer solution was made using an unspecified amount of acetic acid and 0.300 moles of $NaC_2H_3O_2$ in enough water to make 1.500 liters of solution. Its pH was measured as 4.50 on a meter. How many moles of $HC_2H_3O_2$ were used?

a. 0.3100
! b. 0.546
c. 0.610
d. 0.819
e. 1.214

25. The $pK_a$ of acetic acid, $HC_2H_3O_2$, is 4.76. A buffer solution was made using an unspecified amount of $NaC_2H_3O_2$ and 0.300 moles of acetic acid in enough water to make 2.000 liters of solution. Its pH was measured as 4.60 on a meter. How many moles of $NaC_2H_3O_2$ were used?

a. 0.104
! b. 0.208
c. 0.300
d. 0.415
e. 0.434

26. The $pK_a$ of acetic acid, $HC_2H_3O_2$, is 4.76. A buffer solution was made using an unspecified amount of $NaC_2H_3O_2$ and 0.300 moles of acetic acid in enough water to make 1.500 liters of solution. Its pH was measured as 4.55 on a meter. How many moles of $NaC_2H_3O_2$ were used?

a. 0.100
! b. 0.185
c. 0.300
d. 0.370
e. 0.487

27. The $pK_b$ of ammonia solution, $NH_3(aq)$, is 4.76. A buffer solution contains $NH_4Cl$ and $NH_3$, has a pH of 9.45, and a ***total*** ammonia concentration (***all forms***) of 0.400 molar. What is the ammonium ion concentration $[NH_4^+]$ in this buffer?

! a. 0.153
b. 0.186
c. 0.210
d. 0.247
e. 0.300

28. The $pK_b$ of ammonia solution, $NH_3(aq)$, is 4.76. A buffer solution contains $NH_4Cl$ and $NH_3$, has a pH of 9.65, and a ***total*** ammonia concentration (***all forms***) of 0.350 molar. What is the ammonia concentration $[NH_3]$ in this buffer?

a. 0.098
b. 0.146
! c. 0.252
d. 0.389
e. 0.420

29. What is the pH of a 1.00 molar solution of NaCl(*aq*)?

a. 0.00
b. 6.90
! c. 7.00
d. 7.10
e. 13.00

30. What is the pH of a 1.00 molar solution of nitrous acid? The $K_a$ for nitrous acid is $7.1 \times 10^{4}$.

a. 1.57
! b. 1.58
c. 1.60
d. 2.67
e. 3.15

31. Here is a list of some weak acids and their pK's. H*Mac*, $pK_a = 4.46$ H*Tern*, $pK_a = 3.50$ H*Brun*, $pK_a = 5.33$ H*THarn*, $pK_a = 7.33$ H*Pen*, $pK_a = 8.24$ H*Fern*, $pK_a = 6.42$ Which reagents below would you use to make a buffer with pH = 7.00

a. Brunic acid (H*Brun*), and either sodium or potassium brunate
b. Macnic acid (H*Mac*), and either sodium or potassium macnate
c. Pentic acid (H*Pen*), and either sodium or potassium pentate
! d. Tharnic acid (H*THarn*), and either sodium or potassium tharnate
e. Ternic acid (H*Tern*), and either sodium or potassium ternate

32. Here is a list of some weak acids and their pK's. H*Mac*, $pK_a = 4.46$ H*Tern*, $pK_a = 3.50$ H*Brun*, $pK_a = 5.33$ H*THarn*, $pK_a = 7.33$ H*Pen*, $pK_a = 8.24$ H*Fern*, $pK_a = 6.42$ Which reagents below would you use to make a buffer with pH = 5.50

! a. Brunic acid (H*Brun*), and either sodium or potassium brunate
b. Macnic acid (H*Mac*), and either sodium or potassium macnate
c. Pentic acid (H*Pen*), and either sodium or potassium pentate
d. Tharnic acid (H*THarn*), and either sodium or potassium tharnate
e. Fernic acid (H*Fern*), and either sodium or potassium fernate

33. Here is a list of some weak acids and their pK's. H*Mac*, $pK_a = 4.46$ H*Tern*, $pK_a = 3.50$ H*Brun*, $pK_a = 5.33$ H*THarn*, $pK_a = 7.33$ H*Pen*, $pK_a = 8.24$ H*Fern*, $pK_a = 6.42$ Which reagents below would you use to make a buffer with pH = 3.75?

a. Brunic acid (H*Brun*), and either sodium or potassium brunate
b. Fernic acid (H*Fern*), and either sodium or potassium fernate
c. Pentic acid (H*Pen*), and either sodium or potassium pentate
d. Tharnic acid (H*THarn*), and either sodium or potassium tharnate
! e. Ternic acid (H*Tern*), and either sodium or potassium ternate

34. A buffer solution is prepared by taking 0.400 moles of acetic acid (pKa = 4.76) and 0.250 moles of sodium acetate in sufficient water to make 1.400 liters of solution. Calculate the pH of this solution.

a. 4.46
! b. 4.56
c. 4.66
d. 4.86
e. 4.96

35. A buffer solution is prepared by taking 0.250 moles of acetic acid (pKa = 4.76) and 0.400 moles of sodium acetate in sufficient water to make 1.800 liters of solution. Calculate the solution pH.

a. 4.56
b. 4.66
c. 4.86
! d. 4.96
e. 5.06

36 A buffer solution is prepared by taking 0.400 moles of acetic acid (pKa = 4.76) and 0.250 moles of calcium acetate in sufficient water to make 1.400 liters of solution. Calculate the pH of this solution.

a. 4.46
b. 4.56
c. 4.66
! d. 4.86
e. 4.96

37. A buffer solution is prepared by taking 0.250 moles of acetic acid (pKa = 4.76) and 0.400 moles of barium acetate in sufficient water to make 1.400 liters of solution. Calculate the pH of this solution.

a. 4.25
b. 4.45
c. 5.00
! d. 5.27
e. 5.35

38. 40.0 ml of 0.100 M HCl(*aq*) was added to 50.0 ml of 0.100 M NaOH(*aq*) and the mixture was stirred, then tested with a pH meter. What pH should be obtained?

a. 1.95
b. 2.00
c. 7.00
d. 12.00
! e. 12.05

39. 50.0 ml of 0.100 M HCl(*aq*) was added to 40.0 ml of 0.100 M NaOH(*aq*) and the mixture was stirred, then tested with a pH meter. What reading should be obtained for the pH?

! a. 1.95
b. 2.00
c. 7.00
d. 12.00
! e. 12.05

40. 20.0 ml of 0.100 M $H_2SO_4$(*aq*) was added to 50.0 ml of 0.100 M NaOH(*aq*) and the mixture was stirred, then tested with a pH meter. What reading should be obtained for the pH?

a. 1.85
b. 2.00
c. 7.00
d. 12.00
! e. 12.15

41. 40.0 ml of 0.200 M HCl(*aq*) was added to 50.0 ml of 0.200 M NaOH(*aq*) and the mixture was stirred, then tested with a pH meter. What reading should be obtained for the pH?

a. 1.65
b. 2.00
c. 7.00
d. 12.00
! e. 12.35

42. 40.0 ml of 0.200 M HCl(*aq*) was added to 50.0 ml of 0.100 M NaOH(*aq*) and the mixture was stirred, then tested with a pH meter. What pH should be obtained?

a. 0.70
! b. 1.48
c. 1.65
d. 12.00
e. 12.52

43. Given the following: 0.200M $NH_3$(*aq*) with $pK_b$ = 4.76 0.200M $HC_2H_3O_2$(*aq*) with $pK_a$ = 4.76 0.200M NaOH(*aq*) and 0.200M HCl(*aq*). A mixture is made using 50.0 ml of the HCl and 50.0 ml of the NaOH. What is its pH?

a. 3.62
b. 4.57
c. 6.30
! d. 7.00
e. 9.46

44. Given the following: 0.200M $NH_3$(*aq*) with $pK_b$ = 4.76 0.200M $HC_2H_3O_2$(*aq*) with $pK_a$ = 4.76 0.200M NaOH(*aq*) and 0.200M HCl(*aq*). A mixture is made using 50.0 ml of the NaOH and 50.0 ml of water. What is its pH?

a. 1.00
b. 4.55
c. 7.00
! d. 13.00
e. 13.30

45. Given the following: 0.200M $NH_3$(*aq*) with $pK_b$ = 4.76 0.200M $HC_2H_3O_2$(*aq*) with $pK_a$ = 4.76 0.200M NaOH(*aq*) and 0.200M HCl(*aq*). A mixture is made using 50.0 ml of the NaOH and 25.0 ml of the HCl. What is its pH?

a. 7.00
b. 12.52
! c. 12.82
d. 13.00
e. 13.30

46. Given the following: 0.200M $NH_3(aq)$ with $pK_b$ = 4.76 0.200M $HC_2H_3O_2(aq)$ with $pK_a$ = 4.76 0.200M NaOH(*aq*) and 0.200M HCl(*aq*).
What is its pH?

a. 1.18
b. 8.87
c. 8.94
! d. 9.24
e. 9.54

47. Given the following: 0.200M $NH_3(aq)$ with $pK_b$ = 4.76 0.200M $HC_2H_3O_2(aq)$ with $pK_a$ = 4.76 0.200M NaOH(*aq*) and 0.200M HCl(*aq*). A mixture is made using 50.0 ml of the $HC_2H_3O_2$ and 25.0 ml of the NaOH. What is its pH?

a. 2.81
b. 4.46
! c. 4.76
d. 5.06
e. 12.82

48. Given the following: 0.200M $NH_3(aq)$ with $pK_b$ = 4.76 0.200M $HC_2H_3O_2(aq)$ with $pK_a$ = 4.76 0.200M NaOH(*aq*) and 0.200M HCl(*aq*). A mixture is made using 25.0 ml of the HCl and 50.0 ml of the NaOH. What is its pH?

a. 1.18
b. 4.62
c. 7.00
d. 11.38
! e. 12.82

49. Which one of the following is the most appropriate chemical indicator to use in the titration of 0.100 molar HCl(*aq*) by 0.100 molar NaOH(*aq*)? ( the titrant is NaOH )

a. thymol blue, $pK_{ind}$ = 8.8 HInd is yellow and $Ind^-$ is blue
b. bromothymol blue, $pK_{ind}$ = 6.8 HInd is yellow and $Ind^-$ is blue
c. phenolphthalein, $pK_{ind}$ = 9.1 HInd is colorless and $Ind^-$ is pink
d. cresol red, $pK_{ind}$ = 7.9 HInd is yellow and $Ind^-$ is red
! e. These indicators will all work equally well for the titration of a strong acid by a strong base.

50. Which one of the following is the most appropriate chemical indicator to use in the titration of 0.100 molar acetic acid(*aq*) by 0.100 molar NaOH(*aq*)? This is the titration of a weak acid by a strong base. (titrant is NaOH)

a. phenol red, $pK_{ind}$ = 7.3 HInd is yellow and $Ind^-$ is red
b. bromothymol blue, $pK_{ind}$ = 6.8 HInd is yellow and $Ind^-$ is blue
! c. phenolphthalein, $pK_{ind}$ = 9.1 HInd is colorless and $Ind^-$ is pink
d. cresol red, $pK_{ind}$ = 7.9 HInd is yellow and $Ind^-$ is red
e. These indicators will all work equally well for the titration of a weak acid by a strong base.

51. Which one of the following is the most appropriate chemical indicator to use in the titration of 0.100 molar ammonia solution by 0.100 molar HCl(*aq*)? (titrant is HCl)

a. thymol blue, $pK_{ind}$ = 8.8 HInd is yellow and $Ind^-$ is blue
! b. methyl orange, $pK_{ind}$ = 3.8 HInd is red and $Ind^-$ is yellow
c. thymolphthalein, $pK_{ind}$ = 9.9 HInd is colorless and $Ind^-$ is blue
d. cresol red, $pK_{ind}$ = 7.9 HInd is yellow and $Ind^-$ is red
e. These indicators will all work equally well for the titration of a weak acid by a strong base.

52. A mixture contains 0.060 moles of $NaH_2PO_4$ and 0.080 moles of $K_2HPO_4$. It is titrated with 0.500 molar NaOH(*aq*) to neutralize it completely. How many ml of the NaOH solution are required?

a. 280 ml
b. 313.3 ml
c. 372 ml
! d. 400 ml
e. 440 ml

53. A solution is made by taking 100.0 ml of $4.0 \times 10^{-3}$ molar NaOH(*aq*) and mixing in 10.0 ml of $4.0 \times 10^{-2}$ molar HCl(*aq*). The acidity of a small portion was tested using the chemical indicator thymol blue, for which the $pK_{ind}$ = 8.8 and the colors are yellow for Hind and blue for $Ind^-$. What color would be observed as a result of the testing?

a. blue
b. green
! c. yellow
d. orange
e. red

54. A solution is made by taking 100.0 ml of $2.0 \times 10^{-2}$ molar NaOH(*aq*) and mixing in 50.0 ml of $3.0 \times 10^{-2}$ molar HCl(*aq*). The acidity of a small portion was tested using the chemical indicator tyrolian blue, for which the $pK_{ind}$ = 12.5 and the colors are blue for Hind and yellow for $Ind^-$. What color would be observed as a result of the testing?

! a. blue
b. green
c. yellow
d. orange
e. red

55. A solution is made by taking 2.42 grams of NaCl, 6.44 grams of $NaClO_4$, 3.81 grams of $KNO_3$, and 4.64 grams of KBr in enough water to make 500 ml of solution. A portion was treated with thymol blue indicator, for which the $pK_{ind}$ = 8.8 and the colors are yellow for Hind and blue for $Ind^-$. What color should be obtained with the test portion?

a. blue
b. green
! c. yellow
d. orange
e. red

56. A solution is made by mixing 50.0 ml of a nitric acid solution which has a pH of 4.00 with 50.0 ml of a NaOH solution which has a pH of 11.00. What is the pH of the mixture?

a. 7.00
b. 7.50
c. 10.35
! d. 10.65
e. 10.95

57. A solution is made by mixing 50.0 ml of a HCl(*aq*) solution which has a pH of 3.00 with 50.0 ml of a KOH solution which has a pH of 11.00. What is the pH of the mixture?

a. 3.35
b. 4.25
! c. 7.00
d. 7.50
e. 10.65

58. A solution is made by mixing 100.0 ml of a HCl(*aq*) solution which has a pH of 3.00 with 50.0 ml of a KOH solution which has a pOH of 6.00. What is the pH of the mixture?

a. 3.00
! b. 3.185
c. 4.50
d. 7.00
e. 9.25

59. The pH at the equivalence point of the titration of acetic acid solution by sodium hydroxide solution is

a. $< 2.00$
b. $3.50 < pH < 7.00$
c. 7.00
! d. $7.00 < pH < 11.00$
e. $> 11.0$

60. The pH at the equivalence point of the titration of perchloric acid solution by sodium hydroxide solution is

a. $< 2.00$
b. $4.00 < pH < 7.00$
! c. 7.00
d. $7.00 < pH < 11.00$
e. $> 11.0$

61. The pH at the equivalence point of the titration of ammonia solution by hydrochloric acid solution is

a. $< 2.00$
! b. $4.00 < pH < 7.00$
c. 7.00
d. $7.00 < pH < 11.00$
e. $> 11.0$

## Fill in the Blanks

62. The $pK_a$ of a weak acid is 6.50. What is the value of $K_a$ for this acid? ______ (! $3.2 \times 10^{-6}$)

63. A 0.100 molar solution of a moderately strong monoprotic acid in water has a pH = 1.88. Calculate the value of Ka, the acid dissociation constant for this acid. ______ (! $2.00 \times 10^{-3}$)

## True and False

64. A relation involving acids and bases is, $K_a + K_b = K_w$. (!F)

65. A solution of sodium cyanide in water should be basic, not neutral or acid. (!T)

66. The titration of a strong acid by a strong base gives sharper and more well defined end points than titration of a weak acid by a weak base. (!T)

67. A solution of ammonium nitrate in water should be basic, not neutral or acidic. (!F)

68. The ion product constant for water, $K_w$, varies with the temperature of the water. (!T)

69. the $pK_a$ is a reliable measure of the strength of an acid. Strong acids have larger pK values than weak acids. (!F)

## Critical Thinking

70. A student does a titration of $NH_3$(*aq*)(pKb = 4.76) with HCl(*aq*), taking 500.0 ml of 0.300 M $NH_3$(*aq*) and adding 100.0 ml of the HCl solution. He measured the pH of the mixture and found it was 8.94. What was the molarity of the HCl solution? ______ (! 1.00 molar)

71. A student does a titration of $NH_3$(*aq*) (pKb = 4.76) with HCl(*aq*), taking 500.0 ml of 0.300 M $NH_3$(*aq*) and adding 100.0 ml of the HCl solution. He measured the pH of the mixture and found it was 9.54. What was the molarity of the HCl solution? ______ (! 0.500 molar)

72. Given the following: $NH_3$(*aq*) with $pK_b$ = 4.76 $HC_2H_3O_2$(*aq*) with $pK_a$ = 4.76 If a solution is made by mixing 100.0 ml of 0.300 molar $NH_4Cl$(*aq*), 100 ml of 0.300 molar $NH_3$(*aq*), and 100 ml of 0.100 molar NaOH. What should be the pH of the resulting mixture? ______ (! 9.54)

73. Given the following: $NH_3$(*aq*) with $pK_b$ = 4.76 $HC_2H_3O_2$(*aq*) with $pK_a$ = 4.76 If a solution is made by mixing 100.0 ml of 0.300 molar $NaC_2H_3O_2$(*aq*), 100 ml of 0.300 molar $HC_2H_3O_2$(*aq*), and 100 ml of 0.100 molar NaOH. What should be the pH of the resulting mixture? ______ (! 4.46)

74. Given the following: $NH_3$(*aq*) with $pK_b$ = 4.76 $HC_2H_3O_2$(*aq*) with $pK_a$ = 4.76 If a solution is made by mixing 100.0 ml of 0.300 molar $NH_4Cl$(*aq*), 100 ml of 0.300 molar $NH_3$(*aq*), and 100 ml of 0.100 molar HCl(*aq*). What should be the pH of the resulting mixture? ______ (! 8.94)

75. Given the following: $NH_3(aq)$ with $pK_b = 4.76$ $HC_2H_3O_2(aq)$ with $pK_a = 4.76$ If a solution is made by mixing 100.0 ml of 0.300 molar $NaC_2H_3O_2(aq)$, 100 ml of 0.300 molar $HC_2H_3O_2(aq)$, and 100 ml of 0.100 molar HCl. What should be the pH of the resulting mixture? ______ ( ! 4.46)

76. A buffer solution was made by mixing 400.0 ml of 1.00 molar aqueous $NH_3$ (which has a $pK_b$ = 4.76) with 400.0 ml of 1.00 molar aqueous sodium acetate solution. 200.0 ml of 0.500 molar sodium hydroxide solution was added to the mixture. What is the pH of the buffer before addition of the base and after addition of the base? ______, ______ ( ! 9.24, 9.46)

77. A buffer solution was made by mixing 400.0 ml of 1.00 molar aqueous $NH_3$ (which has a $pK_b$ = 4.76) with 400.0 ml of 1.00 molar aqueous sodium acetate solution. 200.0 ml of 0.500 molar hydrochloric acid solution was added to the mixture. What is the pH of the buffer before addition of the base and after addition of the acid? ______, ______ ( ! 9.24, 9.02)

## Chemicals In Use 13: Buffers and Breathing

78. "Respiration" refers to

! a. the transport of the blood gases
b. loss of "acid" from the body
c. breathing carbon dioxide
d. the buffering of blood
e. the conversion of $CO_2(g)$ to $H_2CO_3(aq)$

79. Normally, the blood in the arterial system maintains a $[CO_2]/[HCO_3^-]$ ratio such that the pH of the blood is about

a. 4.2
b. 5.6
! c. 7.4
d. 9.2
e. 10.8

80. When a person hyperventilates, the body involuntarily removes $CO_2$ at an excessive rate. What effect does this have on the pH of the blood?

a. no effect, pH remains at 5.6
! b. increases the pH
c. decreases the pH
d. no effect, pH remains at 7.4
e. instead of remaining steady, the pH fluctuates constantly, causing sweating, then shivering, then back again

# Chapter 17 Solubility and Simultaneous Equilibria

## Multiple Choice

1. For $H_3PO_3$, which is actually a diprotic acid, $K_{a1} = 1.0 \times 10^{-2}$ and $K_{a2} = 2.6 \times 10^{-7}$. Calculate a value for the $[H^+]$ in a 0.500 molar solution of $H_3PO_3$?

a. $1.0 \times 10^{-2}$ mol/L
! b. $6.6 \times 10^{-2}$ mol/L
c. $7.1 \times 10^{-2}$ mol/L
d. $8.5 \times 10^{-3}$ mol/L
e. $3.3 \times 10^{-3}$ mol/L

2. For $H_3PO_3$, which is actually a diprotic acid, $K_{a1} = 1.0 \times 10^{-2}$ and $K_{a2} = 2.6 \times 10^{-7}$. Calculate a value for the $[H_2PO_3^-]$ in a 0.500 molar solution of $H_3PO_3$?

a. $1.0 \times 10^{-2}$ mol/L
! b. $6.6 \times 10^{-2}$ mol/L
c. $7.1 \times 10^{-2}$ mol/L
d. $8.5 \times 10^{-3}$ mol/L
e. $3.3 \times 10^{-3}$ mol/L

3. For $H_3PO_3$, which is actually a diprotic acid, $K_{a1} = 1.0 \times 10^{-2}$ and $K_{a2} = 2.6 \times 10^{-7}$. Calculate a value for the $[HPO_3^{2-}]$ in a 0.500 molar solution of $H_3PO_3$?

! a. $2.6 \times 10^{-7}$ mol/L
b. $6.6 \times 10^{-2}$ mol/L
c. $7.1 \times 10^{-2}$ mol/L
d. $8.5 \times 10^{-5}$ mol/L
e. $3.3 \times 10^{-3}$ mol/L

4. Tenzic acid, $H_2$*Tenz*, is a diprotic acid with dissociation constants: $K_{a1} = 4.0 \times 10^{-4}$, $K_{a2} = 1.5 \times 10^{-9}$. Calculate a value for the $[H^+]$ in a 0.500 molar solution of $H_2$*Tenz*?

! a. $1.4 \times 10^{-2}$ mol/L
b. $1.6 \times 10^{-2}$ mol/L
c. $4.0 \times 10^{-4}$ mol/L
d. $2.0 \times 10^{-2}$ mol/L
e. $4.0 \times 10^{-2}$ mol/L

5. Tenzic acid, $H_2$*Tenz*, is a diprotic acid with dissociation constants: $K_{a1} = 4.0 \times 10^{-4}$, $K_{a2} = 1.5 \times 10^{-9}$. Calculate a value for the [H*Tenz*$^-$] in a 0.500 molar solution of $H_2$*Tenz*?

! a. $1.4 \times 10^{-2}$ mol/L
b. $1.6 \times 10^{-2}$ mol/L
c. $4.0 \times 10^{-4}$ mol/L
d. $2.0 \times 10^{-2}$ mol/L
e. $4.0 \times 10^{-2}$ mol/L

6. Tenzic acid, $H_2$*Tenz*, is a diprotic acid with dissociation constants: $K_{a1} = 4.0 \times 10^{-4}$, $K_{a2} = 1.5 \times 10^{-9}$. Calculate a value for the [$Tenz^{2-}$] in a 0.500 molar solution of $H_2$*Tenz*?

a. $7.7 \times 10^{-7}$ mol/L
b. $2.7 \times 10^{-5}$ mol/L
c. $3.8 \times 10^{-5}$ mol/L
d. $4.0 \times 10^{-4}$ mol/L
! e. $1.5 \times 10^{-9}$ mol/L

7. For $H_3PO_4$, a tripotic acid, $K_{a1} = 7.1 \times 10^{-3}$, $K_{a2} = 6.3 \times 10^{-8}$ and $K_{a3} = 4.5 \times 10^{-13}$ What is the pH of a solution which contains 0.200 moles of $KH_2PO_4$ and 0.300 moles of $Na_3PO_4$ per liter of solution?

a. 7.20
b. 7.34
c. 12.17
d. 12.35
! e. 12.52

8. For $H_3PO_4$, a tripotic acid, $K_{a1} = 7.1 \times 10^{-3}$, $K_{a2} = 6.3 \times 10^{-8}$ and $K_{a3} = 4.5 \times 10^{-13}$ What is the pH of a solution which contains 0.300 moles of $Na_2HPO_4$ and 0.400 moles of $KH_2PO_4$ per liter of solution?

! a. 7.08
b. 7.20
c. 7.33
d. 10.20
e. 12.22

9. Carbonic acid, $H_2CO_3$, is a diprotic acid with dissociation constants: $K_{a1} = 4.5 \times 10^{-7}$, $K_{a2} = 4.7 \times 10^{-11}$. What is the value of the ratio, $[H_2CO_3]/[HCO_3^-]$, in a solution containing both species which is maintained at pH = 7.00 by means of a buffer.

a. 1.00
b. 0.33
c. 0.25
! d. 0.52
e. 1.33

10. The expression for the solubility product of $Ba_3(AsO_4)_2$ is,

! a. $[Ba^{2+}]^3[AsO_4^{3-}]^2$
b. $[3Ba^{2+}]^3[2AsO_4^{3-}]^2$
c. $3[Ba^{2+}] \times 2[AsO_4^{3-}]$
d. $3[Ba^{2+}]^3 + 2[AsO_4^{3-}]^2$
e. $[Ba^{2+}]_3[AsO_4^{3-}]_2$

11. The expression for the solubility product of copper(II) hydroxide is,

a. $[Cu^{2+}][2\ OH^-]$
b. $[Cu^{2+}] \times 2[OH^-]^2$
c. $[Cu^{2+}]^2[OH^-]$
! d. $[Cu^{2+}][OH^-]^2$
e. $[Cu^{2+}] \times \frac{1}{2}[OH^-]^2$

12. The expression for the solubility product of silver oxalate ($Ag_2C_2O_4$) is,

a. $[Ag_2^{2+}][C_2O_4^{2-}]$
b. $[Ag^+][C_2O_4^{2-}]^2$
c. $2[Ag^+][C_2O_4^{2-}]$
! d. $[Ag^+]^2[C_2O_4^{2-}]$
e. $2[Ag^+]^2[C_2O_4^{2-}]$

13. The solubility of $Ag_2NtO_4$, silver nortonate, in pure water is $4.0 \times 10^{-5}$ moles per liter. Calculate the value of $K_{sp}$ for silver nortonate.

a. $1.6 \times 10^{-9}$
b. $6.4 \times 10^{-14}$
! c. $2.6 \times 10^{-13}$
d. $4.0 \times 10^{-5}$
e. $4.0 \times 10^{-15}$

14. Which one of the following salts has the highest solubility in water, expressed in moles per liter?

a. $PbF_2$, $K_{sp} = 3.6 \times 10^{-8}$
b. $Ag_2CrO_4$, $K_{sp} = 1.2 \times 10^{-12}$
c. $CaF_2$, $K_{sp} = 3.9 \times 10^{-11}$
! d. $BaF_2$, $K_{sp} = 1.7 \times 10^{-6}$
e. $PbI_2$, $K_{sp} = 7.9 \times 10^{-9}$

15. The solubility of silver sulfate ($Ag_2SO_4$), in moles per liter, can be expressed in terms of the resulting ion concentrations. Which one of the following relationships is correct?

a. solubility = $2[Ag^+]$
b. solubility = $[Ag^+]$
c. solubility = $[2Ag^+]$
d. solubility = $2[SO_4^{2-}]$
! e. solubility = $[SO_4^{2-}]$

16. The solubility product for $Ag_3PO_4$ is: $K_{sp} = 2.8 \times 10^{-18}$. What is the solubility of $Ag_3PO_4$ in water, in moles per liter?

! a. $1.8 \times 10^{-5}$ M
b. $2.5 \times 10^{-5}$ M
c. $1.7 \times 10^{-91}$ M
d. $3.1 \times 10^{-5}$ M
e. $4.1 \times 10^{-5}$ M

17. The solubility product for $BaSO_4$ is $1.1 \times 10^{-10}$. Calculate the solubility of $BaSO_4$ in pure water, in moles per liter.

a. $5.5 \times 10^{-11}$ mol/L
! b. $1.0 \times 10^{-5}$ mol/L
c. $2.1 \times 10^{-5}$ mol/L
d. $1.1 \times 10^{-10}$ mol/L
e. $2.2 \times 10^{-10}$ mol/L

18. The solubility product for $PbCl_2$ is $1.7 \times 10^{-5}$. What is the solubility of $PbCl_2$ in pure water, in moles per liter?

a. $2.4 \times 10^{-4}$ mol/L
b. $6.2 \times 10^{-2}$ mol/L
c. $7.7 \times 10^{-3}$ mol/L
! d. $1.6 \times 10^{-2}$ mol/L
e. $6.0 \times 10^{-5}$ mol/L

19. Calculate the concentration of chloride ions in a saturated solution of lead(II) chloride. The $K_{sp} = 1.7 \times 10^{-5}$.

a. $2.4 \times 10^{-4}$ M
b. $4.8 \times 10^{-4}$ M
c. $3.9 \times 10^{-2}$ M
d. $1.2 \times 10^{-1}$ M
! e. $3.2 \times 10^{-2}$ M

20. The solubility product of barium fluoride ($BaF_2$) is $1.7 \times 10^{-6}$. Calculate the concentration of fluoride ions in a saturated solution of barium fluoride.

a. $7.6 \times 10^{-3}$ M
! b. $1.5 \times 10^{-2}$ M
c. $3.4 \times 10^{-5}$ M
d. $1.7 \times 10^{-6}$ M
e. $3.4 \times 10^{-6}$ M

21. The solubility product for $Mg_3(PO_4)_2$ is $6.3 \times 10^{-26}$. What is the solubility of $Mg_3(PO_4)_2$ in pure water, in grams per liter?

a. $1.7 \times 10^{-23}$ g/L
b. $3.4 \times 10^{-7}$ g/L
! c. $9.5 \times 10^{-4}$ g/L
d. $1.2 \times 10^{-3}$ g/L
e. $2.4 \times 10^{-3}$ g/L

22. The solubility of lead iodide is 578 mg/L at 25 °C. What is the solubility product for $PbI_2$?

! a. $7.9 \times 10^{-9}$
b. $1.6 \times 10^{-6}$
c. $1.1 \times 10^{-11}$
d. $2.7 \times 10^{-12}$
e. $6.3 \times 10^{-6}$

23. The solubility of barium carbonate is 14.8 mg/L at 30 °C. Calculate the $K_{sp}$ value for $BaCO_3$?

a. $7.5 \times 10^{-5}$
b. $1.5 \times 10^{-4}$
! c. $5.6 \times 10^{-9}$
d. $7.5 \times 10^{-6}$
e. $1.5 \times 10^{-3}$

24. What is the solubility, in moles per liter, of AgCl ($K_{sp} = 1.8 \times 10^{-10}$) in 0.0100 molar aqueous potassium chloride solution?

a. $7.5 \times 10^{-5}$ mol/L
! b. $1.8 \times 10^{-8}$ mol/L
c. $1.3 \times 10^{-6}$ mol/L
d. $3.6 \times 10^{-8}$ mol/L
e. $1.5 \times 10^{-7}$ mol/L

25. The $K_{sp}$ value for $MgF_2$ is $6.9 \times 10^{-9}$. A 0.10 M NaF solution is saturated with magnesium fluoride. How many parts per million (ppm) of magnedium are there in this saturated solution?

a. 0.0013
b. 40
c. 0.040
! d. 0.016
e. 2.65

26. The solubility product of $Ag_2CrO_4$ is $1.2 \times 10^{-12}$. How many parts per million (ppm) of silver are there is a saturated solution of silver chromate in pure water?

a. 0.012
b. 0.24
c. 7.2
! d. 14
e. 1.6

27. The solubility of barium sulfate varies with the composition of the solvent. In which one of the solvent mixtures below would $BaSO_4$ have the lowest solubility?

a. pure water
b. 0.10 M $Na_2SO_4$(aq)
! c. 1.0 M $(NH_4)_2SO_4$(aq)
d. 0.5 M $Ba(NO_3)_2$(aq)
e. 1.0 M HCl(aq)

28. Which one of the following solids would be more soluble in a solution of one of the common strong acids than in pure water?

a. KCl
b. $MgCl_2$
c. $NaNO_3$
d. LiBr
! e. $ZnCO_3$

29. Which one of the following solids would be more soluble in a 1.00 molar solution of one of the common strong acids than in pure water?

a. LiCl
! b. ZnS
c. $NaNO_3$
d. $K_2SO_4$
e. $MgCl_2$

30. $ZnCO_3$, a slightly soluble substance, is least soluble in which one of the following solvents?

a. pure water
! b. 0.20 M $ZnCl_2$
c. 0.10 M $Na_2CO_3$
d. 0.10 M HCl
e. 0.20 M NaCl

31. What is the maximum concentration of $Mg^{2+}$ ion that can co-exist with $F^-$ in a 0.10 M NaF solution without causing any precipitate of magnesium fluoride to form. The $K_{sp}$ of $MgF_2$ is $6.6 \times 10^{-9}$?

a. $1.3 \times 10^{-7}$ M
b. $6.6 \times 10^{-9}$ M
c. $6.6 \times 10^{-8}$ M
d. $1.6 \times 10^{-7}$ M
! e. $6.6 \times 10^{-7}$ M

32. Calculate the minimum concentration of $Ag^+$ ion that must be added to (or built up in) a 0.120 M NaF solution in order to initiate a precipitate of silver chromate. The $K_{sp}$ of $Ag_3PO_4$ is $2.8 \times 10^{-18}$.

a. $4.8 \times 10^{-9}$ mol/L
! b. $2.9 \times 10^{-6}$ mol/L
c. $2.0 \times 10^{-6}$ mol/L
d. $9.5 \times 10^{-7}$ mol/L
e. $1.4 \times 10^{-6}$ mol/L

33. The solubility product for $Ag_3PO_4$ is $2.8 \times 10^{-18}$. What is the solubility of silver phosphate in solution which is also contains 0.10 moles of silver nitrate per liter?

a. $4.4 \times 10^{-4}$ mol/L
b. $4.4 \times 10^{-15}$ mol/L
! c. $2.8 \times 10^{-15}$ mol/L
d. $3.6 \times 10^{-16}$ mol/L
e. $2.8 \times 10^{-13}$ mol/L

34. The solubility product of lead fluoride, ($PbF_2$) is 3.6 x $10^{-8}$. What is its solubility in 0.10 M NaF solution, in grams per liter?

! a. 8.8 x $10^{-4}$ g/L
b. 3.9 x $10^{-4}$ g/L
c. 13 g/L
d. 3.9 x $10^{-3}$ g/L
e. 8.8 x $10^{-5}$ g/L

35. How many parts per million of calcium are there in a saturated solution of $CaF_2$ ($K_{sp}$ = 3.9 x $10^{-11}$) in pure water.

a. 1.4
b. 3.2
! c. 8.6
d. 17
e. 21

36. NaOH is added to sea water containing 0.050 moles per liter of $Mg^{2+}$ ion until the pH reaches 12.0. A certain quantity of $Mg(OH)_2$ precipitates. What is the concentration of the magnesium ions remaining in the solution, in moles per liter? The $K_{sp}$ of $Mg(OH)_2$ is 7.1 x $10^{-12}$.

a. 1.5 x $10^{-7}$ M
b. 5.0 x $10^{-8}$ M
c. 7.1 x $10^{-9}$ M
! d. 7.1 x $10^{-8}$ M
e. 7.1 x $10^{-6}$ M

37. Sea water contains magnesium ions, $Mg^{2+}$. A solution of sea water was treated with sodium hydroxide until its pH was 11.0. If the $K_{sp}$ of $Mg(OH)_2$ is 7.1 x $10^{-12}$, what is the maximum concentration of magnesium ions remaining in the water?

a. 7.1 x $10^{-9}$ mol/L
b. 7.1 x $10^{-19}$ mol/L
c. 5.0 x $10^{-8}$ mol/L
d. 4.4 x $10^{-4}$ mol/L
! e. 7.1 x $10^{-6}$ mol/L

38. Will a precipitate form (yes or no) when 20.0 mL of 1.8 x $10^{-3}$ M $Pb(NO_3)_2$ is added to 30.0 mL of 5.0 x $10^{-4}$ M $Na_2SO_4$? The $K_{sp}$ of ($PbSO_4$) is 6.3 x $10^{-7}$.

! a. no, because the ion product < $K_{sp}$
b. no, because the ion product > $K_{sp}$
c. yes, because the ion product < $K_{sp}$
d. yes, because the ion product > $K_{sp}$
e. no, because the ion product > $K_{sp}$

39. Will a precipitate of $MgF_2$ form when 300 mL of $1.1 \times 10^{-3}$ M $MgCl_2$ solution is added to 500 mL of $1.2 \times 10^{-3}$ M NaF? The $K_{sp}$ of $MgF_2$ is $6.9 \times 10^{-9}$.

a. yes, because the ion product, $Q > K_{sp}$
! b. no, because the ion product, $Q < K_{sp}$
c. no, because the ion product, $Q = K_{sp}$
d. yes, because the ion product, $Q < K_{sp}$
e. no, because the ion product, $Q > K_{sp}$

40. In an experiment, it is planned to add 300 mL of $2.0 \times 10^{-5}$ M $AgNO_3$ to 200 mL of $2.5 \times 10^{-9}$ M NaI. Will a precipitate form? What is the precipitate? $K_{sp}$ (AgI) = $8.3 \times 10^{-17}$.

a. yes, the ppt is $AgNO_3(s)$
b. yes, the ppt is $NaNO_3(s)$
c. yes, the ppt is NaI(s)
! d. yes, the ppt is AgI(s)
e. no

41. For $PbCl_2$, $K_{sp} = 1.7 \times 10^{-5}$. Will a precipitate of $PbCl_2$ form when 200 mL of $3.0 \times 10^{-2}$ M $Pb(NO_3)_2$ solution is added to 300 mL of $5.0 \times 10^{-2}$ M KCl? Choose one of the following.

a. yes, the ion product, $Q_{sp} > K_{sp}$
! b. no, the ion product, $Q_{sp} < K_{sp}$
c. no, the ion product, $Q_{sp} = K_{sp}$
d. yes, thc ion product, $Q_{sp} < K_{sp}$
e. no, because the ion product, $Q_{sp} > K_{sp}$

42. Will a precipitate form when 250 mL of 0.20 M $Na_2CrO_4$ is added to 250 mL of 0.20 M $AgNO_3$? What will be the concentration of the $Ag^+$ ion remaining in solution when equilibrium is reached? The $K_{sp}$ for $Ag_2CrO_4$ is $1.2 \times 10^{-12}$.

! a. yes, $[Ag^+] = 4.9 \times 10^{-6}$ M
b. yes, $[Ag^+] = 2.0 \times 10^{-5}$
c. yes, $[Ag^+] = 2.4 \times 10^{-11}$ M
d. no, $[Ag^+] = 1.0 \times 10^{-3}$ M
e. no, $[Ag^+] = 2.0 \times 10^{3}$ M

43. $PbCO_3$, $PbCl_2$, $PbI_2$, and PbS are all only very slightly soluble in pure water. Which one (ones) would be significantly more soluble in acidic solution than in pure water?

a. $PbI_2$, PbS and $PbCO_3$
! b. only $PbCO_3$ and PbS
c. only $PbCl_2$ and $PbI_2$
d. only $PbCO_3$
e. All four are significantly more soluble in acidic solution.

44. The group, $Zn(NO_3)_2$, $Zn(CO_3)_2$, $ZnCl_2$, and ZnS contains some salts which are only very slightly soluble in pure water. Of these salts, which one(s) would be significantly more soluble in acidic solution than in pure water?

a. $Zn(NO_3)_2$, $Zn(CO_3)_2$, and $ZnCl_2$
b. $Zn(NO_3)_2$, and $Zn(CO_3)_2$
c. $ZnCl_2$, and ZnS
d. $Zn(CO_3)_2$, and $ZnCl_2$
! e. $Zn(CO_3)_2$, and ZnS

45. Zinc carbonate, a slightly soluble substance, is most soluble in which of the following solvents?

a. water
b. 0.1 M $ZnCl_2$
c. 0.1 M NaOH
! d. 0.1 M HCl
e. 0.2 M $Na_2CO_3$

46. Given the following information:

$Ni(OH)_2(s)$ <===> $Ni^{2+}(aq)$ + 2 $OH^-(aq)$ $K_{sp} = 6.0 \times 10^{-16}$
$H_2O(l)$ <===> $H^+(aq)$ + $OH^-(aq)$ $K_w = 1.0 \times 10^{-14}$

Which choice gives the equilibrium constant for the reaction,
$Ni(OH)_2(s)$ + 2 $H^+(aq)$ <===> $Ni^{2+}(aq)$ + 2 $H_2O(l)$

a. $K_{sp}/K_w$
b. $K_{sp}/K_w$
! c. $K_{sp}/K_w^2$
d. $K_w^2/K_{sp}$
e. $2 \times K_w/K_{sp}$

## Fill in the Blanks

47. A precipitate will form when an anion solution and a cation solution are mixed if ______ exceeds the value of ______. (! ion product, solubility product constant)

48. Addition of a common ion to a solution of a slightly soluble salt will ______ the solubility of the slightly soluble salt. (! decrease)

49. Carbonic acid, $H_2CO_3$, is a diprotic acid with dissociation constants: $K_{a1} = 4.5 \times 10^{-7}$, $K_{a2} = 4.7 \times 10^{-11}$. Calculate the pH of a 0.100 molar solution of $Na_2CO_3$, which is the sodium salt of the one of the conjugate bases of this acid. ______ (! 11.65)

50. For $H_3PO_3$, which is actually a diprotic acid, $K_{a1} = 1.0 \times 10^{-2}$ and $K_{a2} = 2.6 \times 10^{-7}$. Calculate the pH of a 0.100 molar solution of $Na_2HPO_3$, which is a salt of one of the conjugate bases of this acid. ______ (! 9.90)

51. The pH of a saturated solution of $Mg(OH)_2$, whose $K_{sp}$ is $7.1 \times 10^{-12}$, is ______ (! 10.08)

52. The pH of a saturated solution of $Ca(OH)_2$ is 12.37. What is the $K_{sp}$ of $Ca(OH)_2$? ______ (! $6.5 \times 10^{-6}$)

## True and False

53. The oxide ion can exist in aqueous solutions provided the pH is higher than 11.00. ( !F)

54. The sulfide ion is too basic to exist as such in aqueous solutions. ( ! T)

## Critical Thinking

55. NaOH is added to sea water containing 0.050 moles per liter of $Mg^{2+}$ ion until the pH reaches 12.0. A certain quantity of $Mg(OH)_2$ precipitates. How many parts per billion of magnesium are left behind in the solution? The $K_{sp}$ of $Mg(OH)_2$ is $7.1 \times 10^{-12}$.

a. 1.2 ppb
b. 12 ppb
c. 0.17 ppb
! d. 1.7 ppb
e. 173 ppb

56. Consider the equilibrium,

$$AgCl(s) + Br^-(aq) \Longleftrightarrow AgBr(s) + Cl^-(aq)$$

whose equilibrium constant is related to $K_{sp}(AgCl) = 1.8 \times 10^{-10}$ and $K_{sp}(AgBr) = 5.0 \times 10^{-13}$. Calculate the equilibrium constant for the reaction, and tell whether AgCl will react with 1.00 molar KBr(aq) to a significant extent. ______, ______ (! $3.6 \times 10^2$, yes)

57. A saturated solution of silver nortonate, $Ag_2NtO_4$, in water contains 72 ppm of silver. What is the $K_{sp}$ of $Ag_2NtO_4$?

a. $1.5 \times 10^{-3}$
! b. $1.5 \times 10^{-10}$
c. $3.7 \times 10^{-11}$
d. $3.0 \times 10^{-10}$
e. $1.2 \times 10^{-9}$

58. A saturated solution of silver nortonate, $Ag_2NtO_4$, in water contains 72 ppm of silver. How many parts per million of silver would there be in a 1.00 molar aqueous solution of $Na_2NtO_4$ which has been saturated with silver nortonate? ______ (! 1.3 ppm)

## Chemicals In Use 14: Limestone Caverns and Coral Reefs

59. A capped, unopened carbonated beverage under a high $CO_2(g)$ pressure has a higher $H^+$ ion concentration than a bottle of the same beverage which has been standing open for some time. ( !T)

60. Limestone consists mainly of

a. $NaHCO_3$
b. $Al_2(CO_3)_3$
c. $H_2CO_3$
d. $MgCO_3 \cdot CaCO_3$
! e. $CaCO_3$

61. The equilibrium for the action of hydrogen ion on calcium carbonate as it occurs within a fissure of a limestone cavern is

a. $CaCO_3(s) + H^+(aq) \Longleftrightarrow Ca^{2+}(aq) + HCO_3^-(aq)$
b. $CaCO_3(s) + H_2O \Longleftrightarrow Ca(OH)^+(aq) + HCO_3^-(aq)$
c. $CaCO_3(s) + H_2O \Longleftrightarrow CaO(s) + H_2CO_3(aq)$
d. $CaCO_3(s) + 2H_2O \Longleftrightarrow Ca(OH)_2(aq) + H_2CO_3(aq)$
! e. $CaCO_3(s) + H_2CO_3(aq) \Longleftrightarrow Ca^{2+}(aq) + 2\,HCO_3^-(aq)$

62. Which one of the following reactions represents the final one in the sequence by which stalactites and stalagmites are formed within limestone caverns?

! a. $Ca(OH)_2(aq) + H_2CO_3(aq) \Longleftrightarrow CaCO_3(s) + 2H_2O$
b. $Ca^{2+}(aq) + 2\,HCO_3^-(aq) \Longleftrightarrow CaCO_3(s) + H_2CO_3(aq)$
c. $CaO(s) + H_2CO_3(aq) \Longleftrightarrow CaCO_3(s) + H_2O$
d. $Ca(OH)^+(aq) + HCO_3^-(aq) \Longleftrightarrow CaCO_3(s) + H_2O$
e. $Ca^{2+}(aq) + HCO_3^-(aq) \Longleftrightarrow CaCO_3(s) + H^+(aq)$

## Chapter 18 Electrochemistry

### Multiple Choice

1. Electrolysis is

a. the splitting atomic nuclei by electrical energy
b. the splitting of atoms by electrical energy
c. the passage of electrical energy through a split-field armature
! d. the chemical reaction which accompanies passage of electrical energy through a liquid electrolyte
e. the chemical reaction which accompanies passage of electrical energy through a metallic liquid

2. An electrolyte is (the best answer, please)

a. a solute which conducts electrical energy
b. an inert electrode which conducts electrical energy
c. a metal which conducts electrical energy through a solution
! d. a pure liquid or solution, containing ions, which conducts electrical energy
b. a solution which conducts electricity

3. Anions

! a. are charged ions which move toward the anode of an electrolysis cell
b. are charged ions which move toward the negative electrode of an electrolysis cell
c. are charged ions which move toward the north pole of a magnetic field
d. are positively charged ions which result from electrical discharge in a liquid solution
e. are ions which attach themselves to any electrode to react chemically during electrolysis

4. Which statement below about electroplating is an incorrect statement?

a. Nickel sulfate solution is the electrolyte used in nickel electroplating.
b. Chromic acid ($H_2CrO_4$) solution is the electrolyte used in chromium electroplating.
! c. Silver nitrate solution is the electrolyte used in silver electroplating.
d. Cyanide baths with gold complexes are the electrolyte used in gold electroplating.
e. Cyanide baths with cadmium complexes are the electrolyte used in cadmium electroplating.

5. Which one of the following metals is the only one which can be prepared by electrolysis of an aqueous solution of one of its salts?

a. aluminum
! b. copper
c. magnesium
d. potassium
e. sodium

6. The products of the electrolysis of molten magnesium chloride using platinum electrodes (an inert electrode) will yield

a. hydrogen gas and chlorine gas
! b. magnesium metal and chlorine gas
c. magnesium metal and oxygen gas
d. magnesium metal and hydroxide ions
e. chlorine gas and platinum-magnesium alloy

7. The products of the electrolysis of aqueous magnesium chloride using platinum electrodes (an inert electrode) will yield

a. magnesium metal and chlorine gas
b. magnesium metal and oxygen gas
c. magnesium metal and hydroxide ions
! d. hydrogen gas and chlorine gas
e. chlorine gas and platinum-magnesium alloy

8. When an aqueous solution of sodium chloride is electrolyzed, a gas is observed to form at the anode. The gas is

! a. chlorine
b. hydrogen
c. hydrogen peroxide
d. oxygen
e. sodium

9. When an aqueous solution of copper sulfate is electrolyzed, a gas is observed to form at the anode. The gas is

a. hydrogen
b. hydrogen sulfide
c. hydrogen peroxide
! d. oxygen
e. sulfur dioxide

10. When an aqueous solution of sodium sulfate is electrolyzed, a gas is observed to form at the anode. The gas is

a. hydrogen
b. hydrogen sulfide
c. hydrogen peroxide
! d. oxygen
e. sulfur dioxide

11. When an aqueous solution of sodium sulfate is electrolyzed, a gas is observed to form at the cathode. The gas is

! a. hydrogen
b. hydrogen sulfide
c. hydrogen peroxide
d. oxygen
e. sulfur dioxide

12. When an aqueous solution of magnesium sulfate is electrolyzed, what product is formed at the cathode?

! a. hydrogen
b. hydrogen sulfide
c. magnesium
d. oxygen
e. sulfur dioxide

13. When an aqueous solution of nickel sulfate is electrolyzed, what product is formed at the anode?

a. hydrogen
b. hydrogen sulfide
c. nickel
! d. oxygen
e. sulfur dioxide

14. When fused sodium chloride is electrolyzed, which statement below correctly describes what occurs?

a. gaseous chlorine is formed at the cathode
b. hydrogen gas is formed at the cathode
! c. liquid sodium is formed at the cathode
d. liquid chlorine is formed at the anode
e. solid sodium is formed at the anode

15. Which statement below is true?

a. Electrolysis cells generate electrical energy when the terminals are reversed
b. Electrolysis was discovered by Lewis Latimer
! c. Galvanic cells generate electrical energy rather than consuming it
d. Galvanic cells were invented by Thomas Edison
e. The Laws of Electrolysis were discovered by Granville Woods

16. When an aqueous solution of $AgNO_3$ is electrolyzed, a gas is observed to form at the anode. The gas is:

a. dinitrogen tetroxide
b. hydrogen
c. mononitrogen monoxide
d. nitrogen dioxide
! e. oxygen

17. When an aqueous solution of sodium chloride is electrolyzed, hydrogen gas is evolved at the cathode. The solution near the cathode becomes:

a. acidic
! b. basic
c. bubbly
d. colored
e. viscous

18. When an aqueous solution of sodium sulfate is electrolyzed, a gas is evolved at the anode. The solution near the anode becomes:

! a. acidic
b. basic
c. bubbly
d. colored
e. viscous

19. When copper is refined using the electrolysis technique, which statement below is true?

! a. Impure copper goes into solution at the anode, and pure copper plates out on the cathode.
b. Impure copper goes into solution at the cathode, and pure copper plates out on the anode.
c. Pure copper goes into solution from the anode and falls to the bottom of the tank .
d. Pure copper goes into solution from the cathode and falls to the bottom of the tank.
e. Pure copper on the bottom of the tank goes into solution and plates out on the cathode.

20. The half-reaction that occurs at the cathode during electrolysis of aqueous sodium iodide solution is:

! a. $2\ H_2O + 2\ e^- \longrightarrow H_2(g) + 2\ OH^-(aq)$
b. $I_2(aq) + 2\ e^- \longrightarrow 2\ I^-(aq)$
c. $2\ I^-(aq) \longrightarrow I_2(aq) + 2\ e^-$
d. $Na^+(aq) + e^- \longrightarrow Na(s)$
e. $Na(s) \longrightarrow Na^+(aq) + e^-$

21. The half-reaction that occurs at the cathode during electrolysis of aqueous $CuCl_2$ solution is,

a. $Cl_2(g) + 2\ e^- \longrightarrow 2\ Cl^-(aq)$
b. $2\ Cl^-(aq) \longrightarrow Cl_2(g) + 2\ e^-$
! c. $Cu^{2+}(aq) + 2\ e^- \longrightarrow Cu(s)$
d. $Cu^+(aq) + e^- \longrightarrow Cu(s)$
e. $2\ H_2O + 2\ e^- \longrightarrow H_2(g) + 2\ OH^-(aq)$

22. The half-reaction that should occur at the anode during electrolysis of aqueous potassium bromide solution is:

a. $Br_2(g) + 2\ e^- \longrightarrow 2\ Br^-(aq)$
! b. $2\ Br^-(aq) \longrightarrow Br_2(l) + 2\ e^-$
c. $2\ H_2O \longrightarrow O_2(g) + 4\ H^+(aq) + 4\ e^-$
d. $2\ H^+(aq) + e^- \longrightarrow H_2(g)$
e. $Na^+(aq) + e^- \longrightarrow Na(s)$

23. The unit of electrical energy used in our work is the

a. ampere
b. coulomb
! c. joule
d. volt
e. watt

24. The unit of electrical quantity used in our work is the

a. ampere
! b. coulomb
d. joule
c. volt
e. watt

25. The unit of electrical current used in our work is the

! a. ampere
b. coulomb
d. joule
c. volt
e. watt

26. The Faraday is a unit of

a. capacitance
b. current
c. power
d. pressure
! e. quantity of electrical charge

27. The quantity of electrical charge can be calculated as

! a. the product of current in amperes by time in seconds
b. the product of potential, in volts, by time in seconds
c. the product of power, in watts, by time in seconds
d. the product of energy, in joules, by time in seconds
e. the product of volts times coulombs

28. The quantity of electrical charge in one equivalent of electrical energy is

a. 4.184 joules
b. 3,600 coulombs
c. 23,060 joules
! d. 96,485 coulombs
e. 3.47 x $10^8$ coulombs

29. How many Faradays of charge are required to cause reduction of 0.200 mole of $Cu^{2+}$ ion to metallic copper?

a. 0.067 Faradays
b. 0.20 Faradays
! c. 0.60 Faradays
d. 0.20 x 96,485 Faradays
e. 0.40 x 96,485 Faradays

30. How many coulombs of charge are required for the reduction of 0.20 mol of $Cr^{3+}$ ions to Cr metal?

a. 0.60 coulombs
b. 3.0 coulombs
c. $2.9 \times 10^4$ coulombs
! d. $5.8 \times 10^4$ coulombs
e. $9.65 \times 10^4$ coulombs

31. How many Faradays of electrical charge are transferred in an electrolysis cell when a current of 2.0 amperes flows for 12 hours?

a. $6.2 \times 10^{-3}$ Faraday
! b. 0.90 Faraday
c. 1.1 Faradays
d. 24 Faradays
e. $8.6 \times 10^4$ Faradays

32. How many Faradays of electrical charge are transferred in an electrolysis cell when a current of 3.0 amperes flows for exactly 4 hours and 30 minutes?

! a. $5.0 \times 10^{-1}$
b. $8.4 \times 10^{-3}$
c. $1.4 \times 10^{-4}$
d. 13.5
e. 17.9

33. What electrical current, in amperes, is required to produce 2.00 Faradays of electrical charge over the course of an 8.000 hour period in an electrolysis cell?

a. 1.57 amperes
! b. 6.70 amperes
c. 107 amperes
d. $6.03 \times 10^3$ amperes
e. $2.41 \times 10^4$ amperes

34. What electrical current, in amperes, is required to produce 2.00 Faradays of electrical charge over the course of 4 hour and 30 minutes and 30 seconds period in an electrolysis cell?

a. 0.0224 amperes
b. 0.0841 amperes
c. 0.336 amperes
d. 11.6 amperes
! e. 11.9 amperes

35. A metal object is to be gold-plated by an electrolytic procedure using aqueous $Au(CN)_4^-$, a gold(III) complex ion, as the electrolyte. Calculate the number of grams of gold deposited in 3.00 min by a constant current of 10.0 A?

a. 0.00689 gram
! b. 1.23 gram
c. 1.32 gram
d. 1.77 gram
e. 3.55 gram

36. How long will it take to produce 78.0 g of Al metal by the reduction of $Al^{3+}$ ions in an electrolysis cell with a current of 2.00 A?

a. 419 seconds
b. 4.34 hours
c. 13.0 hours
! d. 116 hours
e. $6.98 \times 10^3$ hours

37. How many grams of nickel would be electroplated by passing a constant current of 7.2 A through a solution of $NiSO_4$ for 90.0 min?

a. 0.20 g
b. 0.40 g
! c. 12 g
d. 24 g
e. 47 g

38. How many coulombs (q) of electrical charge must pass through an electrolysis cell to reduce 0.44 mol of $Ca^{2+}$ ion to calcium metal?

a. 0.88 coulomb
b. 21,000 coulomb
c. 42,500 coulomb
! d. 85,000 coulomb
e. 193,000 coulomb

39. How many grams of chromium would be electroplated by passing a constant current of 5.2 amperes through a solution containing chromium(III) sulfate for 45.0 min?

a. $9.3 \times 10^{-4}$ g
b. 0.042 g
c. 24 g
! d. 2.5 g
e. $2.3 \times 10^{10}$ g

40. How many coulombs would be required to electroplate 35.0 grams of chromium by passing an electrical current through a solution containing aqueous $CrCl_3$?

a. $0.16 \times 10^4$ coulomb
b. $6.40 \times 10^4$ coulomb
c. $6.50 \times 10^4$ coulomb
! d. $1.95 \times 10^5$ coulomb
e. $1.01 \times 10^7$ coulomb

41. How many minutes would be required to electroplate 25.0 grams of chromium by passing a constant current of 4.8 amperes through a solution containing $CrCl_3$ ?

a. 161 min
b. 322 min
! c. 483 min
d. $1.11 \times 10^4$ min
e. $2.01 \times 10^4$ min

42. The reaction, $Au(CN)_4^-(aq) + 3e^- \longrightarrow Au(s) + 4\ CN^-$, is being used in electrolysis to plate 45.00 grams of gold on a medallion. How many amperes should the current be, if the time is to run from 9:30:00 A.M. to 3:30:00 P.M. the same day?

a. 1.003
b. 1.021
c. 3.010
! d. 3.062
e. 6.020

43. The reaction, $Au(CN)_4^-(aq) + 3\ e^- \longrightarrow Au(s) + 4\ CN^-(aq)$ is being used in electrolysis to plate 18.25 grams of gold on a medallion. If 2.525 amperes will be used starting at 11:30:00 A.M., at what time should it be shut off?

a. 1:10:00 P.M.
! b. 2:27:02 P.M.
c. 3:15:03 P.M.
d. 3:45:06 P.M.
e. 4:35:32 P.M.

44. An electroplating bath containing $K_2CrO_4$ solution is used for chromium plating. If a current of 6.250 amperes was passed through the system for a 6 hr 35 min 20 sec time period, how many grams of Cr would be plated out on the object at the cathode?

! a. 13.32 g
b. 26.63 g
c. 36.16 g
e. 39.95 g
e. 79.89 g

45. Aluminum does not corrode as does iron, because:

! a. A protective layer of $Al_2O_3$ forms on the metal surface.
b. Al does not react with $O_2(g)$.
c. Al is harder to oxidize than is Fe.
d. Fe gives cathodic protection to Al.
e. The electrical circuit cannot be completed on an Al surface.

46. Iron objects such as storage tanks and underground pipelines can be protected from corrosion by connecting them via a wire to a piece of:

a. copper
b. lead
! c. magnesium
d. silver
e. tin

47. Why is cryolite, $Na_3AlF_6$, mixed with alumina prior to electrolysis to produce Al?

a. because sodium is produced which in turn displaces the aluminum
b. the aluminum in cryolite is easier to reduce than that in alumina
c. to provide more fluoride ions
! d. to reduce the required temperature to bring the alumina into solution
e. to shift the equilibrium to the right

48. How long would it take to deposit 500 g of silver metal from a solution containing $Ag^+$ ions using a current of 3.00 amperes?

a. 41.4 minutes
b. 1242 minutes
! c. 2484 minutes
d. 4969 minutes
e. 149000 minutes

49. How many liters of chlorine gas, measured at STP, would be produced by the electrolysis of molten sodium chloride using a current of 3.50 amperes for exactly 46 minutes?

! a. 1.12 liters
b. 2.24 liters
c. 11.2 liters
d. 22.4 liters
e. 44.8 liters

50. When aqueous brine is electrolyzed, the products are:

! a. chlorine gas and sodium hydroxide solution
b. chlorine gas and sodium metal
c. $Na_2O$ and HCl
d. NaCl and sodium metal
e. $NaClO_3$ and $NaClO_2$

51. The Downs cell is used in the electrolytic manufacture of

a. chlorine gas and sodium hydroxide
b. hydrogen gas and chlorine gas
c. hydrogen gas and sodium hydroxide
! d. sodium metal and chlorine gas
e. sodium metal and sodium hydroxide

52. A galvanic cell has two electrodes. Which statement below is correct?

a. oxidation takes place at the anode, which is positively charged
! b. oxidation takes place at the anode, which is negatively charged
c. oxidation takes place at the cathode, which is positively charged
d. oxidation takes place at the anode, which is positively charged
e. oxidation take place at the dynode, which is uncharged

53. A galvanic cell has two electrodes. Which statement below is correct?

a. reduction takes place at the anode, which is positively charged
b. reduction takes place at the anode, which is negatively charged
! c. reduction takes place at the cathode, which is positively charged
d. reduction takes place at the anode, which is positively charged
e. reduction takes place at the dynode, which is uncharged

54. Using these metal ion/metal reaction potentials

| $Cu^{2+}/Cu$ | $Ag^{+}/Ag$ | $Co^{2+}/Co$ | $Nt^{2+}/Nt$ | $Zn^{2+}/Zn$ |
|---|---|---|---|---|
| 0.34 | +0.80 | -0.28 | -1.10 | -0.76 |

Based on the data above, which one of the species below is the best reducing agent?

a. Co(*s*)
! b. Zn(*s*)
c. $Nt^{2+}$ (*aq*)
d. Cu(*s*)
e. $Ag^{+}$ (*aq*)

55. Using these metal ion/metal reaction potentials

| $Cu^{2+}/Cu$ | $Ag^{+}/Ag$ | $Co^{2+}/Co$ | $Nt^{2+}/Nt$ | $Zn^{2+}/Zn$ |
|---|---|---|---|---|
| 0.34 | +0.80 | -0.28 | -1.10 | -0.76 |

Based on the data above, which one of the species below is the best oxidizing agent?

a. Co(*s*)
b. Zn(*s*)
c. $Nt^{2+}$ (*aq*)
d. Cu(*s*)
! e. $Ag^{+}$ (*aq*)

56. Using these metal/metal ion reduction potentials.

| $Cd^{2+}/Cd$ | $Zn^{2+}/Zn$ | $Ni^{2+}/Ni$ | $Xp^{+}/Xp$ | $Cu^{2+}/Cu$ |
|---|---|---|---|---|
| -0.40 v | -0.76 v | -0.25 v | +0.62 v | +0.34 v |

Based on the data above, which one of the species below is the best reducing agent?

! a. Cd(*s*)
b. $Zn^{2+}$(*aq*)
c. Ni(*s*)
d. $Xp^{+}$(*aq*)
e. Cu(*s*)

57. Using these metal/metal ion reduction potentials.

| $Cd^{2+}/Cd$ | $Zn^{2+}/Zn$ | $Ni^{2+}/Ni$ | $Xp^{+}/Xp$ | $Cu^{2+}/Cu$ |
|---|---|---|---|---|
| -0.40 v | -0.76 v | -0.25 v | +0.62 v | +0.34 v |

Based on the data above, which one of the species below is the best oxidizing agent?

a. Cd(*s*)
b. $Zn^{2+}$(*aq*)
c. Ni(*s*)
! d. $Xp^{+}$(*aq*)
e. Cu(*s*)

58. Using these metal ion/metal standard reduction potentials.

| $Cd^{2+}/Cd$ | $Zn^{2+}/Zn$ | $Ni^{2+}/Ni$ | $Xp^{+}/Xp$ | $Cu^{2+}/Cu$ |
|---|---|---|---|---|
| -0.40 v | -0.76 v | -0.25 v | +0.62 v | +0.34 v |

Based on the data above, which one of the species below is the the best oxidizing agent?

a. Xp(*s*)
b. Cu(*s*)
c. $Cd^{2+}$(*aq*)
d. $Zn^{2+}$(*aq*)
! e. $Ni^{2+}$(*aq*)

59. Using these metal ion/metal standard reduction potentials

| $Co^{2+}/Co$ | $Cu^{2+}/Cu$ | $Ag^{+}/Ag$ | $Zn^{2+}/Zn$ | $Ni^{2+}/Ni$ | $Cd^{2+}/Cd$ |
|---|---|---|---|---|---|
| -0.28 v | +0.34 v | +0.80 v | -0.76 v | -0.25 v | -0.40 v |

Calculate the standard cell potential for the cell whose net reaction is

$$Cu^{2+}(aq) + Cd(s) \longrightarrow Cd^{2+}(aq) + Cu(s)$$

a. +0.76 volt
b. +0.06 volt
c. -0.06 volt
! d. +0.74 volt
e. +0.20 volt

60. Using these metal ion/metal reaction potentials

| $Cu^{2+}/Cu$ | $Ag^+/Ag$ | $Co^{2+}/Co$ | $Nt^{2+}/Nt$ | $Zn^{2+}/Zn$ |
|---|---|---|---|---|
| 0.34 | +0.80 | -0.28 | -1.10 | -0.76 |

Calculate the standard cell potential for the cell whose net reaction is

$$Nt(s) + Ag^+(aq) \longrightarrow Nt^{2+}(aq) + Ag(s)$$

a. -0.300 volt
b. +0.300 volt
c. -1.90 volt
! d. +1.90 volt
e. +2.70 volt

61. The anode reaction in the galvanic cell below is

$$Nt(s) + Ag^+(aq) \longrightarrow Nt^{2+}(aq) + Ag(s)$$

a. $Nt^{2+}(aq) + 2e^- \longrightarrow Nt(s)$
b. $Ag^+(aq) + e^- \longrightarrow Ag(s)$
! c. $Nt(s) \longrightarrow Nt^{2+}(aq) + e^-$
d. $Ag(s) \longrightarrow Ag^+(aq) + e^-$
e. $2H^+(aq) + 2e^- \longrightarrow H_2(g)$

62. Using the standard reduction potentials,

$Au^{3+}(aq) + 3e^- \Longleftrightarrow Au(s)$ +1.42 volt

$Ca^{2+}(aq) + 2e^- \Longleftrightarrow Ca(s)$ -2.76 volt

Calculate the value of $E^o_{(cell)}$ for the cell reaction:

$$2\,Au(s) + 3\,Ca^{2+}(aq) \Longleftrightarrow 2\,Au^{3+}(aq) + 3\,Ca(s)$$

a. -1.34 volt
b. +1.34 volt
! c. -4.18 volt
d. +4.18 volt
e. -1.34 volt

63. Using the standard reduction potentials,

$Mg^{2+}(aq) + 2e^- \Longleftrightarrow Mg(s)$ -2.37 volt

$NO_3^-(aq) + 4H^+(aq) + 3e^- \Longleftrightarrow NO(g) + 2H_2O$ + 0.96 volt

Calculate the value of $E^o_{(cell)}$ for the cell with the reaction:

$$3\,Mg(s) + NO_3^-(aq) + 8H^+(aq) \Longleftrightarrow 3\,Mg^{2+}(aq) + 2\,NO(g) + 4\,H_2O$$

a. +1.41 volt
b. -1.41 volt
! c. +3.33 volt
d. +8.46 volt
e. -8.46 volt

64. Using the standard reduction potentials,

$Cr^{3+}(aq) + 3e^- \Longleftrightarrow Cr(s)$ -0.40 volt

$Cl_2(g) + 2e^- \Longleftrightarrow 2\,Cl^-(aq)$ +1.36 volt

Calculate the value of $E^o_{(cell)}$ for the cell with the reaction:

$2\,Cr^{3+}(aq) + 3\,Cl_2(g) \Longleftrightarrow 2\,Cr(s) + 6\,Cl^-(aq)$

a. - 0.96 volt
b. + 0.96 volt
! c. +1.76 volt
d. - 1.76 volt
e. + 0.98 volt

65. For the reaction, $2\,Cr^{2+}(aq) + Cl_2(g) \Longleftrightarrow 2\,Cr^{3+}(aq) + 2\,Cl^-(aq)$, the value of $E^o_{(cell)}$ is 1.78 volt. What is the value of $E^o_{(cell)}$ for

$Cr^{3+}(aq) + Cl^-(aq) \Longleftrightarrow Cr^{2+}(aq) + \frac{1}{2}Cl_2(g)$

! a. -1.78 volt
b. +0.89 volt
c. +1.78 volt
d. -0.89 volt
e. -3.56 volt

66. Using the standard reduction potentials,

$Ni^{2+}(aq) + 2e^- \Longleftrightarrow Ni(s)$ -0.25 volt

$Fe^{3+}(aq) + e^- \Longleftrightarrow Fe^{2+}(aq)$ +0.77 volt

Calculate the value of $E^o_{(cell)}$ for the cell with the reaction:

$Ni^{2+}(aq) + 2\,Fe^{2+}(aq) \Longleftrightarrow Ni(s) + 2\,Fe^{3+}(aq)$

a. +0.52 volt
b. -1.02 volt
c. +2.81 volt
d. +1.02 volt
e. -2.81 volt

67. The cell described by the net reaction,

$Zb^{4+}(aq) + Cl_2(g) + 2\,H_2O \Longleftrightarrow ZbO_2^{2+}(aq) + 2\,Cl^-(aq) + 4\,H^+(aq)$

has a standard potential value of 0.32 volt. Using the value for

$Cl_2(g) + 2e^- \Longleftrightarrow 2\,Cl^-(aq)$ +1.36 volt

determine a value for the standard reduction potential of the (Pt)/$ZbO_2(aq)$,$H^+$(aq)//$Zb^{4+}(aq)$

a. -1.04 volt
! b. +1.04 volt
c. -1.68 volt
d. +1.68 volt
e. +2.40 volt

68. The cell described by the net reaction,

$$2\ U(s)\ +\ 3\ Cl_2(g)\ <\!\!\Longrightarrow\ 6\ Cl^-(aq)\ +\ 2\ U^{3+}(aq)$$

has a standard cell potential of 3.16 volts. Using the standard potential value shown for

$$Cl_2(g)\ +\ 2\ e^-\ <\!\!\Longrightarrow\ 2\ Cl^-(aq) \qquad E^o = +1.36\ \text{volt}$$

determine a value for the standard reduction potential of the (Pt)/$U^{3+}(aq)$,U(s) half cell

! a. -1.80 volt
b. +1.80 volt
c. -1.96 volt
d. -4.52 volt
e. +4.52 volt

69. The cell described by the net reaction,

$$2\ Co^{3+}(aq)\ +\ 2\ Cl^-(aq)\ <\!\!\Longrightarrow\ 2\ Co^{2+}(aq)\ +\ Cl_2(g)$$

has a standard potential of 0.46 volt. Using the standard potential value shown for

$$Cl_2(g)\ +\ 2\ e^-\ <\!\!\Longrightarrow\ 2\ Cl^-(aq) \qquad E^o = +1.36\ \text{volt}$$

determine a value for the standard reduction potential of the (Pt)/$Co^{3+}(aq)$,$Co^{2+}(aq)$ half .lh 0.175"

a. -0.90 volt
b. +0.90 volt
c. +0.91 volt
! d. -1.82 volt
e. +1.82 volt

70. Which one of chemical species in the galvanic cell represented as

$$Sn(s)/Sn^{2+}(aq)//NO_3^-(aq)/NO(g)/(Pt)$$

is the species undergoing reduction?

! a. $NO_3^-(aq)$
b. NO(g)
c. Pt(s)
d. Sn(s)
e. $Sn^{2+}(aq)$

71. Which one of the half reactions below is the anode half reaction in the cell represented as

$$Ni(s)/Ni^{2+}(aq)//MnO_4^-(aq),Mn^{2+}(aq),H^+(aq)/(Pt)$$

! a. $Ni(s)\ <\!\!\Longrightarrow\ Ni^{2+}(aq) + 2\ e^-$
b. $MnO_4^-(aq)\ +\ 8\ H^+(aq)\ <\!\!\Longrightarrow\ Mn^{2+}(aq)\ +\ 4\ H_2O\ +\ 5\ e^-$
c. $Ni^{2+}(aq) + 2\ e^-\ <\!\!\Longrightarrow\ Ni(s)$
d. $Mn^{2+}(aq)\ +\ 4\ H_2O\ +\ 5\ e^-\ <\!\!\Longrightarrow\ MnO_4^-(aq)\ +\ 8\ H^+(aq)$
e. $2\ H_2O\ <\!\!\Longrightarrow\ O_2(g)\ +\ 4\ H^+(aq)\ +\ 4\ e^-$

72. Using the standard reduction potentials,

$Q^{2+}(aq) + 2\,e^- \Longleftrightarrow Q(s)$ -0.85 volt

$NtO_4^{2-}(aq) + 8\,H^+(aq) + 6\,e^- \Longleftrightarrow Nt^{2+}(aq) + 4\,H_2O$ +1.20 volt

Calculate the value of $E^o$ for the cell described as $Q(s)/Q^{2+}(aq)//NtO_4^{2-}(aq),Nt^{2+}(aq),H^+(aq)/(Pt)$

a. -0.35 volt
b. +0.35 volt
c. -2.05 volt
! d. +2.05 volt
e. +2.90 volt

73. A galvanic cell in which the net cell reaction is $A^{3+}(aq) + B(s) \Longleftrightarrow B^{3+}(aq) + A(s)$, has a standard cell potential of 0.500 volts. What is the value of the standard free energy change, in kJ for this system?

a. -32.1 kJ
! b. -144.8 kJ
c. +144.8 kJ
d. -289.5 kJ
e. +289.5 kJ

74. A galvanic cell is composed of these two half cells, with the standard reduction potentials shown

$Co^{2+}(aq) + 2\,e^- \Longleftrightarrow Co(s)$ -0.28 volt

$Cd^{2+}(aq) + 2\,e^- \Longleftrightarrow Cd(s)$ -0.40 volt

What is the standard free energy for the cell reaction of this galvanic cell?

a. -11.6 kJ
b. +11.6 kJ
! c. -23.2 kJ
d. +23.2 kJ
e. -46.3 kJ

75. A galvanic cell is composed of these two half cells, with the standard reduction potentials shown

$Co^{2+}(aq) + 2\,e^- \Longleftrightarrow Co(s)$ -0.28 volt

$Cd^{2+}(aq) + 2\,e^- \Longleftrightarrow Cd(s)$ -0.40 volt

The actual concentrations are: $Co^{2+}(aq) = 0.00100$ M, $Cd^{2+} = 0.100$ M. What is the actual potential of this galvanic cell?

! a. +0.06 volt
b. +0.12 volt
c. +0.24 volt
d. +0.33 volt
e. +0.68 volt

76. A galvanic cell is composed of these two half cells, with the standard reduction potentials shown

| | |
|---|---|
| $Co^{2+}(aq) + 2e^- \Longleftrightarrow Co(s)$ | -0.28 volt |
| $Cd^{2+}(aq) + 2e^- \Longleftrightarrow Cd(s)$ | -0.40 volt |

The actual concentrations are: $Co^{2+}(aq)$ = 0.100 M, $Cd^{2+}$ = 0.0100 M. What is the actual potential of this galvanic cell?

a. +0.06 volt
b. +0.09 volt
! c. +0.15 volt
d. +0.18 volt
e. +0.24 volt

77. A galvanic cell is composed of these two half cells, with the standard reduction potentials shown

| | |
|---|---|
| $Co^{2+}(aq) + 2e^- \Longleftrightarrow Co(s)$ | -0.28 volt |
| $Cd^{2+}(aq) + 2e^- \Longleftrightarrow Cd(s)$ | -0.40 volt |

The actual $Co^{2+}(aq)$ concentration is 0.100 M, the $Cd^{2+}(aq)$ concentration is unknown. The actual cell potential was measured as 0.50 volts. Calculate the actual concentration of the $Cd^{2+}(aq)$ ion.

a. 0.0010 M
! b. 0.0040 M
c. 0.010 M
d. 0.020 M
e. 0.040 M

78. A galvanic cell is composed of these two half cells, with the standard reduction potentials shown

| | |
|---|---|
| $Co^{2+}(aq) + 2e^- \Longleftrightarrow Co(s)$ | -0.28 volt |
| $Cd^{2+}(aq) + 2e^- \Longleftrightarrow Cd(s)$ | -0.40 volt |

The actual concentrations are: $Co^{2+}(aq)$ = 0.00100 M, $Cd^{2+}$ = 0.100 M. What is the free energy change associated with this particular design?

a. -5.79 kJ
! b. -11.6 kJ
c. -23.2 kJ
d. -34.7 kJ
e. -46.3 kJ

79. A galvanic cell is composed of these two half cells, with the standard reduction potentials shown

| | |
|---|---|
| $Co^{2+}(aq) + 2e^- \Longleftrightarrow Co(s)$ | -0.28 volt |
| $Cd^{2+}(aq) + 2e^- \Longleftrightarrow Cd(s)$ | -0.40 volt |

The actual concentrations are: $Co^{2+}(aq)$ = 0.100 M, $Cd^{2+}$ = 0.0100 M. What is the free energy change associated with this particular design?

a. -5.79 kJ
b. -11.6 kJ
! c. -28.9 kJ
d. -34.7 kJ
e. -46.3 kJ

80. A galvanic cell is composed of these two half cells, with the standard reduction potentials shown

$Co^{2+}(aq) + 2e^- \Longleftrightarrow Co(s)$ -0.28 volt

$Cr^{3+}(aq) + 3e^- \Longleftrightarrow Cr(s)$ -0.74 volt

What is the standard free energy for the cell reaction of this galvanic cell?

a. -88.8 kJ
b. -177.5 kJ
! c. -266.3 kJ
d. -295.3 kJ
e. -590.5 kJ

81. A galvanic cell is composed of these two half cells, with the standard reduction potentials shown

$Co^{2+}(aq) + 2e^- \Longleftrightarrow Co(s)$ -0.28 volt

$Cr^{3+}(aq) + 3e^- \Longleftrightarrow Cr(s)$ -0.74 volt

The actual concentrations are: $Co^{2+}(aq)$ = 0.0100 M, $Cr^{3+}$ = 0.00100 M. What is the actual potential of this galvanic cell?

a. +0.40 volt
! b. +0.46 volt
c. +0.52 volt
d. +0.54 volt
e. +1.02 volt

82. A galvanic cell is composed of these two half cells, with the standard reduction potentials shown

$Co^{2+}(aq) + 2e^- \Longleftrightarrow Co(s)$ -0.28 volt

$Cr^{3+}(aq) + 3e^- \Longleftrightarrow Cr(s)$ -0.74 volt

The actual concentrations are: $Co^{2+}(aq)$ = 0.00100 M, $Cr^{3+}$ = 0.100 M. What is the actual potential of this galvanic cell?

a. +0.33 volt
! b. +0.39 volt
c. +0.45 volt
d. +0.94 volt
e. +1.61 volt

83. A galvanic cell is composed of these two half cells, with the standard reduction potentials shown

$Co^{2+}(aq) + 2e^- \Longleftrightarrow Co(s)$ -0.28 volt

$Cr^{3+}(aq) + 3e^- \Longleftrightarrow Cr(s)$ -0.74 volt

The actual $Co^{2+}(aq)$ concentration is 0.00100 M, while the $Cr^{3+}(aq)$ concentration is unknown. The actual cell potential measured 0.38 volts. What is the actual concentration of the $Cr^{3+}$ ion in the cell?

a. 0.00200 M
b. 0.00500 M
c. 0.0100 M
d. 0.0400 M
! e. 0.400 M

84. A galvanic cell is composed of these two half cells, with the standard reduction potentials shown

$Co^{2+}(aq) + 2e^- <\!\!=\!\!=\!\!> Co(s)$ -0.28 volt

$Cr^{3+}(aq) + 3e^- <\!\!=\!\!=\!\!> Cr(s)$ -0.74 volt

The actual concentrations are: $Co^{2+}(aq)$ = 0.0100 M, $Cr^{3+}$ = 0.00100 M. What is the free energy change associated with this particular design?

a. -38.6 kJ
b. -44.4 kJ
c. -231.6 kJ
! d. -266.3 kJ
e. -301.0 kJ

85. A galvanic cell is composed of these two half cells, with the standard reduction potentials shown

$Co^{2+}(aq) + 2e^- <\!\!=\!\!=\!\!> Co(s)$ -0.28 volt

$Cr^{3+}(aq) + 3e^- <\!\!=\!\!=\!\!> Cr(s)$ -0.74 volt

The actual concentrations are: $Co^{2+}(aq)$ = 0.00100 M, $Cr^{3+}$ = 0.100 M. What is the free energy change associated with this particular design?

a. -159.2 kJ
b. -188.1 kJ
c. -191.0 kJ
! d. -225.8 kJ
e. -260.5 kJ

86. Which statement is true in regard to a galvanic cell?

a. $E^o$ for the cell is always positive
b. $E^o$ for the cell is always negative
c. The standard reduction potential for the anode reaction is always positive
d. The standard reduction potential for the anode reaction is always negative
e. The standard reduction potential for the cathode reaction is always positive

## Fill in the Blanks

87. In the lead storage battery, the electrolyte is ______. (! sulfuric acid solution)

88. In the zinc-carbon dry cell, the electrode at which oxidation takes place has a ______ charge. (! Negative)

89. In the nickel-cadmium storage cell ( the "nicad" battery) the electrolyte is ______. (! sodium hydroxide solution)

90. In the Hall process for manufacture of aluminum, the anodes are constructed of ________. (! carbon)

## True and False

91. When an electrical current is passed through fused (molten) magnesium chloride, chlorine gas is produces at the negative electrode. ( !F)

92. During the electrolysis of water, oxygen gas is formed at the anode. ( !T)

93. Lithium metal can be produced readily by the electrolysis of aqueous solutions of lithium nitrate, or lithium chloride. ( !F)

94. If the brine bath is stirred during electrolysis, a different set of products are obtained than would be obtained by electrolysis of an unstirred bath. ( !T)

95. The SI unit of charge is the Faraday. ( !F)

## Critical Thinking

96. An old classmate who works for Eagle metals on the night shift was supposed to do this gold plating job wherein he put 4.5000 troy ounces of gold (1 troy ounce = 31.103486 gram) on a medallion using the reaction,

$$Au(CN)_4^-(aq) + 3\,e^- \longrightarrow Au(s) + 4\,CN^-$$

He somehow got 4.8527 troy ounces on the medallion. Please tell him how many hours, minutes, and seconds he should run the cells with the leads reversed, using a current of 1.350 amperes, to rectify the error and bring the medallions down to the correct weight.
______ ( ! 3 hours, 19 minutes, and 2 seconds)

## Chemicals In Use 15 Ion Selective Electrodes

97. Which one of the following equations forms a basis for measuring ion concentrations by the use of appropriately designed galvanic cells?

! a. $CO_2(aq) + H_2O \rightleftharpoons HCO_3^-(aq) + H^+(aq)$
b. $CO(NH_2)_2(aq) + H_2O \xrightarrow{\text{urease}} CO_2(g) + 2\,NH_3(aq)$
c. $NH_3(aq) + H_2O \rightleftharpoons NH_4^+(aq) + OH^-(aq)$
d. $2\,NH_3(aq) + CO_2(g) \xrightarrow{\text{urease}} CO(NH_2)_2(aq) + H_2O$
e. $CaCO_3(s) + H^+(aq) \rightleftharpoons Ca^{2+}(aq) + HCO_3^-(aq)$

98. Which one of the following ions might a researcher in a medical laboratory be able to measure easily using glass electrodes with suitable glass membranes?

a. $Ca^{2+}(aq)$
! b. $K^+(aq)$
c. $Mg^{2+}(aq)$
d. $Sr^{2+}(aq)$
e. $Ra^{2+}(aq)$

99. What type of electrode uses a solid-state membrane constructed with ion sensitive material?

a. $CO_2$ electrode
b. ion-exchange membrane electrode
! c. glass electrode
d. enzyme-substrate electrode
e. reference electrode

100. A liquid ion-exchange membrane is composed of ______ which is capable of transporting some specific ion across the membrane boundary.

a. a water-soluble inorganic liquid
! b. a water-soluble organic liquid
c. a gel in which an enzyme is embedded
d. a $CO_2$ electrode
e. a $LaF_3$ crystal

# Chapter 19 Metal Complexes and Their Equilibria

## Multiple Choice

1. Which statement below about ligands that form coordination complexes with transition metals is false?

! a. A ligand that attaches itself to a metal ion is functioning as a Lewis acid.
b. Neutral molecules can function as ligands and can form coordination complexes with transition metal ions.
c. Anions can function as ligands and can form coordination complexes with transition metal ions.
d. The ligand in a coordination complex contains a particular atom which functions as the donor atom.
e. The bond between the ligand and the transition metal ion in a coordination complex is a coordinate covalent bond.

2. The number of donor atoms in the edta (ethylenediaminetetraacetate) ligand is

a. 2
b. 3
c. 4
! d. 6
e. 8

3. The charge on the metal ion in the complex ion, $[Co(CO_3)_3]^{3-}$ is

a. 0
b. 2+
! c. 3-
d. 3+
e. 6-

4. A coordination complex formed from cobalt(III) sulfate contains <u>six</u> ammonia molecules which function as ligands. The formula of the compound is properly written as

a. $Co_2(SO_4)_3 \cdot 6NH_3$
! b. $[Co(NH_3)_6]_2(SO_4)_3$
c. $Co_3SO_4(NH_3)_6$
d. $Co_2(SO_4)_3(NH_4)_6$
e. $Co_2[(NH_3)_6](SO_4)_3$

5. A coordination complex formed from chromium(III) chloride contains <u>five</u> ammonia molecules and one water molecule which function as ligands. The correct expression for its formula is

! a. $[Cr(NH_3)_5(H_2O)]Cl_3$
b. $Cr_3Cl(NH_3)_5H_2O$
c. $CrCl_3 \cdot 5NH_3 \cdot H_2O$
d. $Cr(NH_3)_5Cl_3 \cdot H_2O$
e. $[Cr(NH_3)_5]Cl_3 \cdot H_2O$

6. A compound contains a transition metal coordination complex. This coordination complex consists of one iron(III) ion with three $C_2O_4^{2-}$ ions attached as ligands. Which formula below correctly describes a compound that would fit the description?

! a. $K_3[Fe(C_2O_4)_3]$
b. $Fe(C_2O_4)_3$
c. $[Fe(C_2O_4)_3]Cl_3$
d. $Fe(C_2O_4)_3](SO_4)_3$
e. $Ca_3[Fe(C_2O_4)_3]$

7. A complex ion is formed by a $Co^{3+}$ ion which is linked up to one carbonate ion and four ammonia molecules. Which formula below would rcorrectly represent a salt of this complex ion?

a. $Na_3[Co(NH_3)_4CO_3]$
b. $Na[Co(NH_3)_4CO_3]$
c. $[Co(NH_3)_4CO_3]CO_3$
d. $[Co(NH_3)_4CO_3]Cl_2$
! e. $[Co(NH_3)_4CO_3]_2SO_4$

8. A compound contains a transition metal coordination complex. This coordination complex consists of one cobalt(III) ion with five ammonia molecules and one cyanide ion attached as ligands. Which formula below correctly describes a compound that would fit the description?

a. $K_4[Co(NH_3)_5CN]$
b. $K_2[Co(NH_3)_5CN]$
c. $[Co(NH_3)_5CN](SO_4)_2$
! d. $[Co(NH_3)_5CN]SO_4$
e. $[Co(NH_3)_5CN]Cl_3$

9. A coordination complex has the general formula, $Cr(NH_3)_5Cl_3$ and has a limited solubility in water. The reaction of a solution of this complex with silver nitrate solution consistently produces two millimoles of silver chloride for every millimole of the $Cr(NH_3)_5Cl_3$. Reaction with aqueous $H_2SO_4$ or HCl clearly shows a lack of free ammonia in the solution. Which formula below would allow for

! a. $[Co(NH_3)_5Cl]Cl_2$
b. $[Co(NH_3)_5Cl_2]Cl$
c. $[Co(NH_3)_5Cl_3]$
d. $[CoCl_3]\cdot 5NH_3$
e. $[Co(NH_3)_3Cl_3]\cdot 2NH_3$

10. The correct name for the complex ion, $[Ni(CN)_4]^{2-}$, is

a. tetracyanonickel(II)
b. nickel(II) tetracyanide
c. tetracyanonickelo(II)
! d. tetracyanonickelate(II)
e. tetracyanonickelite(II)

11. The correct name for the complex ion, $[Co(NH_3)_6]^{3+}$, is

a. cobalt hexaammine
b. cobalt(III) hexaamine
c. hexaamminecobaltate(III)
d. hexaaminocobaltate(III)
! e. hexaamminecobalt(III)

12. The correct name for the complex ion, $[AuCl_4]^-$, is

a. gold tetrachloride
b. gold(II) tetrachloride
c. auric tetrachloride
d. tetrachlorogold(III)
! e. tetrachloroaurate(III)

13. The correct name for the compound, $[Cr(en)_2Cl_2]_2C_2O_4$, is

! a. dichlorobis(ethylenediamine)chromium(III) oxalate
b. bis(ethylenediamine)dichlorochromate(III) oxalate
c. dichlorobis(ethylenediamine)oxalatochromium(III)
d. dichlorobis(ethylenediamine)oxalatochromate(III)
e. bis(dichlorobis(ethylenediamine)chromium(III)) oxalate

14. The compound, $[Co(NH_3)_6][PtCl_4]$, is named

a. cobalt(III) hexaammine tetrachloroplatinate
b. cobalt(II) hexaammine tetrachloroplatinate
c. cobalt(III) hexaammine platinum tetrachloride
! d. hexaamminecobalt(III) tetrachloroplatinate
e. hexaamminecobalt(III) platinum tetrachloride

15. The correct name of neutral coordination complex below is

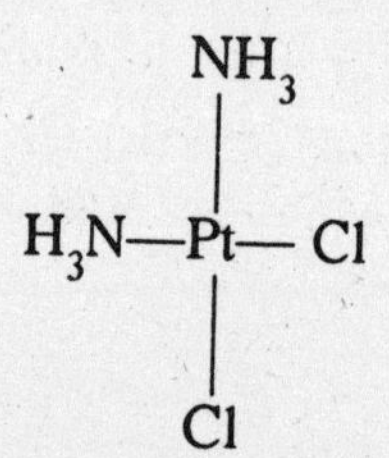


a. *cis*-diaminodichloroplatinate(II)
b. *cis*-diaminodichloroplatinum(II)
! c. *cis*-diamminedichloroplatinum(II)
b. *trans*-diaminodichloroplatinum(II)
c. *trans*-diamminedichloroplatinum(II)

16. The compound, $(NH_4)_3[FeF_6]$, is named

a. ammonium hexafluoroiron(III)
! b. ammonium hexafluoroferrate(III)
c. ammonium iron(III) hexafluoride
d. hexafluoroiron(III) ammonium
e. triammonium hexafluoroiron(III)

17. The coordination complex, $[Co(NH_3)_5(ONO)Cl_2$, is named

a. pentaamminenitrocobalt(III) chloride
b. pentaamminenitritocobalt(II) chloride
c. pentaamminenitrocobalt(II) chloride
! d. pentaamminenitritocobalt(III) chloride
e. nitropentaamminecobalt(II) chloride

18. The coordination number of the chromium ion in the complex ion, $[Cr(dien)_2]^{3+}$ ( dien = $H_2N—CH_2—CH_2—NH—CH_2—CH_2—NH_2$) is

a. 1
b. 2
c. 3
d. 4
! e. 6

19. The coordination number of the platinum ion in the complex ion, $[Pt(C_2O_4)_2]^{2-}$ is

a. 1
b. 2
c. 3
! d. 4
e. 6

20. The coordination number of the nickel ion in the complex ion, $[Ni(en)_3]^{2+}$ is

a. 1
b. 2
c. 3
d. 4
! e. 6

21. A platinum coordination complex has the formula, $[Pt(NH_3)_2BrCl]$ and exhibits a square planar geometry. How many geometric isomers of this compound should exist?

a. 1(no isomers)
! b. 2
c. 3
d. 4
e. 5

22. A platinum coordination complex has the formula, $[Pt(NH_3)ClBrI]^-$ and exhibits a square planar geometry. How many geometric isomers of this compound should exist?

a. 1(no isomers)
b. 2
! c. 3
d. 4
e. 5

23. A coordination complex in which the ligands exhibit an octahedral geometry about the central metal ion has the formula, $[Co(NH_3)_4Cl_2]$. How many geometric isomers of this compound should exist?

a. 1 (no isomers)
! b. 2
c. 3
d. 4
e. 5

24. Which one of the following complex ions is a chiral complex?

a. square planar $[Pt(en)_2]^{2+}$
b. tetrahedral $[NiCl_4]^{2-}$
! c. octahedral $[Co(en)_3]^{3+}$
d. octahedral $[Co(NH_3)_4Cl_2]^+$
e. square planar $[Pt(NH_3)ClBrI]^-$

25. The splitting of the 5-fold degenerate d-orbital energy level of transition metal ions by the field potential of a particular geometry gives rise to the pattern,

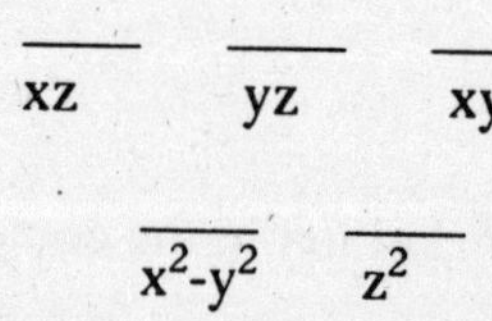


Which one of the following geometries produces the splitting pattern above?

a. linear
b. octahedral
c. planar triangular
d. square planar
! e. tetrahedral

26. The following order is observed in the spectrochemical series for the ligands listed,

$$CN^- > NO_2^- > en > NH_3 > H_2O > C_2O_4^{2-} > OH^- > F^- > Cl^- > Br^- > I^-$$

It was observed that for the two 6-coordinate cobalt complexes, $CoF_6^{3-}$ and $Co(NH_3)_6^{3+}$, one is diamagnetic and one is paramagnetic. How many unpaired electrons are there in the paramagnetic complex ion?

a. 1
b. 2
c. 3
! d. 4
e. 5

27. The following order is observed in the spectrochemical series for the ligands listed,

$$CN^- > NO_2^- > en > NH_3 > H_2O > C_2O_4^{2-} > OH^- > F^- > Cl^- > Br^- > I^-$$

How many unpaired electrons would you expect to find in the complex ion, $Cr(CN)_6^{3-}$ if the value of the splitting parameter is as large as it is in typical cyanide complexes?

a. 1
b. 2
! c. 3
d. 4
e. 5

28. The following order is observed in the spectrochemical series for the ligands listed,

$$CN^- > NO_2^- > en > NH_3 > H_2O > C_2O_4^{2-} > OH^- > F^- > Cl^- > Br^- > I^-$$

$FeF_6^{3-}$ and $Fe(CN)_6^{3-}$ are both negatively charged coordination complexes.

a. $FeF_6^{3-}$ and $Fe(CN)_6^{3-}$ are both diamagnetic.
b. $FeF_6^{3-}$ is diamagnetic while $Fe(CN)_6^{3-}$ is paramagnetic.
c. $FeF_6^{3-}$ is paramagnetic while $Fe(CN)_6^{3-}$ is diamagnetic.
! d. Both complex ions are paramagnetic, but $FeF_6^{3-}$ has a larger number of unpaired electrons than $Fe(CN)_6^{3-}$ hence its paramagnetism is more pronounced.
e. Both complex ions are paramagnetic, but $Fe(CN)_6^{3-}$ has a larger number of unpaired electrons than $FeF_6^{3-}$ hence its paramagnetism is more pronounced.

29. The following order is observed in the spectrochemical series for the ligands listed,

$$CN^- > NO_2^- > en > NH_3 > H_2O > C_2O_4^{2-} > OH^- > F^- > Cl^- > Br^- > I^-$$

$Cr(NH_3)_6^{3+}$ and $Co(NH_3)_6^{3+}$ are both negatively charged coordination complexes.

a. $Cr(NH_3)_6^{3+}$ and $Co(NH_3)_6^{3+}$ are both diamagnetic.
b. $Cr(NH_3)_6^{3+}$ is diamagnetic while $Co(NH_3)_6^{3+}$ is paramagnetic.
! c. $Cr(NH_3)_6^{3+}$ is paramagnetic while $Co(NH_3)_6^{3+}$ is diamagnetic.
d. Both complex ions are paramagnetic, but $Cr(NH_3)_6^{3+}$ has a larger number of unpaired electrons than $Co(NH_3)_6^{3+}$ hence its paramagnetism is more pronounced.
e. Both complex ions are paramagnetic, but $Co(NH_3)_6^{3+}$ has a larger number of unpaired electrons than $Cr(NH_3)_6^{3+}$ hence its paramagnetism is more pronounced.

30. The following order is observed in the spectrochemical series for the ligands listed,

$$CN^- > NO_2^- > en > NH_3 > H_2O > C_2O_4^{2-} > OH^- > F^- > Cl^- > Br^- > I^-$$

$Cr(NH_3)_6^{3+}$ and $Ni(NH_3)_6^{2+}$ are both negatively charged coordination complexes.

a. $Cr(NH_3)_6^{3+}$ and $Ni(NH_3)_6^{2+}$ are both diamagnetic.
b. $Cr(NH_3)_6^{3+}$ is diamagnetic while $Ni(NH_3)_6^{2+}$ is paramagnetic.
c. $Cr(NH_3)_6^{3+}$ is paramagnetic while $Ni(NH_3)_6^{2+}$ is diamagnetic.
! d. Both complex ions are paramagnetic, but $Cr(NH_3)_6^{3+}$ has a larger number of unpaired electrons than $Ni(NH_3)_6^{2+}$ hence its paramagnetism is more pronounced.
e. Both complex ions are paramagnetic, but $Ni(NH_3)_6^{2+}$ has a larger number of unpaired electrons than $Cr(NH_3)_6^{3+}$ hence its paramagnetism is more pronounced.

31. Many, but not all coordination complexes are colored due to absorption of light in the visible region of the spectrum. A complex ion which has its absorption maximum in the blue region of the visible spectrum would most likely appear to be

a. blue
b. green
c. yellow
d. orange
! e. red

32. Many, but not all coordination complexes are colored due to absorption of light in the visible region of the spectrum. Which one of the following coordination complexes of transition metals is definitely *not* expected to be a colored complex?

a. $Co(NH_3)_6^{3+}$
b. $Cr(NH_3)_6^{2+}$
c. $Cu(NH_3)_4^{2+}$
d. $V(H_2O)_6^{3+}$
! e. $Zn(NH_3)_4^{2+}$

33. Which one of the following coordination complexes would you expect to be the most stable (having the highest formation constant)?

a. $Co(NH_3)_6^{3+}$
b. $CoF_6^{3-}$
c. $Co(en)_3^{3+}$
d. $Co(dien)_2^{3+}$ dien = $H_2N—CH_2—CH_2—NH—CH_2—CH_2—NH_2$
! e. $Co(edta)^-$ edta = ethylenediaminetetraacetate

## Fill in the Blanks

34. The ethylenediamine molecule, usually abbreviated as en, functions as a bidentate ligand and forms coordination complexes with platinum which have the formula: $[Pt(en)_2]^{2+}$. Based on this information, assign a coordination number to the platinum(II) ion in this coordination complex.
_____ (! 4)

35. The ethylenediamine molecule, usually abbreviated as en, functions as a bidentate ligand and forms a coordination complex with cobalt which has the formula: $[Co(en)_2(H_2O)_2]^{3+}$. Based on this information, assign a coordination number *and* a charge to the cobalt ion in this coordination complex.
_____ (! 6, 3)

36. The equilibrium constant expression obtained for the process

$$Cu^{2+}(aq) + 4\,NH_3(aq) <\!\!=\!\!=\!\!> Cu(NH_3)_4^{2+}(aq)$$

is $[Cu(NH_3)_4^{2+}]/[Cu^{2+}][NH_3]^4$. This constant is called the _____ constant. (! formation)

37. Which complex ion should exhibit the larger value of the crystal field splitting parameter, $Co(NH_3)_6^{2+}$ or $Co(NH_3)_6^{3+}$? _____ (! $Co(NH_3)_6^{3+}$)

38. Which complex ion should exhibit the larger value of the crystal field splitting parameter, $Co(NH_3)_6^{3+}$ or $Rh(NH_3)_6^{3+}$? _____ ( ! $Rh(NH_3)_6^{3+}$)

39. The formation constant for the diammine silver(I) ion is $1.6 \times 10^7$, while the solubility product constant for silver chloride is $1.8 \times 10^{-10}$. What is the equilibrium constant for the reaction,
$AgCl(s) + 2\,NH_3(aq) \Longleftrightarrow Ag(NH_3)_2^+(aq) + Cl^-(aq)$. _____ ( ! $2.9 \times 10^{-3}$)

40. The formation constant for the bis(thiosulfato)argentate(I) ion is $2.0 \times 10^{13}$, while the solubility product constant for silver bromide is $5.0 \times 10^{-13}$. What is the equilibrium constant for the reaction,
$AgBr(s) + 2\,S_2O_3^{2-}(aq) \Longleftrightarrow Ag(S_2O_3)_2^{3-}(aq) + Br^-(aq)$. _____ ( $1.0 \times 10^1$)

41. The formation constant for the diammine silver(I) ion is $1.6 \times 10^7$, while the solubility product constant for silver chloride is $1.8 \times 10^{-10}$. Given the reaction,
$AgCl(s) + 2\,NH_3(aq) \Longleftrightarrow Ag(NH_3)_2^+(aq) + Cl^-(aq)$,
what is the solubility, in moles per liter, of silver chloride in 1.00 molar ammonia solution?
_____ ( ! $1.8 \times 10^{-2}$)

42. The formation constant for the bis(thiosulfato)argentate(I) ion is $2.0 \times 10^{13}$, while the solubility product constant for silver bromide is $5.0 \times 10^{-13}$. Considering the reaction,
$AgBr(s) + 2\,S_2O_3^{2-}(aq) \Longleftrightarrow Ag(S_2O_3)_2^{3-}(aq) + Br^-(aq)$.
What is the solubility, in moles per liter, of silver bromide in 0.200 molar sodium thiosulfate solution?
_____ ( ! $8.6 \times 10^{-2}$)

## True and False

43. Coordination complexes of a particular metal ion in which water is the ligand usually contain fewer ligands than coordination complexes of the same metal with ethylenediamine as the ligand. ( !F)

44. A polydentate ligand is one which has more than one acceptor atom which can attach itself to the metal ion to form a coordination complex. ( !F)

45. A polydentate ligand is one which has more than one donor atom which can attach itself to the metal ion to form a coordination complex. ( !T)

46. $Na^+$, $K^+$, and $Li^+$, being ions of "A" group metals, have a greater tendency to form coordination complexes with ligands than $Fe^{2+}$, $Co^{2+}$, and $Ni^{2+}$ ions. ( !F)

47. Ethylenediaminetetraacetate, abbreviated $edta^{4-}$, is a ligand whose ability to form coordination complexes has found a wide range of useful applications. ( !T)

48. Monodentate ligands form chelate complexes more readily than polydentate ligands due to the chelate effect. ( !F)

49. Another name for the ferrocyanide, a transition metal complex ion in which the iron is in the 2+ oxidation state, is hexacyanoferrite. ( !F)

50. The following order is observed in the spectrochemical series for the ligands listed,

$$CN^- > NO_2^- > en > NH_3 > H_2O > C_2O_4^{2-} > OH^- > F^- > Cl^- > Br^- > I^-$$

It was observed that for the two 6-coordinate cobalt complexes, $CoF_6^{3-}$ and $Co(NH_3)_6^{3+}$, one is diamagnetic and one is paramagnetic. Which one is the diamagnetic complex? ( ! $Co(NH_3)_6^{3+}$)

51. Even though silver chloride is only very very slightly soluble in water, addition of a reagent which forms complexes with silver increases this solubility. ( !T)

52. The following order is observed in the spectrochemical series for the ligands listed,

$$CN^- > NO_2^- > en > NH_3 > H_2O > C_2O_4^{2-} > OH^- > F^- > Cl^- > Br^- > I^-$$

It was observed that for the two 6-coordinate cobalt complexes, $Co(NH_3)_6^{3+}$ and $Co(H_2O)_6^{2+}$, one is diamagnetic and one is paramagnetic. Which one is the paramagnetic complex? ( ! $Co(H_2O)_6^{2+}$)

## Critical Thinking

53. The formation constant for the bis(thiosulfato)argentate(I) ion, $[Ag(S_2O_3)_2]^{3-}$, is $2.0 \times 10^{13}$, while the solubility product constant for silver iodide is $8.3 \times 10^{-17}$. A 0.200 molar solution of $Na_2S_2O_3$ is saturated with AgI. What is the concentration of free iodide ion in the saturated solution? ______ ( ! $7.5 \times 10^{-3}$ molar)

54. The formation constant for the tris(ethylenediammine)nickel(II) complex ion, $[Ni(en)_3]^{2+}$, is $4.1 \times 10^{17}$. What is the concentration of the free nickel(II) ion in a solution in which the equilibrium concentration of free ethylenediamine is 0.400 molar, and that of the nickel complex above is 0.0100 molar. ______ ( ! $3.8 \times 10^{-19}$ molar)

55. The solubility product of silver chromate, $Ag_2CrO_4$, is $1.2 \times 10^{-12}$. Silver chromate will dissolve in aqueous ammonia to a slight extent due to complex ion formation. If the formation constant for the $[Ag(NH_3)_2]^+$ ion is $1.6 \times 10^{+7}$, calculate the solubility of $Ag_2CrO_4$ in 0.500 molar ammonia solution, in moles per liter.

a. $1.1 \times 10^{-3}$
b. $3.9 \times 10^{-3}$
! c. $6.3 \times 10^{-3}$
d. $9.9 \times 10^{-3}$
e. $1.2 \times 10^{-2}$

56. Even though there are two isomers of the 4-coordinate planar complex $[Pt(NH_3)_2Cl_2]$ there are no isomers for a related complex, $[Pt(en)Cl_2]$. Suggest a reason why. ____________________
( ! The "bite" distance of the donor atoms in the chelating ligand, ethylenediamine, is too small to coordinate to trans positions around the $Pt^{2+}$ ion.)

## Chapter 20 Nuclear Reactions and Their Role in Chemistry

### Multiple Choice

1. The scientists of the "Manhattan Project" realized that the conversion of mass to energy becomes significant only in

a. chemiluminescent transformations
b. exothermic reactions
c. explosive chemical reactions
d. spontaneous processes
! c. nuclear reactions

2. Which one of the following nuclides has the maximum binding energy per nucleon, that is, it lies at or near the maximum in the binding energy per nucleon curve?

a. $^{251}Cf$
b. $^{197}Au$
! c. $^{56}Fe$
d. $^{1}H$
e. $^{4}He$

3. The rest mass of a proton is 1.00727252 u and that of a neutron is 1.008665 u. The $^{4}He$ nucleus weighs 4.00150286 u. Calculate the total binding energy of the nucleus.

a. $1.66 \times 10^{-7}$ joule
b. $2.27 \times 10^{-12}$ joule
c. $3.86 \times 10^{-11}$ joule
! d. $4.53 \times 10^{-12}$ joule
e. $4.53 \times 10^{-10}$ joule

4. The rest mass of a proton is 1.00727252 u and that of a neutron is 1.008665 u. The $^{19}F$ nucleus weighs 18.9934659 u. Calculate the binding energy per nucleon for this nucleus.

a. $1.246 \times 10^{-12}$ joule
! b. $2.368 \times 10^{-11}$ joule
c. $4.735 \times 10^{-11}$ joule
d. $6.230 \times 10^{-13}$ joule
e. $8.307 \times 10^{-13}$ joule

5. The rest mass of a proton is 1.00727252 u and that of a neutron is 1.008665 u. The $^{75}As$ nucleus weighs 74.903487 u. Calculate the total binding energy of the nucleus.

a. $1.464 \times 10^{-7}$ joule
! b. $1.045 \times 10^{-10}$ joule
c. $2.090 \times 10^{-10}$ joule
d. $6.235 \times 10^{-12}$ joule
e. $3.453 \times 10^{-10}$ joule

6. The rest mass of a proton is 1.00727252 u and that of a neutron is 1.008665 u. The $^{103}$Rh nucleus weighs 102.880814 u. Calculate the binding energy per nucleon for this nucleus.

! a. $2.442 \times 10^{-12}$ joule
b. $4.884 \times 10^{-12}$ joule
c. $7.534 \times 10^{-11}$ joule
d. $3.602 \times 10^{-13}$ joule
e. $7.803 \times 10^{-13}$ joule

7. The rest mass of a proton is 1.00727252 u and that of a neutron is 1.008665 u. The $^{31}$P nucleus weighs 30.96553 u. Calculate the total binding energy of the nucleus.

a. $2.662 \times 10^{-11}$ joule
b. $2.108 \times 10^{-11}$ joule
! c. $4.216 \times 10^{-11}$ joule
d. $6.585 \times 10^{-12}$ joule
e. $5.885 \times 10^{-11}$ joule

8. The number of naturally occurring nuclides which are radioactive is limited and they are concentrated mostly in one region of the periodic table. Approximately how many naturally occurring radioactive nuclides are there?

a. 10
! b. 50
c. 100
d. 250
e. 500

9. The radiation involving particles with the greatest charge of those in the set below is

! a. alpha radiation
b. beta radiation
c. gamma radiation
d. neutrons
e. x-rays

10. The radiation with the least penetrating characteristics from the set below is

! a. alpha radiation
b. beta radiation
c. gamma radiation
d. neutrons
e. x-rays

11. The radiation involving particles with the greatest mass of those in the set below is

! a. alpha radiation
b. beta radiation
c. gamma radiation
d. neutrons
e. x-rays

12. The rest mass of a proton is 1.00727252 u and that of a neutron is 1.008665 u. The $^{197}Au$ nucleus weighs 196.92320 u. Calculate the binding energy per nucleon for this nucleus.

a. $4.118 \times 10^{-12}$ joule
b. $3.282 \times 10^{-12}$ joule
! c. $1.268 \times 10^{-12}$ joule
d. $2.536 \times 10^{-12}$ joule
e. $7.366 \times 10^{-13}$ joule

13. Identify the missing species in the nuclear equation given

$$^{155}_{26}Fe + ^{0}_{-1}e \longrightarrow ????$$

a. $^{53}_{27}Fe$
! b. $^{55}_{25}Mn$
c. $^{55}_{27}Co$
d. $^{55}_{27}Fe$
e. $^{55}_{25}Fe$

14. Identify the missing species in the nuclear equation given

$$^{211}_{82}Pb + ^{0}_{-1}e \longrightarrow ????$$

a. $^{211}_{81}Tl$
! b. $^{211}_{83}Bi$
c. $^{212}_{82}Pb$
d. $^{210}_{82}Pb$
e. $^{210}_{83}Bi$

15. When a nuclide emits a beta particle, the atomic number of the particle produced

a. is the same as that of the original nuclide
! b. increases by one mass unit
c. decreases by one mass unit
d. increases by two mass units
e. decreases by two mass units

16. Identify the missing species in the nuclear equation given

$$^{245}_{96}Cm + ^{4}_{2}\alpha \longrightarrow ????$$

a. $^{249}_{94}Pu$
! b. $^{241}_{94}Pu$
c. $^{294}_{98}Cf$
d. $^{241}_{98}Cf$
e. $^{245}_{94}Pu$

17. Identify the missing species in the nuclear equation given

$$^{226}_{88}\text{Cm} \longrightarrow {}^{4}_{2}\alpha + ????$$

! a. $^{226}_{86}\text{Rn}$
b. $^{226}_{90}\text{Th}$
c. $^{230}_{90}\text{Th}$
d. $^{230}_{86}\text{Rn}$
e. $^{230}_{88}\text{Ra}$

18. Identify the missing species in the nuclear equation given

$$^{19}_{10}\text{Ne} \longrightarrow {}^{0}_{1}e + ????$$

a. $^{19}_{11}\text{F}$
b. $^{19}_{11}\text{Na}$
! c. $^{19}_{9}\text{F}$
d. $^{19}_{11}\text{Ne}$
e. $^{210}_{9}\text{F}$

19. When $^{54}_{27}\text{Co}$ decays an isotope of iron is formed. What other products are also formed?

! a. a positron + neutrino
b. a positron only
c. beta rays and gamma rays
d. alpha particles and gamma rays
e. gamma rays only

20. Identify the missing species in the nuclear equation given

$$^{87}_{36}\text{Kr} \longrightarrow {}^{86}_{36}\text{Kr} + ????$$

a. $^{0}_{1}n$
b. $^{0}_{1}e$
! c. $^{1}_{0}n$
d. $^{1}_{0}e$
e. $^{1}_{1}p$

21. One Becquerel is equivalent to the radiation produced by

! a. one disintegration per second
b. $3.7 \times 10^{10}$ disintegration per second
c. a mole of disintegration
d. 100 rads
e. one rem

22. This radiation source can be has caused quite a panic in recent years because it is the most common source of natural radiation and one of the easiest to eliminate.

! a. radon
b. cosmic rays
c. rocks and soils
d. foods
e. water

23. The intensity of of x-ray, gamma ray, or any radiation is

a. directly proportional to the distance from the source
b. directly proportional to the square of the distance from the source
c. inversely proportional to the distance from the source
! d. inversely proportional to the square of the distance from the source.
e. inversely proportional to the cube of the distance from the source.

24. If the half-life of a radioactive element is 30.0 years, how long will it take to decay to the point where the activity is 70.0% of the original value?

a. 5.0 years
b. 12.2 years
! c. 15.4 years
d. 30.8 years
e. 86.1 years

25. If a radioactive sample of a newly discovered element, coronium, decays to the point where the activity is 40.0 percent of its original activity in 44.0 years, what is the half-life of the nuclide?

a. 21.6 years
! b. 33.3 years
c. 58.2 years
d. 80.6 years
e. 89.9 years

26. Rhenium-186 is a ß-emitter with a half life of 90 hours. How long would it take for the activity in a sample of this nuclide to decay to exactly one-third of its original value?

a. 121 hours
! b. 143 hours
c. 158 hours
d. 180 hours
e. 189 hours

27. Ytterbium-175 is a ß-emitter with a half life of 101 hours. How long would it take for the activity in a sample of this nuclide to decay to exactly one-fifth of its original value?

a. 215 hours
b. 225 hours
! c. 235 hours
d. 250 hours
e. 275 hours

28. Terbium-152 is a positron emitter with a half life of 18 hours. How long would it take for the activity in a sample of this nuclide to decay to exactly three-eighths of its original value?

a. 32.8 hours
b. 28.8 hours
c. 23.6 hours
! d. 25.5 hours
e. 48.9 hours

29. Gadolinium-147 is positron emitter with a half life of 35 hours. How long would it take for the activity in a sample of this nuclide to decay to exactly two-sevenths of its original value?

a. 68 hours
! b. 63 hours
! c. 71 hours
d. 78 hours
e. 106 hours

30. How many nucleons are there in the most common isotope of uranium?

a. 234
b. 235
c. 236
! d. 238
e. 239

31. How many nucleons are there in the fissionable isotope of uranium?

a. 234
! b. 235
c. 236
d. 238
e. 239

32. The specific activity of a carbon sample from the bones of a mummy in a secret tomb recently discovered in Egypt is found to be at a level which is 60.0% of that of living plants. Calculate the age of the artifact. ($^{14}_{6}C$: $t_{1/2}$ = 5730 years)

a. 3438 years
b. 2865 years
! c. 4222 years
d. 2111 years
e. 8444 years

33. Which one of the following is *not* a device for determining radiation?

a. Geiger counter
b. Wilson cloud chamber
c. scintillation counter
d. film dosimeter
! e. ionometer

34. Which one of the following is a unit of nuclear disintegrations per unit time?

! a. the curie
b. the rad
c. the gray
d. the sievert
e. the rem

35. Which one of the following is a unit of radiation dose which is the same regardless of type of radiation?

a. the curie
b. the becquerel
! c. the rem
d. the rad
e. the gray

36. What is the mass number of an alpha particle?

a. 0
b. 1
c. 2
d. 3
e. 4

37. Which one of the following nuclear processes not cause a change in the atomic number of the nuclide undergoing the process?

a. emission of $\alpha$-particle
b. emission of ß-particle
! c. emission of a $\Gamma$-ray
d. emission of a positron
e. capture of an electron

38. When radium-226 is transmuted by a nuclear decay to radon-222, the process involved is

! a. emission of $\alpha$-particle
b. emission of ß-particle
c. emission of a $\Gamma$-ray
d. emission of a positron
e. capture of an electron

39. All isotopes of elements beyond ___ in the periodic are unstable and hence radioactive, though some have much longer half lives than others.

a. bismuth
b. platinum
! c. lead
d. francium
e. thallium

40. The nuclear reaction which is depicted below is an example of

$$^{197}_{78}\text{Pt} \longrightarrow ^{197}_{79}\text{Au} + ????$$

a. emission of $\alpha$-particle
! b. emission of ß-particle
c. emission of a $\Gamma$-ray
d. emission of a positron
e. capture of an electron

41. The nuclear reaction which is depicted below is an example of

$$^{200}_{81}\text{Tl} \longrightarrow ^{200}_{80}\text{Au} + ????$$

a. emission of $\alpha$-particle
b. emission of ß-particle
c. emission of a $\Gamma$-ray
! d. emission of a positron
e. capture of an electron

42. Carbon-14 is generated in the atmosphere by the nuclear reaction,

$$^{14}_{7}\text{N} + ^{1}_{0}n \longrightarrow ^{14}_{7}\text{C} + ????$$

The missing species is

a. an $\alpha$-particle
b. a ß-particle
c. a $\Gamma$-ray
d. a positron
! e. a proton

43. The two steps in the reaction which takes place in breeder reactors is

$$^{238}_{92}\text{U} + ^{1}_{0}n \longrightarrow \text{XXXX} + ^{0}_{-1}e$$

$$\text{XXXX} \longrightarrow \text{YYYY} + ^{0}_{-1}e$$

The missing species are

a. $^{239}$Np and $^{239}$Am
b. $^{239}$Np and $^{241}$Pu
! c. $^{239}$Np and $^{239}$Pu
d. $^{239}$U and $^{239}$Np
e. $^{237}$Np and $^{237}$Pu

44. The radiocarbon dating method can be used to determine the age of

a. carbonate rocks
b. silicate rocks
! c. art canvases
d. bronze beakers
e. Roman coins

45. A tokamak is

a. a linear particle accelerator
b. a highly shielded breeder reactor
! c. a fusion generator which uses magnetic fields to contain the nuclei and heat them to the degree needed to start a sustained reaction
d. a third generation fission reactor
e. a plasma driven reactor based on inertial confinement

46. Fission reactors are restrained from going supercritical through

a. use of liquid metal coolants
b. use of heavy water coolant
! c. use of control rods which absorb neutrons
d. use of liquid nitrogen to maintain correct reaction rate
e. proper reactor design

## True and False

47. Tritium is an isotope of hydrogen which is radioactive. ( !T)

48. The mass of a nucleus is exactly equal to the weights of all the protons and neutrons which are in that nucleus, taken separately. ( !F)

49. The emission of a positron produces the same effect as replacement of a proton by a neutron. ( !T)

50. When radiation passes through gaseous matter, ionization results. ( !T)

51. Neutrinos have very high penetrating power. ( !T)

## Fill in the Blanks

52. A neutral particle with the mass of an electron is known as a ______. ( ! neutrino)

53. The energy that an electron receives when accelerated under through a potential difference of one volt is one ______. ( ! electron volt)

54. The energy which holds the nucleons in the nucleus together is called ______. ( ! binding energy)

55. The force which holds the nucleons in the nucleus together is called ______. ( ! nuclear strong force)

56. The change of one nuclide into another is termed ______. ( ! transmutation)

57. A nucleus which results from the combination of a particle and nucleus is termed a ______. ( !compound nucleus)

58. The nucleus of a deuterium atom is called a ______. ( ! deuteron)

59. A ______ measures radiation by detecting the resultant ionization. ( ! Geiger counter)

60. A ______ can detect a subatomic particle by way of visible tracks in a vapor. ( ! cloud chamber)

61. A ______ will detect radiation by way of flashes of light which are emitted when particles strike. ( ! scintillation counter)

62. A common unit of absorbed radiation dose is the ______. ( ! the rad or the gray)

## Critical Thinking

63. Radium 226, which undergoes a first order nuclear decay process to yield $^{222}Rn$, has a half life of 1620 years. A 24.45 milligram sample of $^{226}Ra$ was placed in a time capsule on January 1, 1932 with instructions to open the capsule and weigh it again on July 1, 2276. How many milligrams of the radium should there be in the capsule at that time?

a. 21.85 milligrams
! b. 21.11 milligrams
c. 19.25 milligrams
d. 20.14 milligrams
e. 21.65 milligrams

## Chemicals In Use 16 Positron Emitters and the PET Scan

64. In comparison to the electron, the positron

a. has an equivalent mass and a negative charge
! b. has an equivalent mass and a positive charge
c. has a higher mass and a positive charge
d. has a higher mass and a negative charge
e. has a lower mass and a positibe charge

65. In a PET scan which of the following does not occur?

a. a positron-emitting radionuclide is introduced into the tissue
b. the gamma-emitting radionuclide is introduced into the tissue
c. positrons and electrons are destroyed and replaced by gamma radiation
! d. the radiation is being sent through the body is generated externally
e. the radiation originates within the site being scanned

66. The MRI technique is a less harmful technique (to tissue) than the PET scan. ( !T)

67. The MRI technique is a less harmful technique (to tissue) than the PET scan because

! a. it is noninvasive
b. it involves radiation
c. it doesn't produce electrical impulses
d. it uses glucose to cross the blood-brain interface
e. it does not utilize calcium ions

68. Which type of tissue is transparent to MRI

a. fatty tissue
! b. bone tissue
c. muscle tissue
d. non-malignant tumorous tissue
e. cancerous tissue

69. Assume you have a serious liver disease and you also wear a pacemaker. Which technique would be the least dangerous to you and most likely give an indication as to the extent of the damage to the organ?

! a. PET
b. x-ray scan
c. MRI
d. MRI + x-ray scan
e. CAB

## Chapter 21 Organic Compounds, Polymers and Biochemicals

### Multiple Choice

1. Which one of the compounds whose condensed structural formulas are given below is butane?

a. $CH_3CH_3$
b. $CH_3CH_2CH_3$
! c. $CH_3CH_2CH_2CH_3$
d. $CH_3CH_2CH_2CH_2CH_3$
e. $CH_3CH_2CH_2CH_2CH_2CH_3$

2. Which one of the compounds whose condensed structural formulas are given below is *n*-pentane?

a. $CH_3CH_3$
b. $CH_3CH_2CH_3$
c. $CH_3CH_2CH_2CH_3$
! d. $CH_3CH_2CH_2CH_2CH_3$
e. $CH_3CH_2CH_2CH_2CH_2CH_3$

3. How many isomers of butane, $C_4H_{10}$, can be drawn?

a. 1 (no isomers)
! b. 2
c. 3
d. 4
e. 5

4. How many isomers of pentane, $C_5H_{12}$, can be drawn?

a. 1 (no isomers)
b. 2
! c. 3
d. 4
e. 5

5. $CH_3—CH_2—O—H$ and $CH_3—O—CH_3$ are a pair of compounds which are

a. allotropes
b. allomers
c. isochrones
! d. isomers
e. allotones

6. Which one of the groups below goes by the name, *n*-propyl

a. $CH_3CH_2—$
b. $CH_2{=}CH—$
! c. $CH_3—CH_2—CH_2—$
d. $CH_2{=}CH—CH_2—$
e. $CH_3—CH{=}CH—$

7. The functional group, —OH, is found in which one of these types of organic compounds?

a. alkanes
b. alkenes
c. amines
! d. alcohols
e. ethers

8. Functional groups containing nitrogen at the active site are found in

a. alkenes
b. alcohols
! c. amines
d. carboxylic acids
e. ethers

9. Functional groups containing the C═O group or linkage are found in

a. alkenes
! b. ketones
c. amines
d. alcohols
e. ethers

10. One of the compounds listed below is an organic base and functions as a proton acceptor. Identify this organic base.

a. $CH_3—CH_2—CH_2—O—H$
b. $CH_3—CH═CH_2$
c. $H_3C—\underset{\underset{O}{\|}}{C}—CH_3$
d. $CH_3—CH_2—\underset{\underset{O}{\|}}{C}—O—H$
! e. $CH_3—CH_2—NH_2$

11. Based on the properties of the attached functional group, which one of the compounds below would you expect interact most strongly with water, thereby increasing it solubility?

a. $CH_3—CH_2—Cl$
b. $CH_3—CH_2—F$
c. $CH_3—CH_2—I$
d. $CH_3—CH_2—S—H$
! e. $CH_3—CH_2—O—H$

12. Based on the properties of the attached functional group, which one of the compounds below would you expect interact most strongly with water, thereby increasing it solubility?

a. $CH_3—CH_2—H$
! b. $CH_3—CH_2—NH_2$
c. $CH_3—O—CH_3$
d. $CH_3—CH_2—I$
e. $CH_3—CH_2—S—H$

13. Which one of the following compounds is not an unsaturated compound?

a. $H_2C{=}CH_2$
b. $CH_3—CH{=}CH_2$
c. $H_2C{=}CHCl$
! d. $CH_3—CH_2—O—H$
e. $H_2C{=}CH—O—H$

14. Various hydrocarbons are separated from the crude oil mixture by

a. filtration
b. chromatography
c. precipitation
! d. distillation
e. centrifugation

15. The alkane hydrocarbon whose skeleton is shown immediately below, is regarded as a derivative of

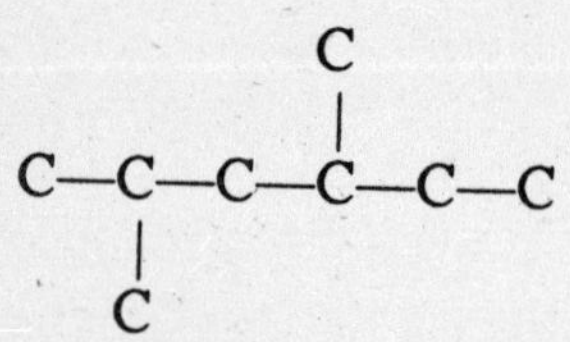

a. butane
! b. hexane
c. heptane
d. octane
e. pentane

16. The alkane hydrocarbon whose skeleton is shown immediately below, is regarded as a derivative of

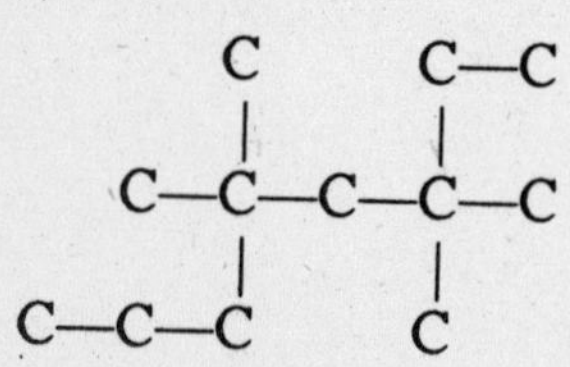

a. dodecane
b. decane
c. heptane
d. pentane
! e. octane

17. The hydrocarbon fragment, —$CH_3$, which has one position open (unfilled) in named

a. methide
b. methonium
c. methal
! d. methyl
e. trihydrocarbide

18. The hydrocarbon fragment shown below is properly named

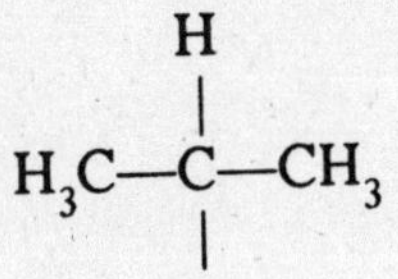


a. propyl
b. propide
c. propanium
! d. isopropyl
e. propanyl

19. The name for the compound, $CH_3$—CH=CH—$CH_3$, is

a. butene-2
b. butene-3
! c. 2-butene
d. 2-butyne
e. 2-butyl

20. If two of the hydrogen atoms in butene are replaced by two chlorine atoms to form dichlorobutene, how many different possible dichlorobutenes will there be?

a. 1
b. 2
! c. 3
d. 4
e. 6

21. Draw a structural formula for cyclohexane, a cyclic saturated hydrocarbon ($C_6H_{12}$). How many σ-bonds are there in cyclohexane?

a. 12
b. 16
c. 17
! d. 18
e. 20

22. Draw a structural formula for benzene. How many σ-bonds are there in a benzene molecule?

a. 6
! b. 12
c. 14
d. 18
e. 20

23. Draw a structural formula for cyclohexane, a cyclic saturated hydrocarbon ($C_6H_{12}$). How many $\pi$-bonds are there in cyclohexane?

! a. 0
b. 2
c. 3
d. 4
e. 6

24. The general structural formula for an ether can be expressed as

a. R—C—H
‖
O

b. R—C—O—H
‖
O

c. R—C—O—R'
‖
O

d. R—C—R'
‖
O

! e. R—O—R'

25. Which one of the compounds below has the general structural formula for an alcohol?

a. R—C—H
‖
O

b. R—C—O—H
‖
O

c. R—C—O—R'
‖
O

d. R—C—R'
‖
O

! e. R—CH—O—H
|
R'

26. Draw a structural formula for benzene. How many $\pi$-bonds are there in a benzene molecule?

a. 0
b. 2
! c. 3
d. 4
e. 6

27. Which one of the following is a product of the oxidation of $CH_3{-}CH_2{-}CH_2{-}O{-}H$?

a. $H_3C{-}\underset{\underset{O}{\|}}{C}{-}CH_3$

b. $H_3C{-}\underset{\underset{O}{\|}}{C}{-}O{-}CH_3$

c. $CH_3{-}CH_2{-}CH_3$

! d. $CH_3{-}CH_2{-}\underset{\underset{O}{\|}}{C}{-}O{-}H$

e. $CH_3{-}CH_2{-}O{-}CH_3$

28. Which one of the species below is a prodcut of the oxidation of $CH_3{-}\underset{\underset{O-H}{|}}{CH}{-}CH_3$

! a. $H_3C{-}\underset{\underset{O}{\|}}{C}{-}CH_3$

b. $H_3C{-}\underset{\underset{O}{\|}}{C}{-}O{-}CH_3$

c. $CH_3{-}CH_2{-}CH_3$

d. $CH_3{-}CH_2{-}\underset{\underset{O}{\|}}{C}{-}O{-}H$

e. $CH_3{-}CH_2{-}O{-}CH_3$

29. When $CH_3{-}CH{=}CH_2$ is treated with chlorine, a compound is formed in the $Cl_2$ adds across the double bond. How many $\sigma$-bonds are ther in the product of this addition reaction?

a. 7
b. 8
c. 9
! d. 10
e. 11

30. When $CH_3$—CH═$CH_2$ is reacted with water in the presence of a catalytic amount of acid, a new compound is formed. Which one of the set below might be the product of this reaction?

a. $CH_3$—CH═$CH_2$

b. $CH_3$—C═CH—O—H
|
H—O—H

! c. $CH_3$—CH—$CH_3$
|
H—O

d. $H_3C$—C—$CH_3$
‖
O

e. $CH_3$—$CH_2$—C—H
‖
O

31. The reaction of $(CH_3)_2$—$CH_2$—$CH_2$—O—H with concentrated HBr using controlled heating yields which one of the products below?

a. $CH_3$—$CH_2$—$CH_2$—Br
! b. $(CH_3)_2$—CH—$CH_2$—Br
c. $(CH_3)_2$—CH—$CH_2$—O—Br
d. $(CH_3)_2$—CH—$CH_4^+$ $Br^-$
e. $(CH_3)_2$—CH—$CH_3$

32. The compound, trimethylamine is a ______ and has the formula ______.

a. base, $(CH_3NH_2)_3$
b. base, $(CH_3NH)_2$
c. base, $(CH_3)_3NH_2$
! d. base, $(CH_3)_3N$
e. base, $(CH_3)_3(NH_2)_2$

33. Dodecylamine, $CH_3(CH_2)_{10}CH_2NH_2$, is absolutely insoluble in water. Yet, it can be converted to a water soluble form. Which of the species below represents the water soluble form of this compound?

a. $CH_3(CH_2)_{10}CH_2NH_2$—O—H
b. $CH_3(CH_2)_{10}CH_2NH_2$—Cl
c. $CH_3(CH_2)_{10}CH_2NH$—$CH_3$
! d. $CH_3(CH_2)_{10}CH_2NH_3^+$ $Cl^-$
e. $CH_3(CH_2)_{10}CH_2NH_2$—O—$CH_3$

34. The compound below is classified as which type of compound?

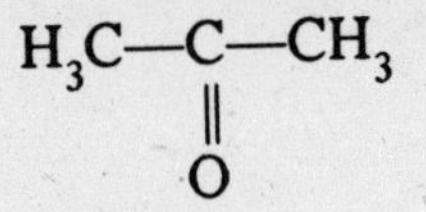


a. aldehyde
! b. ketone
c. acid
d. ester
e. amine

35. The compound below is classified as which type of compound?

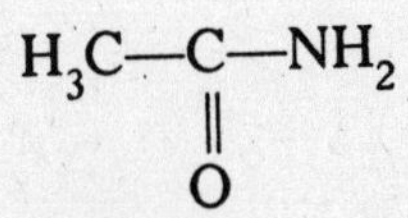


a. aldehyde
b. amine
c. ester
! d. amide
e. amino acid

36. The compound below is classified as which type of compound?

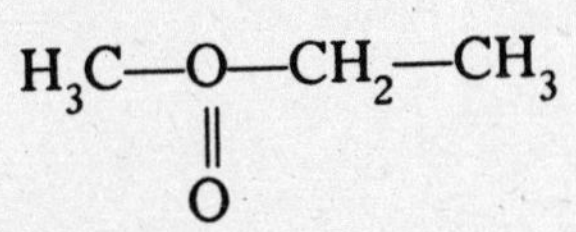


a. ketone
b. aldehyde
! c. ester
d. acid
e. amide

37. The compound below is correctly named as

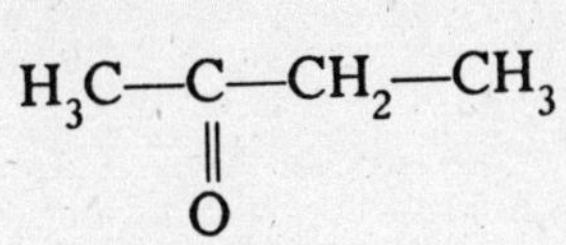


a. butanamine
b. butanamide
c. butanketone
! d. 2-butanone
e. 2-butanal

38. The compound below is correctly named

$$H_3C{-}CH_2{-}CH_2{-}CH_2{-}\underset{\underset{\displaystyle O}{\|}}{C}{-}H$$

a. pentaketone
b. pentaldehyde
! c. pentanal
d. pentanone
e. pentanoic acid

39. The compound below is correctly named

$$H_3C{-}CH_2{-}CH_2{-}CH_2{-}\underset{\underset{\displaystyle O}{\|}}{C}{-}NH_2$$

a. pentanamine
b. pentamine
c. 1-pentanamide
! d. pentanamide
e. pentaketoneamine

40. The ester which is prepared by heating 1-pentanol with acetic in the presence of an acidic catalyst is named

a. acetic pentanoate
b. pentanoic acetate
! c. 1-pentyl acetate
d. acetyl 1-pentanoate
e. acetyl pentanol

41. Linear polyethylene is a straight chain polymer. The repeating unit in linear polyethylene is

! a. $-CH_2-CH_2-$
b. $-CH{=}CH-$
c. $-CH{\equiv}CH-$
d. $-C{\equiv}C-$
e. $-CH_2{=}CH_2-$

42. Which one of the following sets contains the proper compounds from which an addition copolymer can be formulated?

a. $H_2C{=}CH_2$ and $H_2C{=}CH_2$
b. $F_2C{=}CF_2$ and $F_2C{=}CF_2$
c. $C_6H_5CH{=}CCl_2$ and $H_2N{-}(CH_2)_6{-}NH_2$
! d. $C_6H_5CH{=}CCl_2$ and $H_2C{=}CHCl$
e. $C_6H_5CH{=}CCl_2$ and $H{-}O{-}\underset{\underset{\displaystyle O}{\|}}{C}{-}(CH_2)_4{-}\underset{\underset{\displaystyle O}{\|}}{C}{-}O{-}H$

43. Which one of the following sets contains the proper compounds from which a condensation can be formulated?

a. $H_2N-(CH_2)_6-NH_2$ and $H_2C=CCl_2$
b. $H_2C=CHCl$and $F_2C=CF_2$
c. $C_6H_5CH=CCl_2$ and $H_2N-(CH_2)_6-NH_2$
d. $C_6H_5CH=CCl_2$ and $H_2C=CHCl$
! e. $H_2N-(CH_2)_6-NH_2$ and $H-O-\underset{O}{\underset{\|}{C}}-(CH_2)_4-\underset{O}{\underset{\|}{C}}-O-H$

44. The basic repeating unit in the rubber made from naturally occurring material is

! a. $-CH_2-\underset{CH_3}{\underset{|}{C}}=CH-CH_2-$

b. $-H_2C-CH_2-$

c. $-\underset{CH_3}{\underset{|}{HC}}-CH_2-$

d. $-H-\underset{CH_3}{\underset{|}{C}}-\underset{CH_3}{\underset{|}{C}}-H-$

e. $-H-\underset{C_6H_5}{\underset{|}{C}}-C-H-$

45. The basic repeating unit in polystyrene is

a. $-CH_2-\underset{CH_3}{\underset{|}{C}}=CH-CH_2-$

b. $-H_2C-CH_2-$

c. $-\underset{CH_3}{\underset{|}{HC}}-CH_2-$

d. $-H-\underset{CH_3}{\underset{|}{C}}-\underset{CH_3}{\underset{|}{C}}-H-$

! e. $-H-\underset{C_6H_5}{\underset{|}{C}}-C-H-$

46. Which one of the functional groups below is not usually found in carbohydrates?

a. aldehyde
! b. amide
c. ether
d. hydroxy
e. ketone

47. Lipids, by definition, are naturally occurring compounds which

a. contain ester groups
b. contain cholesterol
c. contain fatty acids as structural units
! d. are water insoluble but soluble in nonpolar solvents
e. are unsaturated

48. Triglycerides are compounds which contain (combined)

a. fatty acids and choline
! b. fatty acids and glycerin
c. fatty acids and phospholipids
d. cholesterol and other steroids
e. lecithin and choline

49. The chemical bond which links glycerin to a fatty acid is an example of what type of linkage?

a. ether
b. peptide
! c. ester
d. ionic
e. hydrogen bond

50. When lard is hydrolyzed, the products are

a. amino acids
b. alcohols and lipids
c. glycerin and lipids
d. carbohydrates
! e. glycerin and fatty acids

51. DNA is a(n)

a. enzyme
! b. nucleic acid
c. protein
d. peptide
e. steroid

52. Which one of the following compounds is not a lipid?

a. andosterone
! b. glucose
c. cholesterol
d. testosterone
e. triglycerides

## Fill in the Blanks

53. The sheer number of organic compounds, greater than all other kinds put together, is made possible by the ability of ______ to ________________. ( ! carbon, form long chains and still bond to other atoms)

54. The hydrocarbon group formed by removal of one hydrogen from methane is called ______ ( ! methyl)

55. ______ are the major source of hydrocarbon compounds. ( ! fossil fuels)

56. The name of the alkane hydrocarbon whose skeleton is shown below is ______.

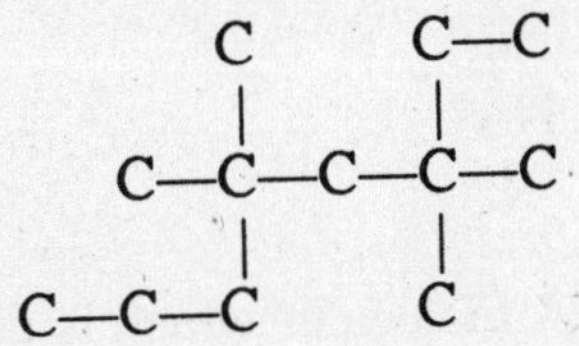

( ! 3,3,5,5-tetramethyloctane)

57. The amide formed when formic acid is heated with ammonia and the product is separated from the reaction mixture is ____________ (name and formula). ( ! formamide, $H{-}C({=}O){-}NH_2$)

58. When the ester, isopropyl butanoate is hydrolyzed in the presence of catalytic amounts of a strong base, two compounds are obtained. Their names are ______ and ______. ( ! isopropyl alcohol (2-propanol) and butanoic acid)

59. In a copolymer, there are ______ kinds of repeating units. ( ! Two)

## True and False

60. A reactive group, called a functional group, which is attached to the hydrocarbon portion of the molecule, influences or determines the kind of reactions a molecule undergoes. ( !T)

61. Hydrocarbons can be either gases, liquids or solids. ( !T)

## Critical Thinking

62. A carbonyl compound with 6 carbons was hydrogenated in a high pressure apparatus at 300°C using a hydrogenation catalyst. The product had the formula,

$$CH_3{-}CH_2{-}\underset{\substack{\| \\ O}}{CH}{-}CH_2{-}CH_3$$

The original carbonyl compound was therefore ______. ( ! 3-pentanone)

63. An ester with 6 carbons was hydrolyzed in the presence of catalytic amounts of a base to yield two compounds, one of which was a 4 carbon acid with a branched chain. What is the name of the *other* product of the reaction? ______ ( ! ethanol)

64. A very important compound, used as a fertilizer, can be prepared by the reaction of
H—O—C—O—H with ammonia under the proper conditions. This important compound is a(n)
‖
O

! a. amide
b. ammonium salt
c. acid salt
d. amine
e. ketone

## Chemicals In Use 17 DNA and "Genetic Fingerprinting"

65. Which one of the following statements is not true about DNA?

a. Genes are made up of DNA.
b. Every cell in the body has the entire set of DNA molecules which is unique to the person
c. DNA is composed of a double helix.
! d. DNA is a virus that attacks bacteria.
e. DNA fingerprinting is used as evidence in criminal cases

66. The acronym, RFLP, used in connection with use of DNA for identification and matching purposes has been gotten increased emphasis recently. What does the acronym stand for?

! a. restriction-fragment length polymorphism
b. restriction-fragile length polymorphism
c. radiation fragment length polymorphism
d. radiation fraction long polymorphism
e. real felons love police

67. Being polymorphic in length means that RFLPs

a. cause an electrical pulse
b. are not unique to a particular DNA sample
! c. are polymorphic in molecular mass
d. have no variations in length
e. are not polymorphic in molecular mass

68. Electrophoresis refers to

a. a supporting medium designed to carry sample RFLPs
! b. the movement of ions in an electrical field caused by an electrical force
c. lighter ions moving more rapidly than heavier ions
d. the transfer of the invisible RFLPs to a nylon membrane
e. the base pairing in DNA

## Supplementary Chapter 1 Simple Molecules and Ions of Nonmetals: Part I

### Multiple Choice

1. The chemical elements of which the matter in the universe is composed are

! a. primarily hydrogen and helium ( > 99%)
b. primarily nitrogen and oxygen, with hydrogen and helium present in a lesser amount
c. primarily calcium, magnesium and carbon in planetary rocks and stellar interiors
d. primarily ferrous metals like iron, cobalt, and nickel
e. probably primarily some unknown and maybe even hitherto undiscovered elements when all the matter in the millions of galaxies and universes which exist in the cosmos are considered

2. In the earth's environmental region (atmosphere, crust) the largest percent of ***all*** the atoms in existence are found as

a. nitrogen atoms
! b. oxygen atoms
c. hydrogen atoms
d. calcium atoms
e. carbon atoms

3. The boiling point of hydrogen, $H_2$, is about

a. -272 °C
! b. -252 °C
c. -200 °C
d. -100 °C
e. - 30 °C

4. The XYZ corporation is one of the largest producers of hydrogen gas in the Western region. Based on your knowledge of the methodology being used currently in the industry, which method listed below are they probably using?

a. electrolysis of water
b. reaction of sodium with water
c. reaction of metals with acid
d. isolation from natural gas
! e. reaction of hydrocarbons with steam

5. A two person chemistry team from XYZ corporation are doing a laboratory demonstration at an eight grade class in a local school, in which they are going to prepare some hydrogen gas, then observe and demonstrate some properties of hydrogen. What is the common method used in the field for generating small amounts of the gas?

! a. reaction of zinc metal with a strong acid, like hydrochloric acid
b. reaction of zinc metal with hot water
c. reaction of zinc metal with concentrated sodium hydroxide solution
d. reaction of zinc metal with superheated steam
e. reaction of zinc metal with alcohol

6. Sally Mae, a graduate student, needs to carry out an NMR measurement on the compound she just synthesized. Of the available solvents she decides to use $D_2O$ to dissolve her compound. $D_2O$, also known as "heavy water", contains two deuterium atoms, where deuterium is the second most abundant isotope of hydrogen. How many nucleons are there in each deuterium atom?

a. 1
! b. 2
c. 2.5
d. 3
e. 4.

7. The molecular weight of $D_2O$ is most nearly, in g/mole,

a. 16
b. 17
! c. 18
d. 19
e. 20

8. Concerning tritium, the radioactive isotope of hydrogen, how many nucleons are there in its nucleus?

a. 1
b. 2
c. 2.5
! d. 3
e. 4

9. Ionic hydrides react with water to liberate $H_2(g)$. In turn, these ionic hydrides are formed when metals of groups IA and IIA are reacted with

a. superheated steam
! b. hydrogen gas
c. strong acids
d. weak acids
e. water

10. The electronic configuration of the hydride ion is the same as that of

a. hydrogen
! b. helium
c. lithium
d. deuterium
e. tritium

11. All of the following are properties of the hydride ion except

a. strong base
b. reducing agent
c. reactive towards water
d. reacts with acids to produce $H_2$ gas
! e. proton donor

12. Which one of the following hydrides is the one which reacts ***violently*** with water? (a good point to remember when stressing the point that all chemical fires aren't best handled with water)

a. $NH_3$, a nitrogen hydride
b. $H_2S$, a sulfur hydride
c. $SiH_4$, a silicon hydride
d. $CH_4$, a carbon hydride
! e. NaH, a sodium hydride

13. Hydrogenated oils are being used in place of animal fats for health reasons in the food industry. Which of the following is one of the effects that hydrogenation has on a property of vegetable oil?

a. lowers its fat content
b. increases the number of double bonds
! c. increases the melting point
d. gives it a yellow color
e. converts it to soap

14. The air that we breathe every day is composed mainly of nitrogen and oxygen. What is the percent (V/V) concentration of oxygen in air?

! a. 21
b. 78
c. 1
d. 50
e. 25

15. A tank of pure oxygen such as one used in a laboratory was probably obtained from industrial sources, where it was made by

a. heating potassium chlorate
b. electrolysis of water
! c. distillation of liquid air
d. pyrolysis of superheated steam
e. passing superheated steam of heated metal oxides

16. Dave is preparing his pre-lab for next week's lab experiment. In this experiment, he will be heating a mixture of potassium chlorate and manganese dioxide. What would be one of the chief products of this reaction?

a. potassium permanganate
! b. potassium chloride
c. potassium hypochlorite
d. manganese chlorate
e. manganese chloride

17. Professor Morris will be doing a lecture demonstration for his class--the preparation of ozone. Which method will he be most likely to use.

a. heating $O_2$ to a high temperature
b. distilling liquid air
c. heating chlorate salts
! d. passing an electric discharge through $O_2$
e. electrolysis of water

18. The oxide ion reacts rapidly with water to form

a. ozone
b. oxygen
! c. hydroxide ion
d. peroxide ion
e. hydrogen peroxide

19. Really heavy water could be prepared if we use the heaviest isotope of oxygen with the non-radioactive heavy isotope of hydrogen. This would be "heavier" than regular water. How many percent heavier would these molecules be?

a. 10 %
b. 20 %
! c. 22 %
d. 25 %
e. 33 %

20. The major use of the oxygen produced by manufacturing processes worldwide is for

a. waste water treatment
b. manufacture of hydrogen peroxide
c. rocket fuel, in the form of liquid oxygen
! d. production of steel and other metals
e. preparation of oxygen based bleaches as an environmentally preferred alternative to to chlorine based bleaches

21. Ozone is a modification of oxygen. It is called

a. an isotope of oxygen
b. an isomer of oxygen
! c. an allotrope of oxygen
d. a heliotrope of oxygen
e. an allomer of oxygen

22. An element which forms more compounds with more elements than any other is

a. fluorine
b. chlorine
c. sodium
! d. oxygen
e. carbon

23. The refractory oxides are binary oxides with elements from

a. Group IA
! b. Group IIA
c. Group VIA
d. Group IB
e. Group VIB

24. An amphoteric oxide is one (in) which

a. can be employed as either a cathode or an anode
b. can function as an oxidizing agent or a reducing agent
! c. can neutralize either an acid or a base
d. can exist as more than one isomer
e. the other element can form more than one compound with oxygen, and these different compounds have the other element in different oxidation states.

25. When a given transition metal forms several oxides having different oxidation states, it is observed that

a. the higher the oxidation state, the more amphoteric the oxide
b. the lower the oxidation state, the more amphoteric the oxide
! c. the higher the oxidation state, the more acidic the oxide
d. the higher the oxidation state, the more basic the oxide
e. the higher the oxidation state, the more powerful the oxide

26. Which statement about $H_2O_2$ is definitely not true?

a. It has a higher boiling point than water.
b. It has a higher density than water.
c. It has a higher molecular weight than water.
! d. It has a higher chemical stability than water.
e. It has a higher basicity than water.

27. Chemically, hydrogen peroxide is well recognized as

a. a strong reducing agent
! b. a strong oxidizing agent
c. a strong acid
d. a strong base
e. a good coordinating ligand

28. Alkali metals of group IA react with oxygen. Which statement below is true about the reaction?

a. The metals, M, always react with $O_2$ to form compounds with the general formula $M_2O$.
b. The metals, M, always react with $O_2$ to form compounds with the general formula MO.
c. The metals, M, always react with $O_2$ to form compounds with the general formula $MO_2$.
d. The metals, M, react with $O_2$ to form varied products, depending on the metal, with the generic formulas MO and $M_2O$.
! e. The metals, M, react with $O_2$ to form varied products, depending on the metal, with the generic formulas MO, $MO_2$ and $M_2O$.

29. Nitrogen gas is an extremely important industrial chemical, and some 30 million tons are produced for use yearly. How is nitrogen for industrial use produced?

a. electrolysis of concentrated ammonia solution
b. the reaction, $4LiNH_2 + 4CaO \longrightarrow 4LiOH + 4CaH + 2N_2$
! c. reaction of hydrocarbons with ammonia gas
d. passing ammonia gas over iron shavings at 650 °C
e. the fractional distillation of liquid air

30. The workers in Bob's Custodial Services use commercial strength ammonia solutions as a cleaning agent to clean the floors of mansions. How is this ammonia manufactured in the first place?

! a. reacting $H_2$ with $N_2$ in the presence of a catalyst
b. reduction of $NO_2$ with sodium or hydrogen peroxide
c. heating ammonium salts
d. extraction from ammonium compounds in the soil
e. treating ammonium salts with strong base

31. Ammonia is a very important derivative compound of nitrogen and, in terms of quantity manufactured, is one of the top 10 industrial chemicals. The major use of ammonia is

a. manufacture of nitric acid
! b. manufacture of chemical fertilizers
c. manufacture of industrial cleaning chemicals
d. as a refrigerant in the refrigeration industry
e. in the manufacture of nylon

32. Sodium amide, a product of the reaction of ammonia with sodium, is used chiefly as a

a. reducing agent
b. source of ammonium
c. cleaning agent
d. bleach
! e. strong base

33. When $NH_4Cl$ is heated above 500 °C it sublimes. The actual products are

a. $NH_3 + Cl_2$
! b. $NH_3 + HCl$
c. $H_2 + NH_2Cl$
d. $N_2 + HCl$
e. $NH_3 + H_2 + Cl_2$

34. Amy just started working for Maids-R-Us. She missed one of the training sessions, the one on safety--mixtures to avoid, so Amy didn't know that a dangerous and toxic substance can be produced when her detergent containing chlorine bleach is mixed with ammonia. This dangerous substance is

a. $NO_2$
b. $N_2O$
! c. $N_2H_4$
d. $NH_4Cl$
e. NO

35. The oxidation number of nitrogen in ionic metal nitrides is

! a. -3
b. -2
c. 0
d. +1
e. +2

36. The correct name for the compound, $NaN_3$ is

a. sodium nitride
! b. sodium azide
c. sodium nitrazide
d. sodium nitrazine
e. sodium trinitride

37. The correct name for the compound, $Na_3N$ is

! a. sodium nitride
b. sodium azide
c. sodium nitrazide
d. sodium nitrazine
e. sodium trinitride

38. The XYZ Corporation is going to raze its old abandoned offices to make way for a new building which will accomodate their needs based on their five year plan. Which of the following compounds would the demolition crew be most likely to use in their explosives to implode the old structure and cause it to collapse?

a. $NaN_3$
! b. $Pb(N_3)_2$
c. $Mg_3N_2$
d. $Na^+NH_2^-$
e. $Na_3N$

39. What chemical substance is manufactured using the Ostwald process?

a. hydrogen peroxide
b. hydrazine
c. ammonium nitrate
! d. nitric acid
e. dynamite

40. The person who invented the process for manufacturing dynamite is

a. Fritz Haber
b. Wilhelm Ostwald
! c. Alfred Nobel
d. Carl Wilhelm Scheele
e. Antoine Laurent Lavoisier

41. Nitric acid is one of the top 20 chemicals manufactured worldwide. Its principal use is in

a. the manufacture of nitroglycerin
b. the manufacture of nitrites for the food industry
c. the manufacture of gunpowder
d. the plastics recycling industry
! e. the manufacture of ammonium nitrate for the fertilizer industry

42. One major difference in properties of hydrochloric acid and those of nitric acid resides in the fact that

a. HCl(*aq*) contains a spectator ion while $HNO_3$(*aq*) contains no spectator ion.
b. HCl(*aq*) is an oxidizing acid while $HNO_3$(*aq*) is a non-oxidizing acid.
! c. HCl(*aq*) is a non-oxidizing acid while $HNO_3$(*aq*) is an oxidizing acid.
d. HCl(*aq*) is a halo acid while $HNO_3$(*aq*) is an oxo acid.
e. HCl(*aq*) does not expand when heated, while $HNO_3$(*aq*) expands when heated.

43. Kellie, an undergraduate student, complains to her TA that the brown coloration in the old bottle of concentrated nitrc acid means that it is "spoiled". Her teaching assistant explains that the brown coloration in an old bottle of concentrated nitric acid is most likely due to

a. NO
! b. $NO_2$
c. $N_2O$
d. $HNO_2$
e. $N_2O_5$

44. Different oxides of nitrogen in whch the oxidation state of the nitrogen is different are known. Which of the following are known (observed) oxidation states for nitrogen oxides?

a. +1, +3, +4, +5
b. +2, +4, +5
c. +1, +3, +5
! d. +1, +2, +3, +4, +5
e. +2, +3, +5

45. The nitrogen oxide used as an anesthetic has the formula,

a. NO
! b. $N_2O$
c. $N_2O_3$
d. $N_2O_4$
e. $N_2O_5$

46. The nitrogen compound which is used in airbag devices in the personal automobile industry is

a. $NaNH_4$
! b. $NaN_3$
c. $NaNO_2$
d. $NaNH_2$
e. $NH_4OCl$

47. Which one of the following compounds contains nitrogen in a positive rather than a negative oxidation state?

a. $NaNH_2$
b. $NH_3$
c. $N_2H_4$
d. $NH_2OH$
! e. $N_2O$

48. Which one of the following is not a form of elemental carbon?

a. coke
b. charcoal
c. graphite
d. fullerenes
! e. kevlar

49. A particular water filter manufacturer claims its product will absorb impurties from water. Victor learned from the salesman that the filter contains powdered charcoal. The absorbent properties of this powdered charcoal are due mainly to

a. chemical reactivity of the charcoal
b. polar nature of the surface of the charcoal
! c. the high surface area of the charcoal in this form
d. the porous structure of the charcoal
e. the basic (to litmus) properties of the charcoal

50. An oxide of carbon which is used as a fuel in industrial plants is

! a. CO
b. $CO_2$
c. $C_2O_3$
d. $C_2O_4$
e. $CO_3^{2-}$

51. A combustible fuel mixture used commercially which goes by the name "water gas" is

a. $C_{(powder)}$ + $H_2(g)$
b. $C_{(powder)}$ + $H_2O(g)$
c. $CO(g)$ + $H_2O(g)$
! d. $CO(g)$ + $H_2(g)$
e. $CO_2(g)$ + $H_2(g)$

52. An important chemical which is used industrially as an inexpensive base is obtained by the thermal decomposition of a carbonate compound or a bicarbonate compound. The chemical is

! a. CaO (lime)
b. $CO(NH_2)_2$ (urea)
c. $Ca(HCO_3)_2$ (calcium bicarbonate)
d. $H_2C_2O_4$ (oxalic acid)
e. $HC_2H_3O_2$ (acetic acid)

53. Which one of the following is *not* one of the major industrial uses of carbon dioxide?

a. production of carbonated beverages
b. production of solid refrigerants
c. production of the gaseous component in fire containment equipment
d. supercritical extraction of caffeine from coffee, an non-contaminating technique
! e. reducing agent in metallurgy of iron

54. Which one of the following is *not* a gas at ordinary temperatures and pressures?

a. CO
b. $CO_2$
c. HCN
d. $H_2S$
e. $CS_2$

55. Some carbide compounds behave as though they are salts, containing a cation and an anion. Calcium carbide, $CaC_2$, behaves as though it contains the $C_2^{2-}$ ion, a diatomic species. The Lewis structure for the carbide ion reveals

a. 1 $\sigma$-bond, 1 $\pi$-bond and 6 non-bonded valence electrons
b. 2 $\sigma$-bonds, 1 $\pi$-bond and 4 non-bonded valence electrons
! c. 1 $\sigma$-bond, 2 $\pi$-bonds and 4 non-bonded valence electrons
d. 1 $\sigma$-bond, 3 $\pi$-bonds and 2 non-bonded valence electrons
e. 1 $\sigma$-bond, no $\pi$-bonds and 8 non-bonded valence electrons

56. Joe Jackson has been a welder for 15 years, and his son Bill knows a few things about acetylene gas, including where he obtains it. One day at a demonstration lecture in his high school chemistry class Bill was completely amazed when the visiting scientist from the local university generated acetylene gas by mixing some gray rocks with water. What chemical substance was the gray rocks composed of?

a. calcium carbonate
b. silicon carbide
! c. calcium carbide
d. magnetite
e. boron carbide

57. Joe Jackson has been a welder for 15 years, and his son Bill knows a few things about acetylene gas. After doing a demonstration on generating acetylene by mixing water with some gray rocks, the visiting scientist did a second demonstration where he mixed water with some magnesium carbide and generated a combustible gas. Bill didn't think it was acetylene--it didn't smell right and it didn't burn the same way acetylene does. What chemical substance was the gas which is generated from reacting magnesium carbide with water?

a. hydrogen
b. carbon monoxide
! c. methane
d. ethylene
e. propane

58. A covalent network carbide compound which has structural features and a hardness comparable to that of diamond is

a. boron carbide
b. calcium carbide
c. magnesium carbide
! d. silicon carbide
e. antimony carbide

## Fill in the Blanks

59. What is the correct name for the compound: $BaO_2$? ______ ( ! barium peroxide)

60. In an important intermediate step in the Ostwald process, air is used to convert the NO(*g*) produced by the catalytic reaction into ______. ( ! $NO_2(g)$ )

61. When considered as a pair of complementary processes, the net effect of the Haber process and the Ostwald processes combined is to convert ______, ______, and ______ into nitric acid. ( ! natural gas, air, and water)

62. Urea, a very important fertilizer, is made by the reaction of ______ with ______. ( ! $CO_2$ with $NH_3$)

63. Hydrogen cyanide can be readily produced by the reaction of a cyanide salt with ______. ( ! a strong acid)

## True and False

64. Ammonium salts of strong acids, like $NH_4Cl$, $NH_4I$ and $(NH_4)_2SO_4$ dissolve in water to give solutions which are gasic to litmus paper. ( !F)

65. The percent of oxygen in the air is slowly decreasing due to the greenhouse effect. ( !F)

66. The percent of oxygen in the air is slowly decreasing due to excessive combustion of hydrocarbon fossil fuels. ( !F)

67. The physical and chemical properties of $O_3$ and $O_2$ are very similar, as might be suspected from their common components. ( !F)

68. $O_3$ is a better reducing agent than $O_2$. ( !F)

69. Ozone is a dangerous component of polluted atmospheres, to the point where it is subject of special and individual attention in weather reporting and forecasting. ( !T)

70. All metals which form oxides form oxides which are basic anhydrides. ( !F)

71. Transition metal oxides with the metal in a high oxidation states, like $CrO_3$, $Mn_2O_7$ and $V_2O_5$, are acidic oxides rather than basic oxides. ( !T)

72. Ammonia solutions are highly ionized to form primarily a mixture of $NH_4^+$ and $OH^-$. Very little unionized ammonia remains in solution. ( !F)

73. An important intermediate step in the Ostwald process is the conversion of $NO_2$ to $NO_3$ using the oxygen in the air stream. ( !F)

74. Carbon dioxide and carbon disulfide are both stable, non-combustible substances. ( !F)

## Critical Thinking

75. From the equation for the self ionization of liquid ammonia we deduce several important things when we compare it to the related equation in the water solvent system. From these comparisons, which compound below should be the strongest base than can exist in $NH_3(l)$?

a. KOH
b. KCl
c. $NH_4Cl$
! d. $KNH_2$
e. $Ag(NH_3)_2^+$

76. The ionic nitrides which dissolve in liquid ammonia immediately react with that solvent. What equation below describes the true situation?

a. $N^{3-}{}_{(amm)} + NH_3(l) \longrightarrow N_2(g) + 3H^-{}_{(amm)}$
! b. $N^{3-}{}_{(amm)} + 2NH_3(l) \longrightarrow 3NH_2^-{}_{(amm)}$
c. $N^{3-}{}_{(amm)} + 2NH_3(l) \longrightarrow NH_4^+{}_{(amm)} + NH_2^-{}_{(amm)} + \frac{1}{2}N_2(g)$
d. $N^{3-}{}_{(amm)} + 3NH_3(l) \longrightarrow 2N_2(g) + 9H^+{}_{(amm)}$
e. $N^{3-}{}_{(amm)} + 2NH_3(l) \longrightarrow \frac{3}{2}N_2(g) + 3H_2(g)$

## Supplementary Chapter 2 Simple Molecules and Ions of Nonmetals: Part II

### Multiple Choice

1. Roger, a chemistry student has just learned that when a metal sulfide is roasted in air, the products are

a. a metal oxide and hydrogen sulfide
! b. a metal oxide and sulfur dioxide
c. metal sulfate and hydrogen sulfide
d. metal sulfite and hydrogen sulfide
e. metal and sulfur dioxide

2. Some samples taken from sulfur deposits deep in a mine shaft in a sub-Sahara country were analyzed and their isotopic composition was recorded. The data:

| isotope | mass, amu | % abundance |
|---|---|---|
| $^{32}S$ | 31.97207 | 94.996 |
| $^{33}S$ | 32.97146 | 00.762 |
| $^{34}S$ | 33.96786 | 04.226 |
| $^{36}S$ | 35.96709 | 0.0016 |

Calculate the average atomic weight of the sulfur from this cave.

a. 32.062
! b. 32.065
c. 32.066
d. 32.892
e. 33.720

3. Large amounts of sulfuric acid are produced in the United States. The principal use of the sulfuric acid is in

a. the manufacture of nitric acid
b. the manufacture of plastics
! c. the manufacture of fertilizers
d. the manufacture of paper in the paper and pulp industry
e. the manufacture of steel

4. Sulfuric acid has several properties which gives it an advantage over other strong acids for industrial uses and also for laboratory use. Which one on the following listed set is ***not*** one of these advantages?

a. cost
b. can be prepared readily in nearly pure form ( > 98% pure)
c. is chemically stable, and does not decompose easily
d. is not volatile, hence doesn't produce corrosive vapors under ordinary handling conditions
! e. is chemically non-reactive with most metals and hence can be shipped in regular metal containers

5. The bottle filled with chunks of sulfur on the laboratory shelf contains this allotrope of sulfur:

! a. orthorhombic
b. monoclinic sulfur
c. $S_\beta$
d. plastic sulfur
e. red sulfur

6. XYZ Chemicals, like most of the producers in the chemical industry, uses sulfur dioxide chiefly for production of

a. sulfur
! b. sulfuric acid
c. sodium sulfate
d. hydrogen sulfide
e. sodium sulfide

7. Which one of the following is not true about $H_2SO_3$?

! a. It forms sulfuric acid upon warming.
b. It is actually hydrated $SO_2$.
c. It produces $SO_2$ when heated to 150 °C.
d. It produces sulfite ion when treated with excess strong base.
e. The actual molecule has never been detected in $H_2O$.

8. In big cities where acid rain and pollution are problems serious efforts have been made to remove $SO_2$ from vent gases. Which one of these compounds will do it?

a. $CaSO_4$
b. $CaCl_2$
! c. $Ca(OH)_2$
d. NaOH
e. $Na_2SO_3$

9. The local pharmacist has been asked to order sodium bisulfite. However, on his premade order form there is no listing for sodium bisulfite. So he instead circles this equivalent:

a. sodium sulfite
b. sodium sulfide
c. sodium bisulfate
d. sodium thiosulfate
! e. sodium metabisulfite

10. Elkton Corporation is an industrial giant that uses large quantities of sodium sulfite. For what purpose is most of the sodium sulfite which is produced annually used?

a. reduction of ores
! b. manufacture of paper pulp
c. manufacture of soap
d. plastics manufacturing industry
e. food preparation and preservation industry

11. The oxidation number of sulfur in $SO_3$ is

a. 2
b. 3
c. 4
! d. 6
e. 8

12. Sulfur trioxide is present in the output stream of waste gases from a certain industrial process. In compliance with certain regulations, sulfur trioxide is being neutralized by reaction with an excess of sodium carbonate. The products are $CO_2$ and

a. sodium bicarbonate
b. sodium bisulfite
c. sodium bisulfate
d. sodium hydrogen sulfate
! e. sodium sulfate

13. Which of the following is a useful catalyst for facilitating the reaction,

$$SO_2 + O_2 \longrightarrow SO_3?$$

a. iron
b. platinum
c. tetraarsenic heptasulfide
d. bismuth oxides
! e. vanadium pentoxide

14. A landscaping and lawn care company advertises that it uses "superphosphate" to fertilize lawns and gardens. Customers might be interested to know that "superphosphate" is made by treating which one of the following substance with sulfuric acid:

! a. phosphate rock
b. gypsum
c. limestone
d. bauxite
e. silicate rock

15. A student in Dr. James' research group ran out of a drying chemical, called drierite, late Friday evening. The research associate in the group suggested they coule purchase some Epsom salts at the pharmacy section of the local drug store and use that. What is this chemical substance called that can substitute so readily for drierite?

a. hydrated lime
b. sodium bicarbonate
! c. hydrated magnesium sulfate
d. hydrated phosphate of lime
e. sodium carbonate decahydrate

16. When the drying chemical, drierite, is exhausted it can be regenerated by

a. treating it with a dehydrating agent like sulfuric acid
! b. heating it for a prolonged time at temparatures above 500 °C
c. pumping out the compound using a very good vacuum pump
d. drying it out in direct sunlight at temperatures above 70 °C
e. actually, drierite cannot be regenerated

17. In the fiber and fabric industry, which one of the following compounds can be used to remove excess chlorine from bleached fabric?

a. sodium nitrate
b. sodium sulfate
c. sodium bisulfate
! d. sodium thiosulfate
e. sodium carbonate

18. The photographic process uses a chemical, known as "hypo", to remove unexposed silver halide from the film. The chemical name for "hypo" is

a. sodium sulfite
b. sodium sulfate
! c. sodium thiosulfate
d. sodium hyposulfite
e. sodium thiosulfite

19. Though $H_2S$ is used in general chemistry laboratory as an analytical reagent, handling large quantities of this substance is undesirable. $H_2S$ can be and is preferably generated in the reaction solution mixture by using

a. sodium sulfide
! b. thioacetamide
c. disulfur dichloride
d. ammonium thiosulfate
e. sodium persulfide

20. When phosphorus was first discovered, its discoverer obtained it by distilling residues obtained from

a. milk
! b. urine
c. wine barrel sediments
d. cow manure
e. wood shavings

21. The known allotropes of phosphorus are

a. $P_2$ and $P_4$
b. monoclinic phosphorus and orthorhombic phosphorus
c. white phosphorus and gray phosphorus
! d. white phosphorus and red amorphous phosphorus
e. triclinic phosphorus and monoclinic phosphorus

22. Elemental phosporous is manufactured on an industrial scale by reducing calcium phosphate with

! a. carbon
b. iron
c. zinc
d. hydrogen
e. superheated steam

23. Phosphine, the phosphorus analog of ammonia, can be prepared easily by the reaction of

a. $H_3PO_4$ + $H_2$
b. $H_3PO_3$ + $H_2$
c. $Na_3PO_4$ + $H_2$
! d. $Ca_3P_2$ + $H_2O$
e. $NaN_3$ + $H_3PO_3$

24. A student carrying a bottle of $PCl_5(s)$ to her lab from the stockroom dropped it in the hallway. The shattered contents started to give off smoky fumes with an acrid odor, so the Hazmat Team was quickly called in. What chemical substance was responsible for the fumes?

a. $PCl_5(g)$
b. $PCl_3(g)$
c. $H_2PtCl_4(g)$
d. $H_3PO_4(s)$
! e. $HCl(g)$

25. The reaction of $P_4O_{10}$, an oxide of phosphorus, with water produces,

a. $H_3PO_4$ only
b. $HPO_3$ only
c. $H_3PO_4$ and $HPO_3$
d. $H_3PO_4$ and $H_4P_2O_7$
! e. $H_3PO_4$, $HPO_3$ and $H_4P_2O_7$

26. $H_4P_2O_7$ is named

a. tetraphosphoric acid
b. metaphosphoric acid
c. orthophosphoric acid
! d. pyrophosphoric acid
e. paraphosphoric acid

27. The formula for hypophosphoric acid is $H_3PO_2$. Hypophosphoric acid is now known to be

! a. a monoprotic acid
b. a diprotic acid
c. a triprotic acid
d. an amphoteric substance
e. a substance which actually does not exhibit acid properties($K_a$ < $10^{-10}$).

28. Phosphate anions can form species which can be thought of as being dimers and trimers--precursors to polymers. These larger anions contain

a. P—P—P—P— . . . . . . . chains
! b. P—O—P—O—P—O—P—O— . . . . . . . chains
c. P—P—O—P—P—O—P—P—O— . . . . . . . chains
d. P—H—P—H—P—H—P—H— . . . . . . . chains
e. P—H—O—P—H—O—P—H—O— . . . . . . . chains

29. Which one of the compounds listed below, as the product of an acid base neutralization, is actually a ficticious compound because its displayed anion is fictitious?

a. $Na_3PO_4$
b. $Na_2HPO_4$
! c. $Na_3PO_3$
d. $Na_2HPO_3$
e. $NaH_2PO_3$

30. Gallium arsenide is an important arsenic compound and is used in

! a. light emitting diodes
b. bone structure replacement
c. cardiac pacemakers
d. 10-year batteries in implanted devices
e. food preservatives

31. Phosphorus acid, $H_3PO_3$ is

a. a monoprotic acid
! b. a diprotic acid
c. a triprotic acid
d. an amphoteric substance
e. a substance which actually does not exhibit acid properties($K_a < 10^{-10}$).

32. Which one of the oxides of Group V, shown below, is amphoteric?

a. $N_2O_5$
! b. $Bi_4O_6$
c. $P_4O_6$
d. $As_4O_6$
e. $Sb_4O_6$

33. In its normal physical state, fluorine is

a. a green liquid
b. a colorless liquid
c. a pale green gas
! d. a pale yellow gas
e. a colorless gas

34. Which one of the following materials is unsuitable for containers to be used for shipping liquid hydrogen fluoride?

a. alloy steel
b. wax lined metal containers
c. polyethylene
d. teflon (polyperfluoroethylene)
! e. glass

35. Fluorine can be manufactured from fluorine containing compounds on a commercial scale by

a. oxidation using atomic oxygen
b. oxidation using ozone at high pressures
! c. electrolytic reduction of fused fluorine compounds
d. electrolytic reduction of aqueous solutions of fluorine compounds
e. treatment of fluoride salts, like NaF or KF, with chlorine gas at high temperatures

36. Even though hydrofluoric acid is a monoprotic acid it forms an ion named the bifluoride ion. The formula for the bifluoride ion is

! a. $HF_2^-$
b. $H_2F^+$
c. $HFO^{2-}$
d. $H_2FO^-$
e. $H_2F_2O^-$

37. Fluoride ions are added to drinking water to provide a source of fluorine to (for)

a. control of bacterial organisms in drinking water
b. favor enhanced production of disease resistant antibodies in mammalian blood
! c. development of an acid resistant modification of tooth enamel
d. development of a fluoride coating in the mouth which poisons bacteria
e. neutralization of acids produced in the mouth by bacterial metabolism

38. In its normal physical state, chlorine is

a. a green liquid
b. a colorless liquid
! c. a pale green gas
d. a pale yellow gas
e. a colorless gas

39. Chlorine, one of the top 10 manufactured chemicals in the United States in terms of quantity, is produced industrially primarily by

a. electrolysis of fused NaCl
! b. electrolysis of brine (aqueous NaCl)
c. thermal decomposition of sodium hypochlorite (NaOCl)
d. oxidation of saline solutions by ozone
e. oxidation of saline solutions by charcoal-activated oxygen

40. The major use of the chlorine manufactured in the United States is for

a. manufacture of chlorine bleaches
b. manufacture of perchloric acid and hydrogen chloride and hydrochloric acid
! c. manufacture of organochlorine products like PVC (polyvinyl chloride) plastics
d. disinfection of municipal water supplies
e. manufacture of steel

41. Hydrochloric acid is a major strong inorganic acid. It's prime use in industry is for

a. preparation of AgCl for photographic film
b. production of hydrogen by reaction with zinc metal
! c. removal of oxide films from metals in metallurgy and related industries
d. neutralization of alkaline wastes in water treatment plants
e. chlorination of water to remove harmful organisms

42. All of the following are known oxyacids of chlorine, except

a. $HClO_4$
b. $HClO_2$
c. $HClO$
! d. $H_2Cl_2O_9$
e. $HClO_3$

43. Hydrogen chloride is produced commercially by a variety of methods. Which scheme shown below is *not* a method by which hydrogen chloride is produced commercially?

a. direct union of the elements, $H_2(g) + Cl_2(g) \longrightarrow 2HCl(g)$
b. metathesis, $NaCl(s) + H_2SO_4(l) \xrightarrow{heat} HCl(g) + NaHSO_4(s)$
c. by product of organic reaction, $Cl_2(g) + R{-}H \longrightarrow R{-}Cl + HCl(g)$
! d. displacement reaction, $HBr(g) + Cl_2(g) \longrightarrow HCl(g) + BrCl(g)$
e. preparation of vinyl chloride for polyvinyl chloride plastics,

$$H_2C{=}CH_2 + Cl_2(g) \longrightarrow HCl(g) + H_2C{=}CH{-}Cl$$

44. The physical appearance of the element bromine at ordinary temperatures and pressures is

a. a pale green gas
b. a pale yellow liquid
c. a deep green liquid
d. a deep purple liquid
! e. a deep red liquid

45. Bromine is manufactured in industry by which one of the following procedures?

a. oxidation of NaBr with ozone
b. oxidation of HBr with oxygen ($2HBr + \frac{1}{2}O_2 \longrightarrow H_2O + Br_2$)
c. reaction of bromide salts with nitric acid
! d. reaction of bromide containing brines with chlorine gas
e. reaction of bromide containing brines with fluorine gas

46. Periodic acid comes in several modifications, whose common relation can be verified by checking the oxidation number of the iodine. Which one of the following is *not* a modification of periodic acid?

! a. $H_4I_2O_7$
b. $HIO_4$
c. $H_4I_2O_9$
d. $H_5IO_6$
e. $H_3IO_5$

47. Which one of the following sets does not list the species in the correct order for the property listed immediately before it?

a. acid strength: $HF(aq) < HCl(aq) < HBr(aq) < HI(aq)$
b. ionic radius: $F^- < Cl^- < Br^- < I^-$
! c. difficulty of oxidation: $F^- < Cl^- < Br^- < I^-$
d. enthalpy of hydration: $F^- < Cl^- < Br^- < I^-$
e. ease of reduction: $F_2 < Cl_2 < Br_2 < I_2$

48. Helium has some remarkable or outstanding properties. Which one of the following is ***not*** one of these remarkable properties?

a. becomes a superfluid (loses all viscosity) at very low temperatures
b. has lowest boiling point of any known substance
c. cannot be changed to a solid at atmospheric pressure (no solid state exists at 1 atmosphere pressure at *any* temperature)
d. has no triple point on its phase diagram
! e. becomes a superconductor at a few hundredths of a degree above absolute zero

49. Which set below from among the "noble gases" actually do form compounds?

a. He Ne Ar
b. He Ne Ar Kr
c. Ne Ar Kr Xe
d. Ar Kr Xe
! e. Kr Xe

50. The first compound of a "noble gas" ever synthesized was a compound of which element?

a. He
b. Ne
c. Ar
d. Kr
! e. Xe

## Fill in the Blanks

51. The most common mineral family containing phosphorus is the ______ family of minerals.
( ! calcium phosphate, or apatite)

## True and False

52. Sulfur occurs in nature both in the free state (elemental sulfur) as well as in compounds. ( !T)

53. Natural gas wells contain considerable quantities of $SO_2$ in addition to methane and helium.
( !F)

54. Even though the chemical behavior of the various isotopes of sulfur is the same, some bacteria which degrade sulfur containing compounds seem to have a slight preference for the molecules with a particular isotope of sulfur. ( !T)

55. "Superphosphate of lime" is the name listed on a bag of chemical fertilizer. Does this mean the phosphate content is higher than that of ordinary phosphates? ______ ( !No)

56. The final step in the manufacture of sulfuric acid is the reaction,

$$SO_3(g) + H_2O(l) \longrightarrow H_2SO_4(l)$$

which is carried out using vanadium(V) oxide as a catalyst. ( !F)

57. To dry a precipitate separated by filtration at room temperature in a dessicator a chemist might use $P_4O_{10}$. When this substance reacts with the water vapor distilling off the precipitate to saturate the air space in the dessicator, the product is $H_4P_2O_7$. ( !T)

58. The acidity of the oxides of the elements of group V increase as we proceed from $N_2O_5$ down to $Bi_4O_{10}$. ( !F)

59. Fluorine is so reactive that is ***never*** found in the free state ($F_2$) in nature. ( !T)

## Critical Thinking

The data presented here applies to questions 60 to 64 immediately below. The symbols A, B, C, and D represent the first four halogen elements. The series of reactions presented always proceed in the forward direction.

$$A_2(unk) + 2C^-(aq) \longrightarrow C_2(unk) + 2A^-(aq)$$
$$B_2(unk) + 2D^-(aq) \longrightarrow D_2(unk) + 2B^-(aq)$$
$$B_2(unk) + 2C^-(aq) \longrightarrow C_2(unk) + 2B^-(aq)$$
$$D_2(unk) + 2C^-(aq) \longrightarrow C_2(unk) + 2D^-(aq)$$

60. Will $A_2$ oxidize $D^-$ to produce $D_2$? ( ! Yes)

61. Which one of the species is fluorine?

a. A
b. B
c. C
d. D
! e. There is insufficient information presented to determine this.

62. Which one of the species is bromine?

a. A
b. B
c. C
! d. D
e. There is insufficient information presented to determine this.

63. The physical state of $C_2$ at ordinary temperatures and pressures is ______. ( ! solid)

64. The physical state of $B_2$ at ordinary temperatures and pressures is ______. ( ! gas)

## Special Topic 2.1 Working with Concentrated Sulfuric Acid

65. When concentrated sulfuric acid is added to an equal volume of water, one immediate result is

! a. an extreme amount of heat is generated
b. the temperature of the solution falls
c. a solid is formed which slowly dissolves
d. sulfur dioxide gas is produced
e. the water reduces the sulfuric acid to $H_2S$ gas

66. In terms of quantity by mass produced for manufacturing and other uses, the largest volume industrial chemical in the United States is

a. ammonia
b. nitric acid
c. sodium carbonate
d. chlorine
! e. sulfuric acid

## Supplementary Chapter 3 The Properties and Chemistry of Metals

### Multiple Choice

1. The first step of the process by which magnesium is obtained from magnesium compounds which are dissolved in sea water is by treatment of the water with

a. ammonia solution
! b. lime
c. potassium hydroxide
d. sodium carbonate
e. sodium hydroxide

2. The first chemical step in the treatment of sulfide compounds of metal which are found in ores is called

a. electrolyzing
b. liming
c. milling
d. reducing
! e. roasting

3. Iron ore is reduced to iron by reaction with

a. calcium carbonate
! b. carbon
c. carbon monoxide
d. carbon dioxide
e. sulfur dioxide

4. Which one of the following techniques or chemical substances is not commonly used to reduce metal oxides to metals?

a. carbon
b. carbon monoxide
c. electrolysis
d. $H_2$ gas
! e. S

5. In the blast furnace silica, an objectionable impurity, is removed as

a. iron silicate
b. silicon
c. silicon dioxide
! d. slag, formed by reaction of silica (acidic oxide) with lime (basic oxide)
e. sodium silicate

6. Commercially, aluminum is obtained from the purified aluminum oxide (after pretreatment) by

a. electrolysis of aqueous $Al_2(SO_4)_3$
b. electrolysis of aqueous $KAl(SO_4)_2$
! c. electrolysis of a fused mixture of $Al_2O_3$ and $Na_3AlF_6$(cryolite)
d. reduction with coke at high temperatures
e. reduction with sodium metal

7. Copper is usually refined by

a. distillation
! b. electrolysis
c. lithification
d. the oxygen lance
e. smelting

8. Where are alkali elements and their compounds found in nature?

a. primarily in caves
b. primarily in clay soils
c. primarily in closed lakes which have no outlet
d. primarily in soils from desert regions
! e. primarily in sea water

9. Which alkali element is very important in biochemical systems of plants?

a. Cs
! b. K
c. Li
d. Na
e. Rb

10. Which one of the elements below has the lowest melting point temperature, consistent with the group periodic properties?

a. Al
b. Ca
c. Cr
d. Fe
! e. K

11. Sodium metal is prepared commercially by

a. electrolysis of aqueous sodium chloride
! b. electrolysis of fused sodium chloride
c. reduction of sodium carbonate by lime at 600 °C
d. reduction of sodium carbonate by carbon at 2000 °C
e. reduction of sodium oxide by carbon at 2000 °C

12. Lithium metal is prepared commercially by

a. electrolysis of aqueous lithium chloride
! b. electrolysis of fused lithium chloride
c. reduction of lithium chloride by boron at 2000 °C
d. reduction of lithium carbonate by carbon at 2000 °C
e. reduction of lithium oxide by carbon at 2000 °C

13. K, Cs and Rb are often obtained from their compounds by the same general method, which is

a. a displacement reaction using elemental sodium
b. a reduction using the reaction of hot coke with the metal carbonate
c. a reduction using the reaction of finely powdered aluminum with the metal carbonate
d. electrolysis of an aqueous solution of a soluble salt of the metal
! e. electrolysis of the fused chloride salt of the metal

14. Which one of the alkali metals will "burn" in a nitrogen atmosphere as it combines with the nitrogen exothermically?

a. Cs
b. K
! c. Li
d. Na
e. Rb

15. Sodium hydroxide is a very important commercial chemical, ranking in the top 10 in quantity manufactured annually. Sodium hydroxide is produced commercially by

! a. electrolysis of an unstirred solution of brine (aqueous sodium chloride)
b. electrolysis of molten sodium chloride
c. heating sodium carbonate above 1500 °C to decompose it and yield an intermediate which is converted to sodium hydroxide by the action of water
d. reaction of sodium metal with water
e. reaction of sodium oxide with water

16. Sodium carbonate is an important chemical in the chemical industry. Which statement below about sodium carbonate is ***not*** true?

a. Sodium carbonate can be manufactured by a process known as the Solvay process.
b. Sodium carbonate can be obtained directly from a mineral called trona.
! c. Sodium carbonate can be obtained by heating a mix of limestone and sodium chloride.
d. Sodium carbonate is an important industrial base (in the acid-base sense).
e. A form of sodium carbonate was widely utilized and is still known as "washing soda".

17. Alkali metal salts impart color to flames when they are strongly heated in an open non-luminous flame. The color produced by sodium salts is

a. blue
b. green
c. orange
d. red
! e. yellow

18. Alkali metal salts impart color to flames when they are strongly heated in an open non-luminous flame. The color produced by lithium salts is

a. blue
b. green
c. orange
! d. red
e. yellow

19. Which one of the alkali metals is known for its ability to form *covalent* organometallic compounds with hydrocarbon alkyl groups?

a. Cs
b. K
! c. Li
d. Na
e. Rb

20. Even though the more active metals are expected to react with water, some of them in fact are not attacked by cold water even though hot water or steam react with all of them rapidly. Which one of these active metals listed below displays this property?

a. Ba
b. Ca
c. K
! d. Mg
e. Na

21. Which Group IIA element forms oxides which are amphoteric in behavior?

a. Ba
! b. Be
c. Ca
d. Mg
e. Sr

22. The most industrially important calcium compound is

a. calcium bicarbonate
b. calcium carbonate (also occurs in limestone)
c. calcium chloride
! d. calcium oxide (known as lime)
e. calcium sulfate (also occurs in gypsum)

23. An important aluminum compound which is the matrix for several types of gemstones is

! a. $\alpha$-$Al_2O_3$
b. $\Gamma$-$Al_2O_3$
c. $Al_2(SO_4)_3$
d. $Al(OH)_3$
e. $KAl(SO_4)_2 \cdot 12H_2O$

24. An important use of $Al_2(SO_3)_3 \cdot 18H_2O$, also known as alum, is

a. as an abrasive in "sandpaper"
b. in cosmetics
c. as a gemstone matrix
d. reaction with coke to produce aluminum metal
! e. in municipal and other water treatment plants

25. Chrome yellow, a pigment formerly used in paints, is

a. $As_2S_3$
b. CdS
c. $Na_2CrO_4$
! d. $PbCrO_4$
e. PbS

26. The most abundant metallic element in the earth's crust is

! a. aluminum
b. iron
c. potassium
d. silicon
e. sodium

27. A *metallic* element used in the free (elemental) state in preparation of solid propellants for the large vehicles used in the space program in the United States is

! a. aluminum
b. cadmium
c. lithium
d. potassium
e. sodium

28. Which one of the oxides below is not amphoteric?

a. $Al_2O_3$
! b. $Fe_2O_3$
c. PbO
d. SnO
e. ZnO

29. A major component of many gemstones is

! a. $Al_2O_3$
b. $CaSiO_3$
c. $K_2CrO_4$
d. $NaAl(SO_4)_2 \cdot 12H_2O$
e. $ZnCO_3$

30. Which type of substance would be most strongly attracted by a magnetic field?

! a. a solid ferromagnetic metal or alloy
b. a solid paramagnetic metal or alloy
c. a solid diamagnetic metal or alloy
d. a molten ferromagnetic metal or alloy
e. a molten paramagnetic metal or alloy

31. Which type of substance would be least strongly attracted by a magnetic field?

a. a solid ferromagnetic metal or alloy
b. a solid paramagnetic metal or alloy
! c. a solid diamagnetic metal or alloy
d. a molten ferromagnetic metal or alloy
e. a molten paramagnetic metal or alloy

32. The principal use for vanadium is

a. as a catalyst in the oxidation of $SO_2$ to $SO_3$ in the manufacture of sulfuric acid
! b. manufacture of special steel alloys and cast iron
c. in paint pigments, because it forms intensely colored chemically stable compounds
d. in corrosion resistant coatings for metals used in construction of bridges over salt water
e. as a component of material for special non-sparking hammers which can be safely used in construction in areas with combustible gases.

33. All of the following are industrial uses to which chromium is put except

a. prepation of decorative or protective coatings on metals, usually deposited on the surfaces by electrolysis
b. as a component in stainless steel alloys
c. in heater elements for toasters, irons and such
d. in paint pigments which contain chromate ion and other chromium ions
! e. water treatment chemicals

34. Vanadium forms an oxycation, which is very uncommon for elements in that region of the periodic table. The oxidation state of vanadium in the vanadyl cation is

a. +2
b. +3
c. +4
! d. +5
e. +6

35. The chemical reaction,

$$3\,MnO_4^{2+}(aq) + 4\,H^+(aq) \longrightarrow 2\,MnO_4^-(aq) + MnO_2(s) + 2\,H_2O$$

is an example of

a. acid base neutralization
! b. disproportionation
c. double displacement
d. protonation
e. single displacement

36. Which statement below about iron and its compounds is *not* true?

a. FeO is a known oxide of iron.
b. $Fe_2O_3$ is a known oxide of iron.
c. $Fe_3O_4$ is a known oxide of iron.
d. Solutions of $Fe^{3+}$ salts are acidic due to cation hydrolysis.
! e. Compounds with iron in the +5 oxidation are very important in chemistry of iron.

37. Nickel is a versatile and useful metal. Which one of the following is *not* true about nickel?

a. It is used in anti-corrosion coatings.
b. It is used in manufacture of high impact steel.
c. It is used in manufacture of stainless steel.
d. It is used in manufacture of monel metal, a corrosion resistant alloy with very high strength.
! e. It is the second most abundant metallic element in the earth's crust.

38. Which one of the following metals will *not* react with and dissolve in non-oxidizing acids?

a. Co
! b. Cu
c. Fe
d, Mn
e. Ni

39. Which one of the following elements has been found existing in natural sources in the free (elemental) state?

a. Co
b. Cr
! c. Cu
d. Fe
e. Ni

40. Even though they are transition elements, some oxidation states are very rare if existent at all. Which one of the following a either a very uncommon or a non existent oxidation state?

! a. Au(II)
b. Co(II)
c. Cu(II)
d. Fe(II)
e. Ni(II)

41. Which metallic element has the highest thermal and electrical conductivity of all known metallic elements?

a. copper
b. gold
c. iron
d. nickel
! e. silver

42. Which group of elements is most resistant to chemical acids, as a group?

a. the Ca - Ba - Sr group
b. the Cr - Mo - W group
! c. the Cu - Ag - Au group
d. the Fe - Ru - Os group
e. the Zn - Cd - Hg group

43. Which group of elements is least resistant to chemical acids, as a group?

! a. the Ca - Ba - Sr group
b. the Cr - Mo - W group
c. the Cu - Ag - Au group
d. the Fe - Ru - Os group
e. the Zn - Cd - Hg group

44. Which reagent below does *not* react with metallic copper to any significant extent?

a. cold dilute (1M) nitric acid
b. hot dilute (6M) nitric acid
c. hot concentrated (15M) nitric acid
! d. cold dilute (6M) sulfuric acid
e. hot concentrated (18M) sulfuric acid

45. Which reagent below does *not* react with metallic copper to any significant extent?

a. cold dilute (1M) nitric acid
b. hot dilute (6M) nitric acid
c. hot concentrated (15M) nitric acid
! d. hot concentrated (12M) hydrochloric acid
e. hot concentrated (18M) sulfuric acid

46. Which one of the following reagents will react with metallic silver?

b. cold dilute (3M) acetic acid
e. cold dilute (3M) ammonia
c. cold dilute (3M) hydrochloric acid
! a. cold dilute (3M) nitric acid
d. cold dilute (3M) sulfuric acid

47. A metal which dissolves in 3M $HCl(aq)$, 3M $H_2SO_4(aq)$, 3M $HNO_3(aq)$ and 3M $NaOH(aq)$ is

a. Cd
b. Co
c. Fe
d. Hg
! e. Zn

48. A metal which dissolves in 3M $HCl(aq)$, 3M $H_2SO_4(aq)$, 3M $HNO_3(aq)$ and 3M $NaOH(aq)$ is

! a. Al
b. Ag
c. Cd
d. Co
e. Fe

49. A metal oxide which, in a finely divided state, will *not* dissolve in 3M $HCl(aq)$, 3M $H_2SO_4(aq)$, 3M $HNO_3(aq)$ and 3M $NaOH(aq)$ is

a. aluminum oxide
! b. iron (III) oxide
c. lead (II) oxide
d. tin (II) oxide
e. zinc oxide

50. All these elements or their compounds are used in the manufacture of commercial electrical batteries except

a. Ag
! b. Cr
c. Hg
d. Ni
e. Zn

51. All these elements or their compounds are used in the manufacture of commercial electrical batteries except

a. Cd
b. Hg
c. Mn
d. Pb
! e. Sn

52. Which element below is not attacked by (does not react with) any of the following reagents: 6M $HNO_3$ 15M (conc) $HNO_3$ 18M (conc) $H_2SO_4$ 12M (conc) HCl 6M HF

a. cadmium
! b. gold
c. iron
d. mercury
e. silver

53. An element used in the preparation of galvanized steel or iron is

a. Cd
b. Cr
c. Co
d. Ni
! e. Zn

54. A metal with a property such that structures made of this metal crumble and disintegrate when temperatures hover around 0 °C for extended periods (weeks and months).

a. aluminum
b. copper
c. lead
! d. tin
e. zinc

## Fill in the Blanks

55. The rock, sand, and other unwanted material present in an ore when it is mined are collectively referred to as ______. ( ! gangue)

56. The thermite process involves the reaction of iron(III) oxide with ______. ( ! aluminum)

57. A compound of barium which can be safely ingested (and excreted), leading to its use in internal medicine even though soluble barium is a toxic element. ______ ( ! barium sulfate)

58. An alloy with a very low melting point which is used in the switching element in sprinkler systems is ________. (! Wood's metal)

59. A qualitative test for the presence of $Fe^{3+}$ ions in solution (which is sensitive enough for use in water quality testing) is the reaction of the $Fe^{3+}$ ions with the ______ ion. (! $SCN^-$)

## True and False

60. Beryllium, though it is in Group II of the periodic table, is different in many ways from the remaining members of the group. ( !T)

61. $SnCl_4$ has a melting point of -33 °C and a boiling point of 114 °C. These suggest that $SnCl_4$ is a covalent compound with discrete molecules and just average intermolecular forces. ( !T)

62. Even though atomic radii tend to show a regular increase within a group or family as the atomic number increases, post lanthanide elements do not appear to follow this pattern closely. ( !T)

63. The ability to display several different oxidation states is much more pronounced among metallic members of transition elements than among non-transition metals. ( !T)

64. The most common oxidation states of chromium in its compounds are the +2, +4, and +5 states. ( !F)

## Critical Thinking

65. Which one of the following compounds does *not* belong to the class of compounds called "alums"?

a. $NaAl(SO_4)_2 \cdot 12H_2O$
b. $KCr(SO_4)_2 \cdot 12H_2O$
c. $NH_4Fe(SO_4)_2 \cdot 12H_2O$
d. $KV(SO_4)_2 \cdot 12H_2O$
e. $NH_4Zn(SO_4)_2 \cdot 12H_2O$

66. Which one of the following is not a property of some post lanthanide transition metals which is linked to the "lanthanide contraction"?

a. high mass density of the neutral element in the solid state
b. high resistance to oxidation of the neutral element
c. comparative chemical inertness
d. higher than expected ionization energies
! e. formation of oxides which are much more stable than those in the previous row